The Catholic Church and Democracy in Chile and Peru

A TITLE FROM THE HELEN KELLOGG INSTITUTE FOR INTERNATIONAL STUDIES

Kwan S. Kim and David F. Ruccio, eds.
Debt and Development in Latin America (1985)

Scott Mainwaring and Alexander Wilde, eds.
The Progressive Church in Latin America (1989)

Bruce Nichols and Gil Loescher, eds.
The Moral Nation: Humanitarianism and U.S. Foreign Policy Today (1989)

Edward L. Cleary, O.P., ed.
Born of the Poor: The Latin American Church since Medellín (1990)

Roberto DaMatta
Carnivals, Rogues, and Heroes: An Interpretation of the Brazilian Dilemma (1991)

Antonio Kandir
The Dynamics of Inflation (1991)

Luis E. González
Political Structures and Democracy in Uruguay (1991)

Scott Mainwaring, Guillermo O'Donnell, and J. Samuel Valenzuela, eds.
Issues in Democratic Consolidation: The New South American Democracies in Comparative Perspective (1992)

Roberto Bouzas and Jaime Ros, eds.
Economic Integration in the Western Hemisphere (1994)

Mark P. Jones
Electoral Laws and the Survival of Presidential Democracies (1995)

Dimitri Sotiropolous
Populism and Bureaucracy: The Case of Greece under PASOK, 1981–1989 (1996)

Peter Lester Reich
Mexico's Hidden Revolution: The Catholic Church in Law and Politics since 1925 (1996)

Michael Fleet and Brian H. Smith
The Catholic Church and Democracy in Chile and Peru (1997)

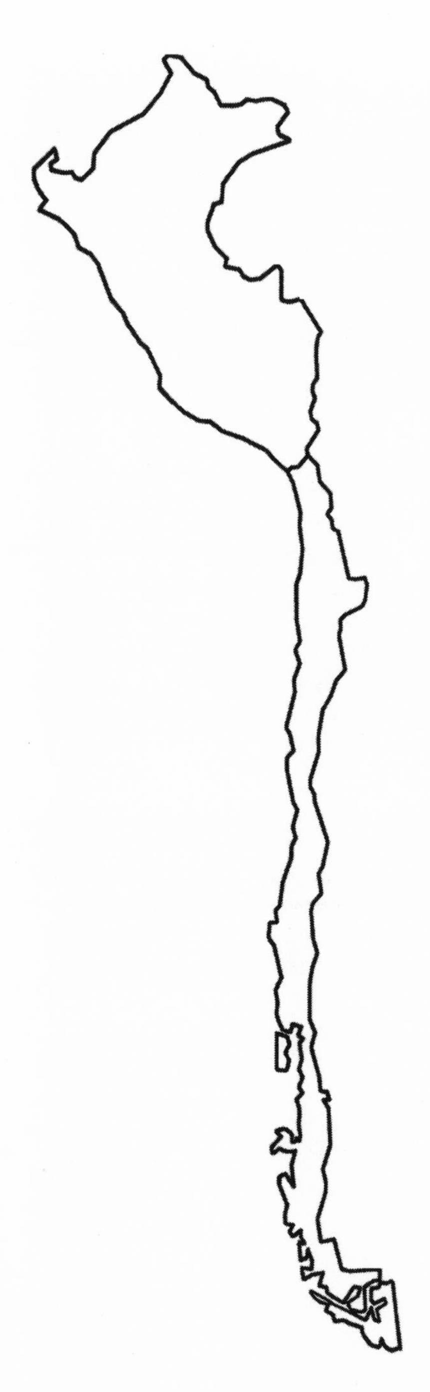

The Catholic Church and Democracy in Chile and Peru

MICHAEL FLEET

and

BRIAN H. SMITH

University of Notre Dame Press

Notre Dame, Indiana

Manufactured in the United States of America

Library of Congress Cataloging-in-Publication Data
Fleet, Michael.
The Catholic Church and democracy in twentieth-century Chile and Peru / Michael Fleet and Brian H. Smith.
p. cm.—(A title from the Helen Kellogg Institute for International Studies.)
Includes bibliographical references and index.
ISBN 0-268-00818-3 (alk. paper)
1. Catholic Church—Chile—History—20th century. 2. Catholic Church—Peru—History—20th century. 3. Democracy—Religious aspects—Catholic Church—History—20th century. 4. Church and social problems—Chile—History—20th century. 5. Church and social problems—Peru—History—20th century. 6. Chile—Church history—20th century. 7. Peru—Church history—20th century.
I. Smith, Brian H., 1940– . II. Title. III. Series.
BX1468.2.F57 1996
282′.83′0904—dc20 96-28967
CIP

The paper used in this publication meets the minimum requirements of the American National Standard for Information Sciences—Permanence of Paper for Printed Library Materials, ANSI Z39.48-1984.

Book design by Will H. Powers
Set in Minion and Amerigo type by Stanton Publication Services, Inc., St. Paul

Contents

Preface

There are many whom we wish to thank for their help in the writing of this book. Our research began in late 1986, when Michael Fleet received a Howard Heinz Foundation grant to study Christian-Marxist relations in Latin America from a base in Santiago, Chile. It enabled him to do attitudinal surveys and to interview Christian and Marxist elites in Chile and Peru the following year. Neither survey could have been carried out without the help of the Centro Belarmino's Center for Sociocultural Research (CISOC) in Santiago, the Catholic University of Peru's Faculty of Social Science, and the good offices of dozens of priests, nuns, and lay Catholic leaders in both Lima and Santiago. With additional support from Marquette University and from the Organization of American States, Fleet spent most of 1987 in Santiago and Lima.

By late 1987, however, Christian-Marxist relations were beginning to lose their intellectual and topical political appeal. Marxists and Christians were working together fluidly to restore or strengthen democracy in both countries. But most Marxists were in the throes of ideological or political crisis, and few Christians felt the need to pursue or reflect on Christian-Marxist relations as such. In this context, moreover, both of us realized that the more important and interesting story unfolding was the range and complexity of the Catholic Church's political influence, and we decided to tackle this phenomenon jointly, using Fleet's data and materials.

The analytical framework presented in chapter 1 was the initial fruit of our collaboration. It was born of a series of conversations that we had in late 1987, and by the following spring we had agreed upon a structure for the overall project. Smith then wrote initial drafts of the Introduction and chapter 1, which surveys the literature and then develops our analytical framework. These drafts were subsequently revised and refined—several times—by both of us. Chapter 2, on the Chilean Church through the early 1980s, was a joint undertaking, drawing on work that each of us had done previously (Smith 1982 and Fleet

1985). Fleet returned to the field in the summers of 1988 and 1990, gathering materials on the Church, interviewing additional Catholic elites (bishops, priests, and nuns) and laypeople (the latter of varying degrees of political and Church involvement), and monitoring political developments in both countries.

This field research, and Smith's work in the summers of 1990 and 1991, were funded by a 1990 grant from the United States Institute of Peace. Fleet spent the fall of 1990 at Notre Dame's Kellogg Institute for International Affairs, working closely with the noted Church scholar Phillip Berryman (who was also a Residential Fellow), and exploiting the bountiful resources of the Hesburgh Library. While at the Kellogg and during the following spring while on sabbatical leave from Marquette, Fleet analyzed the survey and interview data and wrote drafts of chapters 3 through 7. These chapters were revised and refined by both of us over the next several years, as we sought to produce a text with which we were each satisfied, both stylistically and in terms of content. Fleet returned, for the last time, to both Peru and Chile in 1992 to complete the elite interviews and to look for materials dealing with each country's transition and consolidation processes. During the fall of 1993, while on sabbatical leave from Ripon, Smith worked on the drafts of chapters 3 through 7. Fleet wrote the concluding chapter (8) in the summer of 1995, presenting it to a panel on religion and politics at LASA's Washington, D.C., meeting later that year.

The study is, thus, a genuinely collaborative undertaking, although each of us has had primary responsibility for certain aspects and sections. The analytical framework, for example, relies heavily of Smith's previous work (1982), and he is largely responsible for the Introduction and chapter 1, and for the hypotheses at the conclusion to chapters 2 and 3. The bulk of chapter 2, on the other hand, was written jointly, and Fleet, whose field work put him in closer touch with the basic material, wrote the initial drafts for all other chapters. Once in draft form, however, all chapters underwent substantial revision and rewriting by both of us, with Fleet usually doing the final rewrite.

In addition to the several sponsoring institutions mentioned above, we wish to thank the computer centers at Marquette and Ripon College for their assistance in data analysis and in facilitating the electronic exchange of chapter drafts with minimum distortion or loss. We also would like to acknowledge the Slinger Inn in Slinger, Wisconsin, which offered us a hospitable environment (and delicious apple pie!) when we met, as we did many times, half-way between Milwaukee and Ripon. And we want to extend special thanks to Jim Langford of the University of Notre Dame Press for his support of an initially cumbersome manuscript, and to our editors, Ann Rice and John McCudden, for helping to make it less so.

Finally, we wish to thank our wives, Jean Fleet and Mary Kaye Smith, and children—Maria Elena, Sara, Rachel, and Katie Fleet, and Sean and Katie Smith—for their enduring patience over the course of the project, and especially when one or the other of us was in the throes of the anxiety or grumpiness that seem an inescapable part of these enterprises.

The Catholic Church
and Democracy
in Chile and Peru

Introduction

The "modern age" has been a source of continuing difficulty for Roman Catholicism. Modern thinkers and movements have been cutting away at the Church's temporal and spiritual power since the late sixteenth century. For most of this period, Catholic authorities strenuously resisted the "modern world." They opposed forces that were pressing for freedom, equality, democracy, and individual rights, defending, instead, the monarchical regimes which these forces were challenging. In the 1860s, when Pope Pius IX rejected outright the possible separation of church and state, he termed progress, liberalism, and modern civilization "the principal errors of our time," with which it was impossible for the Roman Pontiff "(to) reconcile himself and come to terms."[1]

Shortly thereafter, however, the Church began to rethink its opposition to modernity. Some Catholics concluded, reluctantly, that the Church would have to accommodate modern values if it wanted to regain its waning influence and appeal. Others embraced these values and sought to reconcile them with traditional Catholic beliefs and concerns. All agreed that the Church would never attract the emerging middle classes or win back popular-sector groups if it remained closely aligned with socioeconomic and political elites and/or state authorities. The efforts of these pragmatic and liberal Catholics led to a gradual, uneven, but ultimately substantial "modernization" of the Catholic Church over the first six decades of this century. It culminated in the documents of the Second Vatican Council, held between 1962 and 1965, in which the Church blessed and embraced "the joys and hopes, the griefs and the anxieties of the men of this age."

Part of the Church's accommodation with modernity has been a gradual evolution of its attitude toward democracy. As traditional monarchies gave way to constitutional and democratic regimes, the Vatican came to accept elected representative government as morally legitimate. In 1881, Pope Leo XIII gave tentative support for representative government, claiming that leaders "may in certain cases be chosen by the will and decision of the multitude, without opposition to or impugning of the Catholic doctrine," and that the "people are not hindered from choosing for themselves that form of government which suits best . . . their own disposition."[2]

The Church's experience under communism and fascism in the early- and mid-twentieth century called forth papal support for democratic forms of government as World War II was ending. Pope Pius XII in his Christmas message of 1944 acknowledged that the "bitter experience" of "dictatorial power" was making the peoples of the world "call for a system of government more in keeping with the dignity and liberty of citizens," one in which persons could not be "compelled without being heard" and where they could "express their own views of the duties and sacrifices" asked by rulers.[3]

Clearer affirmations of the positive aspects of democracy and the rights associated with it were made by Pope John XXIII and by the bishops of the Second Vatican Council. In his 1963 encyclical, *Peace on Earth* (*Pacem in Terris*), Pope John XXIII laid out a list of inalienable rights of human beings, including the right to life, to bodily integrity, and to assembly and association; the right to express one's opinions freely, to take an active part in public life, to choose those who are to rule, and to select the form of government in which authority is to be exercised. In 1965 "The Pastoral Constitution on the Church in the Modern World" ("Gaudium et Spes") of Vatican II reiterated this right to political participation, stating that "the choice of government and the method of selecting leaders is left to the free will of citizens." In another document, "Declaration on Religious Freedom" ("Dignitatis Humanae"), the Council Fathers upheld the freedom of conscience and stated that all persons are "to be immune from coercion" and that no one "is to be forced to act in a manner contrary to his own beliefs." Finally, Vatican II also legitimized greater participation in decision-making within the Church itself by encouraging more collegial exercise of authority among pope, bishops, priests, and nuns and more direct forms of cooperation for the laity in the apostolate of the Church.[4]

In evolving toward an acceptance of democratic values and processes, however, official Catholic teaching has continued to affirm that justice, not freedom alone, is the goal of legitimate government, that individual rights must be exercised with concern for the common good, and that a universal moral law exists not subject to change by popular consensus and protected by the voice of the Church. Pope Leo

XIII endorsed the legitimacy of any form of government chosen by the people "provided only it be just, and that it tend to the common advantage." He also insisted that "the laws of nature and of the Gospel which by right are superior to all human contingencies, are necessarily independent of all modifications by civil government." Pius XII claimed that popular sovereignty is always subject to a higher realm and positive law is only inviolable when it conforms—or at least is not opposed—to the absolute order set up by the Creator." He stated, in fact, that it is the mission of the Church to discern and teach that "divinely-established order of beings and ends which is the ultimate foundation . . . of every democracy." The bishops at Vatican II reaffirmed this when they proclaimed the Church's "right to pass moral judgments, even on matters touching the political order, whenever basic personal rights or the salvation of souls make such judgments necessary."[5]

The Church's stance on such core elements of modernity as democracy, the rights of the individual, and popular sovereignty has thus evolved from open hostility to cautious acceptance. It has embraced these values and processes with the qualification that an immutable moral realm underlies them, a realm it must continue to articulate and protect. Its willingness to modernize and accommodate change in the twentieth century has made it a more relevant institution, and a more credible ally of democracy, in an increasingly secular world. Less hierarchical, although better organized, it has grown more sensitive to the needs and requirements of local churches. Its higher authorities have begun to share space and functions with priests, nuns, and laypeople who are closer to the lives and needs of ordinary Catholics. Its leaders became as concerned with serving and empowering their communities as they once were directing and disciplining them.

The Church's reforms and innovations helped it to play a progressive role in support of social change and democratic politics in much of Latin America during the last thirty years. In the late 1950s and early 1960s, Latin American bishops endorsed agrarian and tax reforms, expanded suffrage, and greater government spending in health and education. With financial help from churches in Europe and the United States, they initiated programs in literacy training, production and marketing cooperatives, credit unions, and health and nutrition projects. In these ways, the Church sought to promote peaceful change and thereby head off violent revolution. In the process, it gave additional impulse to modernization and reformist forces within its own ranks.

During the late 1960s and 1970s, the Church emerged as a critic and antagonist of repressive military regimes in several countries. Catholic bishops became champions of human rights and popular interests. They denounced state violence, demanded respect for due process and the rule of law, and called for policies that were more responsive to the needs of the poor. Local churches became havens for the persecuted, providing human warmth, material and legal assistance, and logistical support for those facing or fleeing from repression. Some served as staging grounds for nonviolent resistance to military authorities, and Catholic activists were prominent in movements pressing for a return to civilian rule.

In the early and mid 1980s, Church leaders and activists helped to persuade a number of military governments to relinquish power to civilian successors. In some instances (Brazil and Chile), they facilitated or strengthened compromise agreements between military and civilian leaders. In others (Nicaragua and El Salvador), they served as informal mediators between governments and antigovernment guerilla forces.

The Church thus played a generally progressive role in most of Latin America during the last thirty years. With one or two notable exceptions (Argentina, and Guatemala through the mid 1980s), it has consistently supported peaceful resolution of conflict within a broader commitment to social justice, human rights, and democratic politics. Its accommodation of modernity and democracy has not been without attendant costs, however. Its reforms and initiatives have produced tensions, divisions, and fragmentation within its ranks. As some Catholics have been won back to the fold, others have been lost or alienated. The new collegiality among bishops, greater independence and organizational development among priests and sisters, and the expanded dignity and role of lay men and women, all have helped to generate greater vitality and commitment at all levels of the Church. But they have come at the expense of institutional authority and coherence, as increasing numbers of Catholics have been making moral decisions based on the dictates of their own consciences.

The costs and sacrifices associated with modernization have spawned tensions between radical and conservative Church groups. At issue in most countries is the apparent radicalization of Catholic activists inspired by liberation theology, a new type of religious thinking that reads scripture from the vantage point of the poor. During the 1970s and 1980s, Catholics committed to liberation theology and active in local Christian communities were in the forefront of radical political movements in many countries. They frequently ran afoul of moderate and conservative superiors, many of whom wanted the Church to distance itself from such "politics" and to reaffirm traditional institutional authority and prerogatives. Pope John Paul II, members of the Roman Curia, and most of the bishops named to head vacant dioceses since 1978, the year in which John Paul assumed the papacy, are leading this "restoration" movement.

Since becoming pope, John Paul has reaffirmed Church support for social justice, human rights, and the rule of law, but also has moved to limit the political involvement of local churches. He has replaced retiring bishops (many of them progressives named by Paul VI) with politically and theologically conservative successors. He has publicly criticized the "sectarian" tendencies of some Christian communities and has insisted on stricter compliance with the instructions and teachings of Church authorities. Finally, he has approved the issuance of written warnings against liberation theology, the interrogation of several of its leading exponents, and the temporary silencing of the Brazilian theologian (and former priest) Leonardo Boff.

These moves have encouraged conservative Catholics within most Latin American countries. Some bishops have moved liberation-oriented priests out of their "popular" sector parishes, and sent progressive foreign missionaries home. Liberal seminaries have been closed, and the faculties and curricula of others, along with training programs for lay leaders, have been restructured to emphasize prayer, biblical scholarship, Church history, canon law, and individually oriented pastoral counseling. Loyalty to official Church teachings is being stressed, and pastoral agents (priests, nuns, and lay leaders) are being urged to compete with evangelical Protestants to prevent nominal Catholics from drifting into Pentecostal churches.

The long-term effects of these efforts at retrenchment and restoration are not yet clear; nor are their effects on the support that the Church has given to social justice and democratic politics over the last thirty years. Those who study the Church in Latin America differ in their assessments of this pullback and its impact on the Church's internal structure and its future role in Latin American society. According to the literature, the Latin American Church could be heading in any number of directions during the next several years.

One possibility would be a continued pullback from social and political involvement. Church conservatives are alarmed over the "independence" of the laity, lower clergy, and local communities and they attribute the tensions between laity and hierarchy to the politicization of local Church groups. Many are convinced that the inroads being made by evangelical Protestantism among popular-sector Catholics are partly due to the appeal of the otherworldly spirituality from which the Catholic Church has moved away (wrongly) since Vatican II. The impact of further retreat from social and political involvement would be significant. Occupied with "internal" affairs, and pursuing a primarily spiritual agenda, the Church would no longer serve as moderator and mediator among contending political factions, or as defender of constitutionalism against possible resurgent military intervention or violent revolution. By default, if not intent, it would revert to a position of defending the interests of dominant social and political elites.[6]

A second possibility would be an increasingly polarized Church moving toward de facto, if not formal, schism. This scenario envisions a new reformation in Latin America, with grassroots Catholics creating a Church of the poor whose agenda is informed by liberation theology. The priests, religious women, and laypeople who have taken up new, semiautonomous roles during the 1970s and 1980s will resist those who would have them abandon or reduce their social and political activities on behalf of the poor. Were this to occur, it could have a profound political impact. Instead of a united, but silent Church, there would be two separate churches, one a conventional spiritually oriented church ministering to all classes, and the other a popular church committed to intense social and political involvement on behalf of the poor. Such an outcome might occur under conditions of chronic poverty and the imposed austerity of neoliberal economic policies now being pursued by democratic governments. A radicalized local church might be more inclined to support a resurgence of revolutionary movements or populist forms of military rule than liberal democratic regimes that promise continued economic hardship.[7] In any event, local Church structures are not expected to disappear or to return to the spiritual agenda of preconciliar and pre-Medellín Catholicism. At least a part of the Church, these analysts seem to be saying, has been permanently captured by the poor.

A third and final possibility would be a period of internal adjustment, in which the Church placed renewed emphasis on its primary religious mission but without abandoning social issues. The Church would continue to affirm the connection between religious faith and justice but would insist that greater emphasis be given to spiritual concerns to counterbalance the heavy social and political activism during the 1970s and 1980s. Such emphases would include more attention to spiritual formation of clergy, religious, and laity, maintaining a preferential option for the poor without excluding other classes from the Church's mission, and insisting that all official representatives of the Church (i.e., bishops, priests, nuns, deacons, and advisors and animators of Christian communities) avoid partisan political involvement of any kind. Such moves would affect the social and political impact of the Latin American Church only marginally. They would not alter its generally progressive political impact of recent years but would moderate its pace and timing in line with conditions in each country. The Church would be less involved in politicking, but would continue to provide moral support for social equity and democratic procedures whenever (and by whomever) these values were threatened.[8]

We believe that an adequate assessment of the meaning and long-term impact of the Vatican's efforts at restoration requires a more comprehensive approach to religious and political change than is offered in any of these scenarios. In the first and second, overwhelming emphasis is placed on the Church's domination by class or social forces. In the third, the changes introduced by Church leaders are viewed as being independent of social forces and interests. None of them fully captures the state of the Latin American Church as it is today; nor do they explain how it has come to this point or where it is likely to go from here.

We believe that both societal and religious factors must be considered in analyzing the Catholic Church's political role. We think that to understand its evolution to date and its likely future direction, we must look at social forces, the changing emphases in the Church's sense of its mission, and the formative social experiences of Church groups (that new pastoral strategies have helped to generate) as parts of an interconnected whole.

The interplay among these societal and ecclesial factors can best be seen in the Church's response to social and political movements (such as secularization, social reform, Marxism, and authoritarianism) that have challenged its teachings and institutional prerogatives during the twentieth century. In the chapters that follow, we will be particularly concerned with identifying the ecclesial and societal factors that affect the Church's response to these challenges, its successes or failures, and their long-term effects on internal Church structures and dynamics.

We argue that the new strategies forged by the Church in the face of these modern challenges have made it a supporter of democratic processes. We also believe, however, that perennial religious concerns continue to shape the Church's stance and that not all of these (belief in a common good that takes precedence over individual interests and prerogatives, commitment to universal moral laws articulated by the Church and not subject to popular consensus, and so on) are compatible with the tenets and requirements of classical liberal democracy.

Scope and Method of This Book

The internal dynamics of national Catholic churches, their relationships with civil society, and the impact of their pastoral activities in recent years are complicated matters that are crucial to the future development of most Latin American countries. Chile and Peru are appropriate countries in which to analyze the possibilities in these regards for various reasons.

First, the Church in each country enjoys notoriety and prestige beyond its own boundaries. In particular, each has well-developed local organizations actively involved in social and political issues and supportive of democratic processes. Peru has been a center of liberation theology in Latin America, while the Chilean Church has been one of the most active in human-rights advocacy and defense.

Second, lay Catholics in the two countries have been active politically, playing important roles in leftist parties and movements of national significance. Both countries have faced serious political challenges in which these forces have played important parts. Peru is currently trying to rebuild its political system as it restructures economically and politically, and Chile is trying to reestablish civilian supremacy and greater socioeconomic equity after a generation of socially insensitive military rule. In each case, the Church has attempted to play a moderating role, endorsing and supporting democratic and socially responsible solutions to national problems.

Finally, the appointment of conservative bishops in recent years has undercut liberal dominance of the hierarchies of both countries and has led to their adoption of more cautious religious and political strategies. The two countries are thus ideal contexts in which to examine how rank-and-file and elite Catholics think about important religious and political matters, how Catholic activists are likely to respond to the leads of more conservative authorities, and how these matters might affect the Church's role as promoter and defender of democracy.

Our methods and procedures of analysis include extended interviews of elite and rank-and-file activists, attitudinal survey research, and examination of documentary materials. During 1987 and 1990, one of us interviewed more than sixty Catholic intellectuals and national and local leaders in Peru, and more than ninety in Chile. The interviews covered religious background, beliefs, and attitudes, general ideological and political questions, extent of political involvement, and attitudes toward the left. Interviewees were chosen at random from lists developed with the help of local informants. They included "organizational" Catholics who attended mass regularly and belonged to at least one Church-sponsored organization, "sacramental" Catholics whose only contact with the Church was with its ritual life, and "cultural" Catholics who retained values and sentiments from their Catholic upbringing, but who might or might not "believe," and were neither sacramentally nor organizationally involved in the Church. The interviews were taped and transcribed, and run from 45 to 120 minutes.

In addition, with the help of local research centers (the Bellarmine Center's CISOC and the Catholic University of Lima's social science faculty), Fleet developed and administered to 518 local-level Catholics in Santiago and 484 in Lima attitudinal surveys (of 147 and 63 variables respectively) that included unusually explicit religious and political material. The questionnaires enable us to distinguish among organizational, sacramental, and cultural Catholics whose religious and political attitudes we believe might vary. Based as they are on purposive samples, they do not permit us to characterize the Catholic or Catholic activist populations of either country, but they do generate important insights into the relationships between religious belief and political attitudes and involvement.

These materials provide the bases for our assessment of the Church's internal character, its relationship to civil society, and its role in mediating conflict and assisting in the development of democratic institutions and practices. Chapter 1 lays out a framework for looking at ecclesial and societal factors as they condition the Church's relationship to society. In it we identify major challenges to Church interests over the past two centuries, the Church's various responses, and the multiple, and occasionally conflicting, models of church on which these responses rest.

Chapters 2 and 3 provide an overview of the historical evolution of the Chilean and Peruvian churches during the twentieth century. In each case, we emphasize the pluralism of religious and political perspectives in both the formal Church and the broader Catholic community, and the diffuse, but increasingly progressive, impact of Catholicism on political life.

Chapters 4 and 5 deal with the Chilean Church's impact on the transition from military to civilian rule, and on the consolidation of democratic institutions and practices in the postmilitary period. Chapter 4 discusses the Chilean Church's role in helping to legitimate public criticism of military rule and to facilitate the development of a credible civilian alternative. Chapter 5 analyzes the influence of Church leaders and activists in the more demanding and conflictive challenges of the consolidation process—justice for victims of past human-rights abuses, elimination of remaining authoritarian vestiges in government, economic growth with greater equity for the poor, and continuing social and cultural democratization.

Chapters 6 and 7 cover the Church's impact on these processes in Peru. The Peruvian bishops never enjoyed the national prestige or influence of their Chilean counterparts, and as the years of military rule came to an end, their once dominant, moderately progressive leadership began to retreat. Under both military and civilian rule, their own ideological, ecclesial, and theological divisions prevented them from taking clear positions or providing moral or religious leadership around which the country's social and political forces might unite in support of democracy.

In Chapter 8, we return to the alternative scenarios laid out above, none of which seem likely to eventualize as described. Neither a primarily spiritual nor schismatic Church is probable, but accommodation between higher and lower Church levels will not be easy. Continued social and political involvement by the Church in the years ahead appears unavoidable, especially if the neoliberal economic policies being pursued by newly restored democracies prolong or intensify the difficult socioeconomic conditions in which most Latin Americans live. However, our data also show substantial differences between the bishops and rank-and-file Catholics on abortion, divorce, birth control, the role of women in the Church, and relations with Protestantism. Tensions are likely to persist and may intensify around these issues. The Catholic bishops may lose some of the moral authority they acquired during the years of authoritarian rule. At the same time, the newly constituted civilian governments of Latin America may be prevented from adopting more consensual policies in areas of personal and public morality, something that characterizes most liberal democratic regimes.

1

Church and Society in Theoretical Perspective

The Roman Catholic Church is a large, complex organization firmly rooted in its traditions. Its cautious adaptability has helped to make it the oldest continuous institution in Western culture. After 2,000 years of existence, and despite the powerful secularizing trends of the last 300 years, it remains a significant national institution in virtually every society (European, Latin American, and African) in which it is the principal Christian Church.

Its political influence has been decidedly conservative for most of its history. This was particularly true of the centuries immediately following the Reformation, during which it reaffirmed its hierarchical control and opposed the liberal currents of secular change that were to shape the emergence of the modern world. Since the late nineteenth century, and especially in the last fifty years, however, the Church has undergone significant internal change, expanding the scope of its mission to include the promotion of social justice and human rights in the secular world, decentralizing responsibilities for its various ministries, and affording individual members greater freedom of moral choice. In the process, its impact has become more progressive, particularly in countries (in Latin America, for example) experiencing chronic poverty and human rights abuse.

Changes in as large and complex an institution as the Roman Catholic Church are the result of forces affecting it from within and from without. Internal changes in this century have come in response to secular forces that have challenged the Church's legal prerogatives and its religious or moral credibility. These external threats helped to legitimate new religious emphases and styles of ministry that had arisen earlier but had not yet become normative, and in some cases were actually condemned by Church authorities.

In the end, these new religious ideas and strategies gained acceptance because of their potential for countering external threats to the Church's credibility and influence. They have helped the Church to recapture its capacity for influence on secular society. Their impact has been greater or lesser depending upon the social and political configurations of the national contexts in which the Church interacts with other social forces (attempting to promote or blunt change in these contexts). It has been greatest where secular ideologies, structures, and attitudes have been moving in the same direction and are susceptible to reinforcement by new religious and moral values. The Church's role has also been important when other social and political institutions are stalemated, enabling or obliging it to act as a surrogate political force.

Secular dynamics thus impinge on the Church's pursuit of religious goals, and these objectives change over time, as do the strategies designed to achieve them. As the Church changes, however, it must remain true to certain core or perennial concerns. Its new emphases and styles of ministry have to be justified religiously, i.e., they had to be shown to be consistent with the institution's traditional character and distinguishing characteristics. Religious ideas, structures, and strategies are thus influenced by societal dynamics but are not simply a reflection of them. Similarly, religious values are seldom the primary causes of change in society but can have important reinforcing or legitimating effects at certain moments, especially if aided by secular "carriers."

In this chapter we identify and discuss the interacting religious and secular forces that affected the Church's recent evolution and are likely to condition its future development. These include: (1) *traditional core features or concerns* that are central to its religious mission, providing it with flexibility yet limiting its adaptive capacities; (2) *the historical dominance of an institutional model* of Church and its political implications; (3) *secular challenges* that the Church has faced during the past century and its pastoral responses; (4) *new political roles* (moral tutor, social leaven, and surrogate social and political actor) that the Church has taken on in connection with these responses; and (5) *hypotheses* on how the Church can exercise these new political roles effectively without sacrificing its perennial core features.

Traditional Core Features of the Roman Church

Four organizational features have been central to Roman Catholicism since the fourth century. They are: (1) its hierarchical structure of authority, flowing from the pope, through the bishops, to priests, religious men and women, and finally lay men and women, at the local level; (2) the universal scope of its membership, allowing for uneven allegiances among its constituents; (3) the varying specificities and binding forces of its religious and moral teachings; and (4) its transnational character, with peripheral structures, personnel, finances, and teachings coordinated by a single center, the Vatican.[1] These core features have provided the Church with continuity and adaptability over time. They also limit its capacity for change and its impact on other social forces.

The Roman Catholic Church has been *a hierarchically structured institution* in which religious and moral authority has rested with a pope (the bishop of Rome) and is shared with bishops directly accountable to him. Over the centuries, Rome has defended such a structure as essential for the preservation of unity, doctrinal orthodoxy, and the structure of apostolic succession, on which the integrity of the Church's sacramental life is held to rest.

The chain of command from pope to bishop to priest can facilitate institutional and other changes once they have been embraced by hierarchical authorities. Similarly, key people at or near the top can generate dramatic changes in style and orientation throughout the organization, as occurred during the papacy of Pope John XXIII (1958–63). Local clergy and religious (nuns and monks), on the other hand, have only modest decision-making powers within the Church, although their daily fulfillment of sacramental, teaching, and administrative responsibilities gives them significant de facto authority and influence. Along with Church theologians, they are sources of new ideas and strategies that can rise up the chain of command and may be endorsed by formal Church authorities.

The Church's authority structures limit the amount of change to which its leadership will accede. Since the Reformation, its leaders have viewed challenges to episcopal and papal authority with alarm, and have frequently taken disciplinary action against the "offenders." Changes in the distribution of responsibilities across the chain of command do occur, and greater discretion can be given to local Church leaders from time to time. But any movement from below that threatens vertical authority will be seen by the pope and by most bishops as a threat to the Church's nature and mission, and therefore will be resisted whatever its social or political implications.

A second core feature of the Church has been the *universal scope of its membership*. In Troeltsch's terms, this membership has constituted a "church" rather than a "sect," an institution insisting that God's grace has been offered to all men and women, whatever their class, race, nationality, level of religious or moral development, or other measure of "worthiness."[2] No one is privileged or excluded from membership on the basis of social position, intensity of faith, level of external observ-ance, or other sectarian criteria. Saints and sinners alike are welcome, and excommunication is invoked only rarely, and for very grave sins, e.g., public apostasy or physical attacks on clerics.

Within the Catholic tradition, there are three levels of (legitimate) Church membership. There are *sacramental Catholics* whose faith is expressed in regular or occasional Mass attendance and reception of sacraments. There are *organizational Catholics* who, apart from their sacramental involvement, participate in Church-sponsored programs in spiritual formation, education, welfare services, and social communication. They are intensely exposed to the socialization process of the Church and represent its values and positions most consistently and integrally in their daily lives. Finally, there are *cultural Catholics* who, although baptized, rarely if at all attend Mass and do not participate in any Catholic organizations. They constitute the largest number of laypeople in most countries. They are formally part of the Church and espouse many of its moral values, even if they do not always live up to them in their personal or social lives.

Catholics thus come in all shapes and sizes. They expose the Church to varied perspectives on leading issues and problems, and they give it a potential for influence in virtually every sector of society. They also set limits on its adaptive capacities, and on its development of coherent and consistent moral or political positions. Catholic authorities long have resisted efforts to refashion the Church into an exclusive community; often they have abandoned pastoral programs, initiatives, and decisions that were likely to alienate sizable classes or groups. The Church, after all, must continue to minister effectively to all people.

Its tolerance of varying levels of membership commitment has diluted the Church's impact on the thinking and the activity of lay Catholics. Most nonpracticing Catholics in the "Catholic" countries of Europe and Latin America ignore the admonitions of Church leaders when these are not in accord with their personal moral or social interests. Church leaders may lament this, but they rarely attempt to impose regular practice or attentiveness on "followers." The Church's potential for secular influence thus may be great because of the extent of its membership, but its tolerance of uneven commitment limits the extent of loyalty and obedience it can reasonably expect in return.

Roman Catholicism's third distinguishing feature is the *varying specificities and forces of its religious and moral teachings.* Certain dogmas are specifically worded and considered binding under pain of sin. But these are relatively few in number and limited to theological, rather than moral, matters, e.g., the Trinity, the divinity of Jesus, his virgin birth, Mary's sinlessness from the moment of conception, her assumption into heaven, and the infallibility of the pope when speaking *ex cathedra.*[3] Other teachings are considered authoritative but not "infallible." They are to be taken seriously but unlike the above-mentioned articles of faith do not require unquestioning acceptance. This is because the Church, from early in its history, has used human reason in applying its moral teachings to specific situations, thereby making dissent possible.[4]

In fact, the Church has never claimed infallibility for any of its ethical teachings. When addressing issues of sexual morality (e.g., birth control and abortion), economic or business practices (e.g., the charging of interest on money loaned), and violence (e.g., the killing in war), Church authorities have defended fundamental values and principles but have acknowledged difficulties of application in specific circumstances. Indeed, the Church has modified its teaching in these areas over time precisely because changing contexts have made previous applications of general principles obsolete.[5]

In particular, Church leaders have tried to avoid taking positions on the adequacy of specific economic strategies and political policies. These are contingent matters in which Church leaders have no particular competence, and are best left to lay experts to decide. Accordingly, Church statements on most social questions with moral import are couched in terms general enough to be meaningful in different contexts. This generality and nonbinding force of its moral teachings have enabled the Church to be flexible in approaching particular situations. At the same time, however, they have also diminished its capacity for consistent impact. Its social principles can be interpreted and applied differently by different people in different contexts. For example, wealthy Catholics have often ignored Church social teachings about the obligation to pay workers a just wage, since the Church leaves it up to them to decide what precisely this obligation entails in each particular situation.

The greater the specificity of the Church's social pronouncements, the harder it is for such groups to excuse themselves from compliance.

Given the complexity of contemporary economic and political issues, however, most leaders appear to think that greater specificity will tie the Church to programs and policies to which many Catholics will have legitimate objections, and will adversely affect the Church's unity and moral authority.

The Church's fourth major core feature is its *transnational character.* Catholic dioceses, apostolic territories, parishes, religious congregations, schools, hospitals, and charitable institutions are to be found in every country in the world but are supervised by an administrative and policy-making center in Rome. This facilitates accommodation of distinctive national and cultural concerns within a broader unity of purpose and tradition.

The international character both enhances and limits the Church's capacity for change. The Vatican can do much to nudge a particular national church in new directions through appointment of progressive bishops or by directives for updating from the papal nuncio, the Vatican's official representative to each nation-state. The bishops' ability to move personnel and financial resources from more endowed to less endowed national churches has bolstered official Catholic presence in developing countries. These resources have enabled churches in poor countires to undertake a range of social and economic services in addition to strictly spiritual works.

But Catholicism's transnational character also limits its capacity for absorbing or accommodating change. The Vatican can and will intervene to slow down the pace of change in a particular national church if it believes, for example, that it is not adequately dealing with a threat to a core Catholic concern that change is precipitating. This intervention can take the form of disciplinary action against local clergy, investigations of, or reprimands to, theologians or seminary professors, and the replacement of liberal bishops (usually upon retirement) with those more in accord with Vatican concerns.

Over the centuries, Rome has acted both to stimulate change and to restrain reforms in periphery churches. The Church's transnational character provides the means to bring about change, and the Vatican acts as the guardian of the other three core features (hierarchical authority flows, variegated membership patterns, and differentiated moral and religious teachings). As final arbiter, the papacy has been a central factor in the Church's survival and development over the last two thousand years.

The Dominance of an Institutional Church and Its Political Implications

The Church has defended its core concerns over the years with considerable flexibility. Its ecclesiology, i.e., its understanding of itself as Church, has gone through several paradigmatic shifts, each of which has been justified as either a more faithful expression of its religious mission or a more effective means of fulfilling it.

Five different models of Church can be discerned in Scripture and apostolic practice. In the earliest years, Christians lived as a close-knit group (*community*) of friends who reaffirmed their faith by repeating stories of Jesus and celebrating rituals together. These first believers also preached throughout the Mediterranean, announcing (as *herald*) the beginning of God's kingdom in Jesus' ministry, death, and resurrection. Those who were wealthy served (*servant*) their less fortunate fellow Christians by sharing their resources with them. Following the deaths of the apostles, unity in doctrine and practice was preserved by establishing clerical offices and by defining rules and obligations for members, providing the Church with a juridical dimension (*as institution*). Finally, in each of these modes of existence, the Church sought to be a continuation of Jesus' own life, a sign (*sacrament*) mediating God's grace and pointing to his presence in the world.[6]

To varying degrees and at varying levels, these dimensions have continued to exist throughout Church history. Church as *community* has been more applicable at the local parish level or in smaller monastic groupings. Church as *servant* has been uppermost in the minds of those (clergy and laity) who have staffed the Church's charitable agencies over the years. The Church as gospel *herald* has been represented by the preaching clergy and by nuns and laypeople in Catholic schools. The *sacramental* Church has been sustained and projected through Catholicism's richly symbolic ritual life.

The early apostolic Church functioned in a decentralized, collegial style.[7] As it expanded in numbers and extent of territory covered, however, its institutional aspects grew in importance. Amidst growing persecution, and with the emergence (in the second century) of heresies regarding Jesus' nature, the bishops asserted their authority over the local church and looked to Rome as a centripetal force. Clerical office, central guidance from the successors of St. Peter, and the impositions of sanctions and penalties for major sins came to be used to preserve Church unity and discipline.

The tendency towards *institutionality* was further strengthened in the fourth century with the establishment of Christianity as the official religion of the Roman Empire by Theodosius. The Church expanded rapidly (all citizens of the Empire were obliged to become Christian), and was assigned charitable and educational responsibilities in some regions. Between the fifth and eighth centuries, the Church became a major preserver and transmitter of Western culture, and when the Empire's political power began to decline, Catholic bishops took over the administration of territories and other political functions as well.[8]

These commitments required greater administrative infrastructure and coordination. By the eleventh century the Church established a strong central bureaucracy (the curia) in Rome to assist the pope in overseeing Church affairs and to counter the efforts of secular princes to take back the prerogatives that the Church had been granted earlier. The reforms of Pope Gregory VII (which included a code of canon law) were introduced as a means of curbing abuses that had arisen with greater Church involvement in secular affairs.

Over time, these institutional interests and concerns came to dominate, even displace, the other dimensions. The Protestant Reformation was an attempt to restore a balance among the various styles or models, emphasizing Church as herald of the gospel, a community of believers, and servant in the world. To Roman authorities, however, the reformers threatened the Church's hierarchical structure, transnational linkages, and universal membership. At the Council of Trent (1545–1653), they embraced the *institutional* model even more fully, fashioning a heavily juridical Church determined to maintain its authority, its doctrines and rituals, and its remaining legal privileges.

From the fourth to the twentieth centuries, the dominant model of the Roman Church was thus institutional. The Church conceived of itself as a juridical and clerical entity. Its mission of leading all people to salvation was understood largely in doctrinal and ritualistic terms. The clergy were to teach and safeguard orthodoxy, and to dispense the sacraments as the principal means of obtaining grace and overcoming sin. The laity's role was passive acceptance of doctrinal and moral teachings and reception of the sacraments from bishops and priests. Universality of membership was formally maintained (the sacraments were offered to anyone who came), but most regularly practicing Catholics came from the upper class, which also provided most recruits for the clergy and the money to sustain the institution. Religious and moral teachings were differentiated, but doctrinal issues were considered more important than general ethical norms, and personal morality was stressed over and above social morality. Transnational linkages existed, but distances and difficulties in communication limited the Vatican's ability to control national Church policies.

Its limited capacity for reaching the laity and controlling national churches helps to explain the Church's interest in church-state unity since the 1500s. Lacking adequate resources of its own, the Church used political ties and sanctions to keep its people Catholic, orthodox, moral, and sacramentally serviced. Through treaties and concordats with separate states, the Church and its teachings achieved privileged status. Excommunication and other ecclesial penalties were imposed or reinforced by civil authorities. Civil registers and burial proceedings were controlled by priests to insure that citizens would frequent the sacraments (the primary means for religious salvation). Clerical salaries, the upkeep of Church buildings, and support for its educational and charitable activities were paid wholly or in part by governments.

Under these arrangements, priests, abbots, and bishops played major roles in public life. Clerics, most of whom were aristocrats, aligned themselves with upper-class groups to promote order, stability, and morality in public affairs. With the emergence of constitutional governments in the seventeenth and eighteenth centuries, they became involved in conservative political parties and urged Catholics to support them because of their defense of Church interests and privileges.

This support was not free. Under the *patronato* (first granted to the Spanish king, and later to presidents of individual countries), civil authorities had the right to recommend candidates for bishop and to reject those who might be troublesome for the government or the interests of its supporters. For its part, Rome's access to local churches was through national governments that would "help it" to select local Church leaders and to discipline rebels and nonconformists. Such arrangements militated against a publicly critical or prophetic role for the Church in terms of either government policies or the established social order.

For centuries, popes and bishops accepted these limitations. Preaching and administering the sacraments in all nations were viewed as more important than taking stands against social inequities. The good will of governments and the classes that controlled them guaranteed the Church a public presence and assured Catholic values privileged legal status. Finally, as no other group appeared capable of challenging the interests of either the dominant classes or the Church itself, this strategy was seen as the most, indeed the only, appropriate one.

Under the institutional model of Church, the main resources for Catholic political influence were clerical involvement in conservative political movements, defense of Church interests and privileges by conservative lay Catholic elites, and state policies subsidizing the Church and juridically enforcing its teachings. Reliance on such strategies made the Church a decidedly conservative, and sometimes reactionary, political institution from the early Middle Ages through the early twentieth century. For virtually the entire period, Church representatives accepted unquestioningly monarchical and other authoritarian forms of government, elite domination of policy-making, religious intolerance, and social inequities favoring the rich.

This conservatism came not from Catholicism as such but from its linkages with particular social forces.[9] Under the institutional model, the clergy's ties to the state and the aristocracy made it easier for bishops and priests to defend their interests and privileges. They also made it difficult for the Church to develop religious and social programs that served other classes. Vocations to the priesthood and religious life came from the wealthier families, as did candidates for appointment as bishops, while most Church operations (e.g., places of worship, schools, universities, monasteries, charitable institutions, etc.) were dependent on either government subsidies or the private contributions of the wealthy. In addition, the Church's public and private patrons had a determining, and not always subtle, influence on the way these organizations were run. Until the twentieth century, Catholic schools and universities educated the sons and daughters of socioeconomic, political, and military elites. The best clergy were assigned to well-to-do parishes. Charitable institutions, although assisting poor people, encouraged their docility and seldom opposed the injustices that caused their suffering. In these ways, in fact, the institutional model helped to make the Latin American Church an instrument or appendage of state and dominant-class elites.

There was little the Church could do to counter their influence. Its lay following was drawn disproportionately from these groups and participated passively in rituals directed by the clergy. There were few organizational activities beyond education and charitable services, and these were staffed by priests, monks and nuns. Most baptized Catholics (especially middle- and lower-class elements) attended mass and received the sacraments only rarely or occasionally, and had little contact with the Church.

The Church's Response to Four Modern Challenges

From its inception, the Catholic Church has been of two minds regarding the secular world and authority. It has viewed history as a vale of tears to be endured, with government controlling evil through the imposition of order. But it also has approached history as an opportunity to begin realizing the Kingdom of God and has seen government as a positive force to promote gospel values of peace, justice, and community. For most of its history, the Church has acted in line with the first set of views. As long as its relationship with socioeconomic and political elites was "working," i.e., safeguarding Catholic interests, Church authorities were comfortable with the first view of history and expected little in the way of improvement in the human condition. When these allies began losing ground to class and political adversaries, Church leaders began to look at their policies more critically and to view human history as holding new possibilities not only for the Church but for humankind as a whole.

Over the last one hundred years Catholic values and privileges have been challenged frontally by various tendencies and movements. Among them have been: (1) growing secularization and the attendant demand for separation of church and state; (2) the steady rise of middle- and working-class movements challenging the power of aristocratic and landed interests; (3) the appeal of Marxism among intellectuals, students, and workers; and (4) more frequent recourse to new forms of authoritarianism as a means of solving political and economic crises.

In dealing successively with these challenges, Catholic authorities began to devise strategies and forms of organization of a more progressive character. As they did, they started to rethink their mission and its relation to society and history. They were aided in this by their rediscovery of the early Christian Church, a Church quite different from the centralized, hierarchical, and almost exclusively ritual-oriented Church that most assumed had always existed. Advances in archaeology and ancient-language studies in the late nineteenth and early twentieth centuries brought about a new understanding of the scriptures and of early Church history. This, in turn, reintroduced thinking about the Church as communitarian, service-oriented, and focused on preaching the word as well as celebrating ritual. These features of the early Church became models for the development of new pastoral strategies by the mid-twentieth century.

Without repudiating the institutional conception of the Church (one emphasizing hierarchy, outward conformity, and legal status), Catholic leaders began to restore the communitarian, servant, herald, and sacramental dimensions of Church life within it. As they did, they validated, wittingly or not, many of the concerns of Protestant reformers centuries earlier. Church leaders in some countries began earlier and carried their reforms and initiatives further than those in others, but by the late 1970s, virtually all had come to embrace most of the following positions.

First, with religious pluralism an established fact even in predominantly Catholic countries, they endorsed the principle of religious freedom, and agreed to substantial, though not total, separation of church and state.[10] Initially, some bishops viewed these moves as unfortunate but necessary concessions. In the end, however, even ecclesial and theological conservatives embraced them as measures effectively freeing the Church from harmful external constraints.

Second, they assigned lay Catholics to evangelical work traditionally reserved for clerics (gospel preaching). They trained lay men and women in the Church's moral and social teachings, and then gave them educational and organizational responsibilities. This helped to compensate for the shortage of clerics and strengthened the religious understanding and commitment of many middle- and working-class Catholics. It also developed a Catholic presence among social forces (both professionals and workers) that had drifted away from the Church over the years.

Third, they embraced the idea of pastoral ministry in social justice. They came to see that working for social justice within history and society was part of the salvation process, and that gospel values must be brought to people's attention in concrete and socially relevant terms if they were to appreciate and respond to them positively.

Fourth, they came to view the defense of other universal values, including freedom, equality, and the rights of women, as part of Christian witness and ministry. These values were important to all men and women. Their partial realization on earth was part of God's plan (and a prefiguring of himself), requiring strong defense from the Church, particularly if other forces were unable to provide it.

The Church's appreciation of such modern democratic ideals was strengthened by the Second Vatican Council's (1962–65) reintroduction of the notion of Church as *sacrament.* In the Catholic tradition a sacrament is a ritual that mediates God's grace. It is a sign of intimate union with God and of the unity of all mankind and "an instrument for the achievement of such union and unity." It is an "outward sign" that points to a spiritual reality beyond itself, even as it embodies that reality. Jesus was the perfect sacrament since he pointed to divine power as the source of his works and yet was that power as well. By analogy, signs performed in Jesus' name (water in baptism, bread and wine in the Eucharist, oil in confirmation, ordination, and the last rites, etc.) brought his life into the person receiving them. But from a biblical perspective, humanity itself was sacramental as well; it pointed to God and was an image of God in human flesh. In this light, everything that was humanly good could be viewed as sacramental and as having religious value. The Church was a more perfect sacrament because it mediated God's grace and truth as fully revealed in Jesus. But "the triumphs of the human race are a sign of God's greatness" as well, and also should be celebrated as sacraments.[11]

In fact, according to the bishops, God's presence in the world was to be found, above all, in the human conscience, a person's "most secret core and sanctuary," and in the "authentic freedom . . . an exceptional sign of the divine image" that is also present in all people. People, they said, should be able to denounce "any kind of government that blocks civil or religious liberty" or "multiplies victims of political crimes," "to defend their own rights . . . against any abuse of . . . authority," and to choose their "government and the method of selecting leaders."[12]

These perspectives have helped the Church to play a progressive role in support of social change, human rights, and democracy in Latin America since the early 1960s. The Church's new pastoral emphases have helped it to counter challenges to its survival and relevance. While accepting church-state separation and political neutrality, it has developed new ways of evangelizing middle- and working-class Catholics, and has retained, probably even enhanced, its national presence and influence. The decentralization of its structures, greater use of the Bible, the modification of forms of prayer and worship, and greater lay involvement in its ministries, helped it to offset staffing problems resulting from the decline of priestly vocations in recent years, and may have helped to limit its loss of followers to the Protestant churches. Finally, its defense of the rights and interests of ordinary people has helped the Church to "win back" estranged followers and gain the respect of others, including leftist activists, to whom it has reached out pastorally.

New Political Roles for the Church

The Latin American Church is still a hierarchical institution. The clergy remain at the top, preaching, counseling and administering sacraments. Now, however, nonclerics (i.e., nuns and lay "organizationals") function in ritual and other capacities. The line of decision-making runs up and down, since the hierarchy depends on local churches for information, and since bishops and vicars frequently consult with them in making pastoral decisions.

Both the "sacramental" and "organizational" laity are larger and socially more diverse groups than before. In fact, middle- and lower-class Catholics make up the bulk of those who regularly attend mass and are involved in Church-sponsored social organizations. "Cultural" Catholics are also closer to the Church, since they benefit from many of the new social services provided by the Church.

The political connections of this multidimensional Church are also more complex than those of the Church as institution. Clerics no longer have formal ties to political parties or the state and come primarily from middle-class backgrounds, although the number coming from the working class has increased as well. The Church's more diverse following provides it with linkages to virtually all groups contending for power. Class is still an important factor in Catholic political choice, as reflected by the party preferences of the various lay groups. Parties now have more diverse social bases and are no longer the beneficiaries of clerical directives. The Church's ties are weakest with its former (military and the upper class) allies, some of whom are identified with movements of the nonconstitutional right.[13] The Church has greater access, on the other hand, to former adversaries on the left and in working-class and popular sectors. Liberation theology and the assistance that national and local churches have given to those persecuted by authoritarian governments have greatly enhanced Catholic credibility with constitutional and even some extraconstitutional groups on the left.

With more pastoral decisions being made collegially at national and local levels, the Latin American Church is no longer as tightly controlled by Rome. This is particularly true with respect to missionary and financial assistance to third-world churches whose authorities now deal directly with their European and North American counterparts. The Vatican and the aid-sending churches of the developed countries relate separately to different levels of the Latin American churches. Rome sends nuncios and appoints bishops, while foreign churches provide clergy, nuns, and material assistance to intermediate- and base-level organizations. Neither has to go through local secular authorities in their dealings with a particular Latin American church.

With its extrication from "entangling" alliances, and the social-Christian incubation of a new generation of clergy and laypeople, the Church could reenter the political arena with a markedly different profile. Three new roles are available, each likely to bolster democratic and change-oriented social forces: a *moral tutorial* role to be exercised by clerics committed to broadly progressive and democratic values; the *leavening* action of lay Catholics seeking, in relative autonomy, to apply the Church's social teachings in the political arena; and the role of a *social and political surrogate* that would act in place of other institutions in times of crisis.

In playing a moral tutorial role in Latin America, the clergy can influence political life in less direct and less partisan ways than in the past. Instead of directing the laity how to vote, they could proclaim broad principles (such as respect for the common good, basic human rights, equity, constitutionalism, and popular participation) that would serve as guidelines within which competition, bargaining, and consensus-building could take place. Its lay training programs, on the other hand, can provide the Church with a less direct but nonetheless significant leavening influence in party and governmental circles. With priests and bishops barred from urging or dictating specific options, lay people can provide a Catholic presence in all parties committed to democratic and socially responsible politics.

The reemergence of the herald, communitarian, servant, and sacramental elements has helped Latin American Catholicism to increase its organizational autonomy and "thickness."[14] Now, however, with the help of European and North American churches, it has created a layered network of national, regional, and local organizations that it had neither needed nor been allowed to develop previously. Most of these structures are concerned with broad religious and social matters, and many are staffed by laypeople. Their agencies, offices, and programs are open to all, demanding little in the way of religious orthodoxy or practice. The many people they serve are cultural Catholics closer to the Church now since they benefit from many of the new social services provided by the institution.

Its greater financial independence, its new organizational depth, and the porosity (in terms of membership) of its new local communities enable the Church to act as a surrogate political actor in times of crisis, when citizens are unable to speak or act for themselves. Bishops committed to social justice and human rights are more likely to criticize abuses of public power under a repressive, socially conservative military government, and to provide spaces in which citizens of varying ideological persuasion can gather to defend these values, now that (except in the area of education) they no longer depend on public resources for the fulfillment of their ministries.

Hypotheses on the Church's Exercise of These New Political Roles

The above discussion depicts ideal type linkages between Catholicism and politics in Latin America since the mid-1960s. The relationships portrayed are not manifest to the same degree in all countries. Variations in political context and differences in the development of national churches affect these interactions. Not all political systems over the past generation have afforded the Church the same opportunity for playing a progressive public role, nor have they responded in the same way. Nor have all churches met the "preconditions" for playing new and more progressive roles, namely, extrication from traditional bases of support and development of multiclass lay training programs with a strong social justice component, to the same extent.

Moreover, Vatican efforts to restrain individual Latin American churches during the past decade reflect a fear that their new roles may be jeopardizing one or more of the Church's perennial core concerns. At various times, the actions of priests and bishops as moral tutors, the strategies of Catholic laypeople in the partisan political arena, and the pressures experienced by Church authorities when performing surrogate social and political functions, have posed challenges to the

Church's structure of authority, the universal scope of its membership, the avoidance of partisanship in its social teaching, and its supranational character, especially subordination to Rome.

The Vatican appears to view the social and theological disintegration of some of the Latin American churches as a possibility. Accordingly, it has taken steps to restore traditional vertical structures of authority at international and national levels. If its fears are warranted, recent conservative episcopal appointments and warnings to progressive sectors of clergy and laity may succeed in moving the Church back towards the institutional model and to a concern for strictly ecclesial matters.

We offer a series of hypotheses, therefore, based on the dialectical framework of religion and politics presented in this chapter. These hypotheses will shape our treatment of the historical evolution of the Chilean and Peruvian churches in subsequent chapters as we attempt to assess how successfully the Church has come to play a new role in the politics of Latin America in recent decades without sacrificing its core religious concerns.

Hypotheses relating to extrication and incubation

1. How a national church extricates itself from traditional alliances significantly affects the character of its new values and pastoral orientations.

The assumption here is that more complete extrication will allow the Church greater autonomy, and that in poverty-ridden countries it will use this autonomy to embrace a social Christian approach to its lay training programs and general pastoral agenda. Conversely, with the survival of aspects of church-state union and partisan political ties (e.g., governmental subsidies of Church social programs, lingering ties to conservative parties, etc.), lay programs and movements are likely to remain under tight clerical control and to focus primarily on spiritual, as against social, concerns. This latter pattern can be seen in Colombia, where the Church has been slow to implement the pastoral reforms called for by Vatican II and by the Medellín and Puebla Conferences.[15]

Even when separation from the state and traditional landed interests and the military is complete, however, the conditions under which extrication takes place will have a determining impact on subsequent Church development. In both Mexico and Cuba, the Church's ties with traditional allies were broken, but Church leaders did not turn to social

Christianity. In both cases, extrication was imposed severely by secular forces that had displaced the Church's conservative allies. The lay programs that the Church went on to establish were primarily concerned with removing stringent restrictions that state authorities had imposed on its ministries.[16]

It is where extrication has been substantial and relatively conflict-free, therefore, that the Church is more likely to move in a social Christian direction, to define its pastoral agenda autonomously, apart from party and government influence, and to play an independent political role. We shall test this hypothesis by examining the nature and effects of the extrication processes in Chile and Peru.

2. The rate of incubation of social Catholicism within a national church will also depend on the extent of Vatican support, the level of support and/or tolerance of national church leaders, and the absence of serious threats to hierarchical authority at the time.

The appointment by Rome of social Christian bishops, and Vatican directives supportive of new forms and strategies of Church involvement in politics, will make it much more likely that a national church will develop lay training and involvement programs with a social justice thrust. Without a critical mass of younger, social Christian priests at the national level, however, the incubation of social Catholicism is likely to be slower. Finally, crises of authority in the Church will emerge during incubation if laypeople anxious to play a more autonomous and politically active role as Catholics clash with older-style bishops determined to continue speaking for the Church as a whole. The experiences of the Chilean and Peruvian churches during the 1930s and 1940s will be examined for indications of the interplay of these several factors.

Hypotheses relating to re-entry (as moral tutor, leaven for social justice, and/or social and political surrogate) into the political system

3. For the clergy to play an effective moral tutorial role, the language of their social pronouncements must be sufficiently concrete and contextualized to be relevant but without identifying the Church with partisan strategies or movements. Additionally, secular leaders must be sufficiently well disposed toward the Church to take its moral guidance seriously.

If the public statements on policy issues by bishops and priests remain highly general, failing to link moral values to the specific context at hand, their impact in the political arena will be less. Leaders whose interests are threatened by Church statements will find it easier to avoid public embarrassment if these pronouncements are vague and abstract. Others will find them inadequate as guidelines for policy or action in specific instances. Public reaction to the statements of Chilean and Peruvian bishops on social and political matters over the past half century will be examined for evidence in this regard.

If political statements by clerics are specific and pointed, on the other hand, they are more likely to command greater attention. Their relevance for political strategies or policies will be clearer, and their impact on the attitudes and actions of both elites and citizens greater. However, too much specificity in the Church's social teachings runs the risk of tying it to partisan movements and policies, thus jeopardizing its core interest of openness to all regardless of ideological or political persuasion. If this occurs, both Rome and the national bishops' conferences are likely to take restrictive action to preserve the integrity of the Church's universal mission, regardless of the sometimes chilling political implications involved. Episcopal responses to the ONIS movement in Peru and to the Christians for Socialism movement in Chile will be examined with these issues in mind, as will the Vatican's inquiries into liberation theology in Peru in the 1980s.

Even if the overspecification is avoided, however, the Church's moral tutorial role may not have a decisive impact under all circumstances. Where economic and social conditions are improving, and policy-makers may enjoy significant (albeit minority) political support, the Church's criticisms may have a limited impact on government policy. Church-state interactions in both Chile and Peru in the late 1970s and early 1980s will be examined for lessons in this respect.

4. The Church's capacity for acting as a leaven for moderate social change depends on the availability of secular carriers through which Catholic laypeople can act. Divisions within such carriers and in the wider society may limit the coherence of lay Catholic impact in politics.

Without reform movements in which progressive Catholic laypeople can act, the Church's leavening affect will be minimal. In political systems dominated by landed interests and the military, and challenged by strong leftist movements and/or parties, there is little or no room for social Catholicism. In such contexts, reform-minded Catholics are likely to either abstain from politics (Peru) or to become radicalized out of frustration (Central America).

The political impact of social Catholicism is likely to be greater, for example, where there are viable center or center-left reform parties, as in Chile from the mid-1960s through the early-1970s. Such parties may, but need not, espouse Christian values in their ideology or platform, but cannot be highly critical of the Church. If a country's reformist forces are politically weak, harbor strong anticlerical sentiments, and/or are inconsistent in their commitment to social justice (the case of Peru), on the other hand, they are not likely to attract social Christian Catholics, and Catholicism's political expression will be more diffuse.

The political viability of reformist parties depends on their ability to fashion coalitions among themselves and across social classes. The middle class in Latin America is too small (15 percent or less of the population in most countries) to constitute a sufficient electoral base for centrist parties, which must therefore attract working- and upper-class constituents as well. Such a variegated base could lead to divisions within and across such parties, however, diluting or splitting the political impact of lay Catholics within a single party (the Chilean Christian Democratic Party) or spread across multiple groups (the various MAPUs, and the Izquierda Cristiana)[17] as in Chile in the 1960s.

5. The post–Vatican II Church has some unique institutional capacities to act as a surrogate for other institutions in times of political or economic crisis. However, a prolonged surrogate role for the Church can also create serious internal tensions that can jeopardize the fulfillment of its primary religious mission.

The Church's new commitment to the promotion of justice, its greater organizational depth and role differentiation, and its increased access to international resources have given its national subunits unique capacities for action in times of crisis. A national church can act as a counterweight to political authoritarianism by offering a haven of refuge to enemies of the state, and it can rapidly expand its social activities when public services are cut back. The outreach and porosity of its local communities at such moments also give it opportunities to enhance its moral and religious credibility with groups previously alienated from it.

However, the more that the Church does for the persecuted in times of repression, and the more it attempts to mediate between opposing sides in times of polarization, the greater will be its internal tensions. The energies of priests and nuns turn more to secular than religious tasks, and coordination of expanding social commitments from above become more difficult. Such developments are likely to worry episcopal conservatives and may lead to tension and conflict with progressive priests and nuns at lower levels. Additionally, secular groups and individuals benefiting from these services may appear to be using the Church for their own partisan agendas, thereby alienating ideologically conservative lay Catholics, prompting government attacks on Church personnel, and alarming nervous Church officials concerned with preserving institutional unity and the Church's mission to serve all people.

With the suspension of constitutional guarantees and the imposition of restrictions on other major social and political institutions, the Catholic Church can act as a defender of civic traditions. And, yet, the challenges to its perennial or core concerns are also likely to be greater, and differences between ecclesiastical conservatives and progressives more pronounced. To determine the validity of this proposition, we will examine local-level Church resistance to military rule in Chile between 1973 and 1987 and in Peru in the late 1970s, and resistance to civilian government policies in Peru in the early 1980s.

Hypotheses concerning the Church during the period of redemocratization

6. If sacramental and organizational Catholics have sectarian views or are indifferent to hierarchical authority, conflict between a "popular" Church and the new generation of "Vatican" bishops will dominate Latin American Catholicism for the next generation. This, in turn, is likely to undermine the Church's newly won credibility among cultural Catholics.

The long period of military rule during which lower clergy, religious, and laity enjoyed considerable discretion and responsibility in carrying out new pastoral ministries may have instilled in them a radical ecclesiology more in tune with Protestant than with Roman Catholic traditions, i.e., equality between clergy and laity, democratic decision-making by local communities, and expanded sacramental roles for nonordained Catholics. If this is the case, the conflicting ecclesiologies are likely to produce serious internal tensions, particularly if the Vatican and other Church officials persist in their efforts to restore some aspects of the institutional model of Church. Such top-down efforts may frustrate local-level, radical Catholic activists not wanting to return to that style of Church organization. They may also produce negative reactions from popular-sector cultural Catholics who were direct beneficiaries of the servant, herald, and sacrament models of Church during the period of military rule.

7. If sacramental and organizational Catholics are radicalized as a result of exposure to, and collaboration with, cultural Catholics in times of severe repression, the Church's potential for exerting a moderating impact (as a leaven for social justice and democratic politics) will be short-circuited.

If, as the result of their social involvement, local-level Catholic activists develop radical ecclesiological and ideological views, they will be less of a moderating political force during the period of transition to democracy. They are more likely to identify with hard-line positions (demanding that the wealth accumulated by elites under military rule be redistributed immediately, for example), and they might condone the suspension of civil liberties and due process (and other nonconstitutional actions) to achieve such a goal. In other words, their attachment to liberal democracy will be weaker than their interest in other objectives. And the Church's support for a negotiated settlement with regime officials will be greatly diluted.

The next two chapters offer historical analysis of Chilean and Peruvian Catholicism from the late nineteenth century to the early 1980s. They examine the extent to which the two churches have extricated themselves from traditional alliances, their development of new lay formation and training programs, and the strategies they have used in order to exert a more progressive influence on national politics. In these chapters, we shall test our first five hypotheses, and assess the potential of each church for promoting progressive social change through constitutional means.

In chapters 4 through 7, we look at more recent developments in the two countries, in particular the roles that each church plays in its country's transition from military to civilian rule, and in their subsequent efforts to consolidate democratic institutions and practices. In chapters 5 and 7, we test the validity of hypotheses 6 and 7 regarding the responses of local- and intermediate-level Catholics to the restraints placed on them by the Vatican and the impact of these restraints on the process of reestablishing and consolidating democratic governments.

2

. . . .

The Chilean Church: A Historical Overview

In institutional terms, the Chilean Church is one of Latin America's strongest. It has one of the lowest Catholics-per-priest ratio (4,989) in the region, and a relatively large percentage (68.3 percent) of its priests are Chilean citizens. The vast majority (88 percent) of its parishes have a full-time priest, the number of nuns has grown by 56 percent since 1986, and its hospitals, schools, and other institutions provide social services to a significant portion of the national population.[1] Moreover, the Catholic Church today is one of Chile's most trusted and admired national institutions. For more than thirty years, its bishops have spoken forcefully and influentially on leading social issues, despite their varying ecclesiological and ideological persuasions. At the same time, they have maintained good relations and open lines of communications with their priests, sisters, and local parishes and communities.

And yet, for all the strength of its Church, Chile is among the "least Catholic" of Latin American countries. On the one hand, Chileans are less likely than other Latin Americans to consider themselves Catholics, to attend mass regularly, to "marry in the Church," to have their children baptized, and to participate in religious festivals and processions.[2] On the other hand, avowedly secular parties and movements have made substantial inroads in Chile, and by 1990 Evangelical Protestants accounted for more than 15 percent of its population.

Ironically, the strong secular dimensions of Chilean society and the challenges they have posed over the years have prompted the Chilean Church to strengthen its internal structures earlier and more fully than other Latin American churches. They have also afforded it unique opportunities for exercising the roles of moral tutor, leaven for social justice and democracy, and surrogate social and political actor in times of crisis.

In this chapter, we analyze the Chilean Church's evolution from the mid-nineteenth century through 1982. We do so from the perspective of the hypotheses articulated at the end of the preceding chapter, highlighting the societal factors that have prompted the development of new strategies and roles. We identify four distinct phases or periods of internal Church evolution and political intervention: (1) a period of *extrication* from the state, partial distancing from the Conservative Party, and incubation of progressive Catholic social tendencies (1860–1935); (2) a period of *uneasy coexistence* in the Church between social Catholicism and traditional conservative tendencies (1935–58); (3) an era of rapid development of new pastoral mechanisms in which the Church began to play an active role as *moral tutor and institutional support for social reforms* (1958–73); and (4) a period in which the Church acted as *surrogate opponent and provider of social services* during severe political repression (1973–82).

Extrication from Traditional Allies and Incubation of New Social Tendencies (1860–1935)

The Chilean Church began this period as an institutional Church dependent on state concessions and subsidies, aligned with landed elites whose wealth and power it defended, and the object of indifference or resentment for many middle- and lower-class Chileans. In an effort to escape these circumstances, it severed its ties to the state, took steps to abandon its alliance with the Conservative Party, and took up the "social question," hoping to win back workers, peasants, and others who had left or were leaving its ranks. Among the factors generating and facilitating these developments were: the rise of anticlerical and other secular forces, enlightened leadership from key Church leaders, the emergence of a new generation of priests and lay Catholic activists concerned with social issues, and strong Vatican encouragement for progressive and conciliatory policies at key junctures.

Mounting attack on Church privileges (1860–1920)

Although most Church leaders had backed the Spanish crown, church-state relations were generally harmonious during the decades following independence (1821). The Portalian constitution of 1833 did not formally designate Catholicism the country's official religion, but it affirmed the union of church and state, refused to recognize the civil rights of other religions, and reaffirmed the *patronato*, whereby the government approved the appointment of new bishops and other administrative decisions in exchange for public subsidies. With the arrival of foreigners (especially the British) involved in mining and commerce in the mid-nineteenth century, the Church's monopoly on religion began to erode. Laws were modified in 1844 allowing non-Catholics to marry legally, and in 1865 all denominations were granted the right to worship publicly and establish religious schools.[3]

Church-state relations deteriorated in the 1860s, when Church leaders became defensive and sectarian in response to resurgent anticlericalism. They formed an alliance with the Conservative Party, whose principal bases of support were the traditional landholding families. As mining, commerce, and industry expanded during the late nineteenth and early twentieth centuries, however, a Liberal Alliance of the Radical, Democratic, and Liberal parties arose to challenge the Conservatives, and with them the Church. The industrialists, bankers, professionals, and white- and blue-collar workers who supported these parties favored social reform and separation of church and state. During the 1870s and 1880s, with the Liberals and Radicals dominating the legislature, laws were passed requiring clerics to stand trial in civil courts, making civil marriage mandatory, removing cemeteries from exclusive Church control, and placing civil records in the hands of the state.[4] Church leaders resisted, but failed to prevent, these erosions of privilege. Their Conservative allies had lost their political clout.

Not all Church officials were hostile to the Liberal Alliance, however. Some, like Archbishops Valdivieso (1845–78) and Casanova (1878–1908) of Santiago, thought that the Church's ties to Conservatism were hurting its cause, and sought to make peace with Liberal elements. But most Church leaders continued to identify with Conservatism. During the late nineteenth century, several bishops were official party members, a number of priests held Conservative seats in the legislature, and Church funds were used to support Conservative candidates in elections.

The continuing secularization of society, the growing political power of liberal forces, and emerging threats stemming from the intermeshing of religious and political issues, set the stage for efforts at formal extrication. Actual separation of church from state, which came in 1925, would not have been achieved as amicably, however, had it not been for the leadership of Archbishop Crescente Errázuriz, the political skills of President Arturo Alessandri, and intervention by Rome.

Separation of church and state (1920–25)

In the years following World War I, moderates assumed leadership positions in both church and state in Chile. They seemed to understand and empathize with one another. Even so, an amicable church-state separation would probably not have been possible without Vatican diplomacy.

In 1918, Pope Benedict XV appointed Crescente Errázuriz as archbishop of Santiago. The nephew of former Archbishop Valdivieso and a friend of the Liberal Party leader, Arturo Alessandri, Errázuriz was an aristocrat of liberal sensibilities. He was appalled by his country's poverty, and thought that the Church should be more concerned with people's material well-being. He encouraged the dissemination of the "social" encyclicals of Pope Leo XIII, and promoted educational and social action projects directed at workers and their families.

Errázuriz was personally in favor of continued church-state union, but wanted to avoid a potentially bruising confrontation with anticlerical forces. To do so he was willing to accept formal separation. His friend Alessandri had been elected president in 1920 on a platform that called for disestablishment. Alessandri resigned in 1924, amidst a conflict of powers with the Congress, but agreed to resume his presidency the following year *if* a new constitution reestablished a presidential (as opposed to a parliamentary) system and provided for full religious freedom.

While in "exile," Alessandri met with Vatican secretary of state Pietro Gasparri. Gasparri told Alessandri that the Church would accept separation if certain conditions were met: continued public legal standing for the Chilean Church, indemnification for its confiscated properties, abrogation of the *patronato*, continued religious instruction in public schools, and no constitutional recognition of atheism. Alessandri continued to consult with both the Vatican and Errázuriz in the ensuing months. The constitution's final version met Rome's demands and was forwarded to Gasparri, who quickly cabled his approval to Alessandri and to the papal nuncio in Santiago.[5]

Once the Vatican's position became known, no cleric (bishop or priest) dared to oppose it, and the new constitution was approved overwhelmingly in the September 1925 plebiscite. At the time, separation was something that most Chilean churchmen were willing to accept as the lesser of potential evils. In ensuing years, however, it brought important benefits for the Church's structural development. With the *patronato* set aside, for example, Rome acquired a freer hand in Chilean Church affairs. Eight new dioceses were created between 1925 and 1929, doubling the number of bishops and strengthening the Church's presence in the southern and northern provinces, something that the government had previously prevented it from doing.

Lingering ties with the Conservative Party (1925–35)

The other aspect of extrication—termination of the Church's alliance with the Conservative Party—proved more troublesome. Errázuriz was firmly opposed to clerical involvement in partisan politics, but was in a minority on the issue.[6] Most of the bishops named after 1925 were ideological conservatives and a majority of the country's priests continued to sympathize with the Conservative Party as well.

The controversy persisted for several years. Archbishop Errázuriz died in 1931, and was replaced by José Horacio Campillo (1931–39), a man with strong Conservative sympathies. Deteriorating socioeconomic conditions and the emergence of vigorous Socialist, Communist, and National Socialist parties pushed others into the Conservative camp as a bastion from which to resist renewed anticlericalism.[7] When "noninterventionists" (those opposed to the Church's favoring one party over another) protested, the matter was referred to Rome.

In July 1934, then secretary of state (and later Pope Pius XII) Eugenio Pacelli issued a letter appearing to side with the noninterventionists. Pacelli insisted that no party could claim to represent the Church politically and that Catholics could associate themselves with any party "as long as they (it) offered sufficient assurances to the rights of the Church and its people." He also encouraged the development of Catholic Action programs in Chile, but demanded that they avoid involvement "in the struggles of political parties, even if these be formed by Catholics." Finally, he warned priests and bishops to stay free of partisan politics. They should be helping to develop general criteria regarding political matters, not advocating contingent policies and strategies. But having said this, he left to local bishops the choice of "the most appropriate manner in which to form people's consciences" and further conceded that defense of the Church's rights and of important moral principles could require the union of all Catholics.[8]

These terms enabled pro-Conservative bishops to persist in their partisan ways. They could argue that the proper way to "form consciences" and defend the rights of the Church in Chile was to require Catholic support of a single party. To close this loophole, "noninterventionists" persuaded their colleagues to endorse a pastoral letter explicitly acknowledging the right of Chilean Catholics to belong to different political parties.[9] But individual bishops, priests, and lay Catholics continued to interpret and apply this "policy" in line with their respective sympathies and convictions. A full break with the Conservative Party would not come until the late 1950s, by which time its political standing had diminished greatly and Rome had begun to replace pro-Conservative bishops with prelates of more liberal ideological and ecclesiological views.

The early development of social Christianity (1891–1935)

The separation of church from state in Chile, and the uneasiness caused by the close association of the Church with Conservatism, were both facilitated by the early development of social Christian tendencies within Chilean Catholicism.

The "social question" arose in Chile in the late nineteenth century. Archbishops Valdivieso, Casanova, and Errázuriz worried openly about workers and their families, their low wages and living standards, and their estrangement from a Church that appeared to accept existing conditions, structures, and relationships. In fact, these leaders opposed identification with the Conservative Party in part because it was undermining pastoral efforts among lower-class groups.

Among the first church leaders to begin assisting workers and the unemployed in Chile was the Jesuit priest Fernando Vives. He had come into contact with social Catholicism while studying in Spain and Belgium, and set about organizing Catholic workers and university students when he returned to Chile in 1910. He designed small, parish-based groups of workers who were given catechetical instruction and introduced to Catholic social teaching.[10] Participants would read from the gospel, comment on a particular passage, and then draw conclusions and apply them to the reality they were living. Vives also organized Catholic university students and young professionals in study groups designed to deepen their faith, prepare them to deal with social problems, and carry out projects involving workers and social organizations.[11]

In later years, Vives was joined by other progressive priests. Fellow Jesuit Jorge Fernández, took over Vive's *círculos de estudios* in 1918, developed the well-known Monday gatherings, and later worked with Catholic University students. Father Oscar Larson, a diocesan priest whom Archbishop Errázuriz had sent to Belgium in 1926 to study psychology and Catholic youth movements, returned two years later to become advisor to the National Association of Catholic Students (ANEC), which became very active and influential over the next several years.[12] Confronting (in the late 1920s) an economic depression, a military dictatorship that discouraged partisan political activity, and parties (like the Conservative) that persisted in defending the status quo, they found the priests and their social Christian ideas appealing.

The organizations to which they belonged became the basis of Chilean Catholic Action, which was established in October 1931 to coordinate existing programs of spiritual formation and social assistance and to ensure that socially committed Catholics acted in harmony with Church authorities and policies. Catholic Action was encouraged by the Vatican but appealed particularly to Chilean Church authorities, caught as they were between leftist critics and self-serving Conservative patrons. It offered the Chilean Church new ways to deepen the religious understanding and commitment of Catholics who were still faithful, while attracting workers and other popular elements that had drifted away over the years. And unlike its Conservative Party allies, it provided the Church with activists willing to defend its interests without extracting concessions or alienating the followers of other parties.

Catholic Action units, set up under clerical supervision and control at parish, diocesan, and national levels, were most successful among the young. In its first months of existence the organization enrolled roughly 30,000 members, and by 1936 more than 47,000 young people had joined.[13] Most of the priests and lay Catholics involved either held or soon embraced social Christian values and attitudes. Its activists were the critical mass on which Chile's social Christian movement would be based. During the 1940s and 1950s, social Christians pushed the Church toward greater social involvement and led younger Catholics away from the Conservative Party. Until they became the dominant tendency in the 1950s, however, social Christians coexisted uneasily with traditional Catholics, whose notions of the Church and its proper response to social conditions and processes were quite different.

Uneasy Coexistence of Older and Newer Forms of Catholicism (1935–58)

During the next quarter century, these divergent tendencies were sources of ongoing tension. Many bishops continued to pursue pastoral strategies that appealed to traditional Catholic spirituality and personal morality, even as they encouraged Catholic Action programs and activities. A progressive social Christian party—the Falange Nacional—emerged from the ranks of the Conservative Party, winning the sympathy of a few bishops, a larger number of priests, and a substantial number of Catholic Action activists attracted by its commitment to social justice. Most bishops remained personally loyal to the Conservative Party, however.

During these years, the bishops began to speak more forcefully and cohesively on social justice topics. The general terms of their pronouncements enabled them to do this but also allowed some Catholics to ignore their potentially disagreeable implications. Moreover, most bishops spent more time condemning communism and warning against joint ventures with the left than they did promoting reforms that would benefit the poor. It was only in the late 1950s, with the appointment of additional social Christian bishops and the emergence of a dynamic political party identified with Catholic social teachings, that the seed work of Catholic Action began to bear either ecclesial or political fruit.

Contending pastoral strategies during the 1940s

The Church's pastoral initiatives during this period included both traditional and social Christian programs. They sought to correct deficiencies in Catholic formation and practice. Studies conducted in the late 1930s and early 1940s described appallingly low levels of practice, inadequate numbers of priests and sisters, little or no contact between the Church and "nonsacramental" Catholics, little attachment or loyalty to the Church in the population at large, and high levels of alcoholism, infidelity, and personal corruption among people over whom the Church ought to be exerting influence.[14]

Different approaches were taken. At the mass level, Eucharistic Congresses, ritual celebrations, processions, and other public events were designed to strengthen identification and solidarity with the

Catholic tradition. Additionally, catechisms and religious education programs were redesigned, and programs preparing people to receive the Eucharist and other sacraments were revamped, with an eye to strengthening identification with Catholicism.

Catholic Action activities were the Church's principal instruments of social impact, however. They were strongly Catholic in both makeup and appearance. They were generally, though not always, strongly anti-communist but vocal in their calls for social change and reform. They helped to offset the Church's image as an ally of economic and social elites. Many of the activists involved went on to join political movements, and some of their clerical advisors later became bishops, thereby extending the group's influence within the Church.

Catholic Action enjoyed the support of both progressive and conservative Catholics. Social Christians were happy with the movement's social engagement and with its progressive positions on the issues. These aspects were more troubling to conservatives, although they were hesitant to oppose something backed by the Vatican, and they liked the movement's strong Catholic identity and its hierarchical structure. During the 1940s and early 1950s, however, the Church's conservative image and impact were reinforced by ambiguous episcopal statements on social issues and by recurring conflicts between Church authorities and a social Christian political party known as the Falange Nacional.

Diverse impact of Episcopal statements (1938–49)

Church leaders helped to avert a political crisis following the election of a communist-supported Popular Front government in 1938. Msgr. José Maria Caro, bishop of La Serena and soon to be named archbishop of Santiago, issued a letter that helped to legitimate the new government in the face of rumored coups and cabals. He reminded Catholics of their obligation to obey all duly elected governments and promised the Church's cooperation in promoting the common good. At about the same time, the bishops began to speak collegially on social issues through the Comisión Episcopal Permanente para la Acción Católica. Following the 1938 election, they had unanimously reaffirmed the freedom of Catholics to join any party that respected Church values and interests.[15] During the 1940s, they defended the right of workers to form unions, receive just salaries, and live in adequate housing, and they stressed the obligation of all Catholics to embrace and apply the social teachings of Popes Leo XIII, Pius XI, and Pius XII. Rarely did they offer specific policy recommendations to achieve these goals, however, nor were they critical of prevailing economic structures. They rather attributed social problems to personal moral failings, e.g., the sumptuous life styles of some employers, and called for the conversion of individual hearts as the most effective way of overcoming them.[16]

Such statements were too general to have much of an impact. Moreover, during the mid and late 1940s, they were overshadowed by the Church's more prominent denunciations of communism. Gabriel González Videla, a Radical who had been elected president with Communist Party support in 1946, dismissed his Communist ministers after only five months in office, and later (1948) pushed through a Law of Defense of Democracy that made their party illegal. The Chilean bishops joined the Conservative and Liberal parties in supporting the move. In doing so, they reverted to traditional partisanship, after several years of advocating constitutionalism and ideological toleration. In the context of the emerging Cold War and the Vatican's clear position on Marxism, the Chilean bishops apparently felt compelled to defend anticommunist initiatives unreservedly. Their tilt toward the political right was given added impetus by their conflict with the Falange.

Episcopal conflicts with the Falange Nacional (1947–50)

The Falange Nacional was formed in 1938 by members of the youth branch of the Conservative Party, many of whom were Catholic Action militants who wanted to apply the social encyclicals in Chile's social and political arenas. They had been influenced by Father Oscar Larson, and later by his charismatic successor, the Jesuit sociologist Alberto Hurtado.[17] Hurtado had studied at Louvain University in Belgium, where he wrote on the Chilean Church and came into contact with social Catholicism. He was highly critical of the Conservative Party's opposition to social reform and readily shared his views with others. Many of the young people he guided during the 1940s joined the Falange, leading some to view Catholic Action as a *falangista* training ground. Hurtado quickly became a source of constant anxiety for Conservative Party leaders, and for Church authorities like Caro's auxiliary bishop, Augusto Salinas, who openly sympathized with them. In 1944, with the help of other ecclesial conservatives, Salinas persuaded Caro to remove Hurtado as Catholic Action advisor, and to reassign him to other duties.

The motives and concerns producing Hurtado's ouster resurfaced in 1947. In a letter published in the Conservative newspaper *El Diario Ilustrado*, Bishop Salinas reprimanded the Falange for criticizing the Franco government of Spain and for calling for diplomatic relations with the Soviet Union. He also admonished the party for encouraging Catholics to work within the Marxist-dominated Chilean labor movement and for its electoral alliances with communists and other leftists.[18] Finally, he charged Falange leaders with "corrupting" Catholic Action militants by turning them against the Catholic bishops.

The two organizations were not easy to distinguish. By the late 1940s, the leadership and most Catholic Action activists were falangista sympathizers if not actual party members. In fact, the Falange was a partisan political voice of social Christianity, aspiring to a "third road" alternative to liberal capitalism and Marxist socialism, to the reconciliation of individual and social interests, and to the defense of both freedom and social justice. Their common language and positions caused many people to associate, if not confuse, the two groups.

Falange leaders angrily protested Salinas's harsh words and judgments. Recognizing the Church's authority in "those matters which, according to its own judgment, fall within its (religious) mission," they argued that in questions of contingent politics lay Catholics were free to appraise the facts as they saw fit and were not bound by the directives of Church authorities.

When the dispute became public, and because it involved the judgment of a colleague in a matter believed essential, other bishops came to Salinas's defense, denouncing the Falange as well. The Episcopal Commission for Catholic Action condemned what it termed an offense against "the authority of a member of the hierarchy," and Cardinal Caro endorsed Salinas's views and chided the Falange for its "lack of proper respect" for his auxiliary bishop.

At this point, the party's leaders beat a tactical retreat, offering to dissolve their organization if the hierarchy wished them to do so. They were saved from this fate by the intervention of social Christian Bishops Larraín (of Talca) and Berrios (of San Felipe). They thought that Salinas had overstepped his bounds, but in their public remarks they simply argued that the offense, if any, had been given by individual activists, who should be more attentive to Church authorities, but that there was no need for action against the party itself. Cardinal Caro apparently concurred, and publicly confirmed that the bishops had not asked for the Falange's dissolution.[19]

The dispute diluted the impact of the Church's support for social justice and constitutionalism at the time. In attacking the reform-minded Falange, the official Church gave comfort and support to right-wing parties that opposed reform and pluralism. In 1948, when the Falange opposed the Law of Defense of Democracy and defended the Communist Party's right to compete in elections, conservative politicians and bishops tried to have it removed from the list of parties with which Catholics could affiliate. At this point, however, the Vatican intervened. In a letter sent to the Chilean bishops in 1950, Bishop Dominico Tardini, Vatican Secretary of Extraordinary Affairs, reiterated the position taken by the Pacelli letter of 1934, i.e., that Catholics were free to support any candidates who respected "religion and the doctrine and rights of the Church." The letter also called for unity among Catholics and strong commitment by all to implement the social teachings of the popes.[20] It mentioned no party by name and did little to weaken the Chilean Church's pro-Conservative image at the time. But it implicitly cautioned against either condemning the Falange or losing sight of the need for social reform because of concern for communism. It had a calming effect on the Chilean Church, and left the door open to future collaboration between social Christian Catholics and the Falange.[21]

Social Christian growth and development (1950–58)

During the 1950s the Church's impact on Chilean society grew more progressive. Among the contributing factors were Catholic Action's development of functionally specific organizations, the increased number of local and foreign-born priests serving in working-class parishes, Father Hurtado's death, the appointment of additional social Christian bishops, and the emergence of the Christian Democratic Party as a carrier for social Christian values and perspectives.

Membership of Catholic Action organizations reached an estimated 57,000 in Santiago and 100,000 nationwide during the late 1940s and the 1950s. Increasingly, Catholic Action activists ventured out from their parish bases to factories, farms, secondary schools, universities, and professional circles, where they mingled with the people they wished to influence (secular and lapsed-Catholic activists). Their efforts to adapt their appeals to these environments afforded the Church a more effective presence in areas where its influence had been weak.[22]

During this period, the Church also reached more unchurched workers, urban slum-dwellers, and peasants. The arrival of an estimated 550 foreign missionary priests and nuns between 1954 and 1958 bolstered its presence among popular-sector Chileans. The foreign priests increased the number of clerics working in Santiago by almost 70 percent (from 527 to 891), and lowered the number of Catholics per priest for the country as a whole from 3,101 to 2,666 between 1954 and 1958.[23] Following the example of these foreigners, more Chilean priests volunteered to work among the poor, and priestly training began to incorporate social Christian values. Former Catholic Action chaplains influenced seminary curricula, and some candidates for the priesthood were required to do part-time service in working-class parishes.[24]

Father Hurtado's death, from cancer, in 1952 gave special impetus to the social causes with which he was identified. His work had made him a legend in Chile during his lifetime. Following his dismissal from Catholic Action, he founded the Hogar de Cristo, a production cooperative and refuge for homeless children and adults, and the Asociación Sindical de Chile (ASICH), a social Christian labor federation. His death made it easier for some bishops to embrace his ideas and follow his lead. It also precipitated an increase of vocations to the priesthood among the young men with whom he had worked.

Also affecting the Church were the appointments of new bishops during Msgr. Sebastiano Baggio's tenure (1953–59) as papal nuncio. Baggio was a close friend of Pope Pius XII and an advocate of countering communist influence through the development of Christian Democratic policies and organizations. When he arrived in Chile in 1953, five of the twenty-one Chilean bishops were considered conservatives, two were social Christians, and the remaining fourteen were not identified with either tendency. Between 1955 and 1959, ten episcopal appointments were made on Baggio's recommendations. Of them, six were social Christians, two were conservatives, and two were neutral. By 1959, when Baggio was replaced, social Christian bishops outnumbered the conservatives 7 to 5, with the remaining bishops considered neutral.[25]

Developments within Chilean politics also strengthened the appeal of social Christian ideas. The decline of the Radical Party, and the failure of the maverick Ibáñez government (1952–58) to deliver needed structural reforms in agriculture, industry, and tax laws, created a vacuum in the center of the political spectrum, which falangista Eduardo Frei and the Christian Democratic Party rapidly filled.

Frei had been elected to the Senate in 1949, and emerged as a national political figure during the mid-1950s. He was reelected in 1957 from Santiago, the country's largest senatorial district, and by early 1958 he was a leading candidate in that year's presidential election. His popularity was the foundation on which the new Christian Democratic Party was built. In early 1957, the Falange elected fourteen candidates to the Chamber of Deputies and two to the Senate, its best showing ever. In July of the same year, it joined with social Christian Conservatives and other socially minded Catholics to create the Christian Democratic Party, whose leading figure and political resource was Frei.[26]

In the September 1958 presidential election, Frei won 20 percent of the vote, finishing behind rightist Jorge Alessandri (31.2 percent) and Marxist Salvador Allende (28.5 percent), but ahead of the Radical, Luis Bossay (15 percent). Much of Frei's success was due to his appeal to progressive Catholic voters, whose numbers had grown in recent years. Practicing Catholics still tended to support the right, but those supporting Frei were decidedly more progressive. Seventy-eight percent of them favored the legalization of the Communist Party, compared to 62 percent of those backing Alessandri. Seventy-two percent opted for higher salaries (as against additional investment) as a means of reducing poverty, whereas only 66 percent of those backing Alessandri did so.[27]

At this point, therefore, it hardly could be said that the Church was keeping Catholics conservative. Its impact on organizational Catholics involved in Catholic Action and other social programs was moderately progressive, and by the late 1950s these Catholics had found a political vehicle or carrier in the Christian Democratic Party. The Church had not had much influence on Catholics with whom it had only sacramental contact. But new evangelization efforts, together with changes in the political climate in the early 1960s, would help to move some conservative Catholics to the political center by 1964.

Moral Tutor and Institutional Support for Reform (1958–73)

The period from 1958 to 1973 was one of revitalization for the Chilean Church.[28] Under the leadership of social Christian elements, and with help from European and North American Catholic Churches, it updated its pastoral programs and strategies. As moral tutor and provider of social services, it was deeply involved in urging and shaping social change. Each of these capacities brought the Church enhanced vitality and visibility.

They also politicized its mission, confronting it with new challenges. During the early and mid-1960s, the Church (including many bishops and priests) tied itself closely to the Christian Democratic Party. When Frei was elected president in 1964, many of the Church's lay activists assumed positions with his government. When his Revolution in Liberty stalled in 1967, the Church was caught in the same wrenching debate over the government's performance that plagued the PDC. The resulting divisions and tensions persisted for the remainder of Frei's term, and under the subsequent Allende government as well, making it more difficult for the Church to resist or moderate the polarization of the country as a whole.

These years can be divided into three subperiods: one of pastoral renovation (1958–62); another of close association with Christian Democracy (1962–67); and a third of attempts to maintain internal unity and play a moderating social role (1967–73).

Pastoral renovation (1958–62)

The Chilean Church was revitalized by innovations and structural reforms undertaken during the late 1950s and early 1960s. Set in motion earlier, they were given additional urgency by the advances of Marxism in Latin America and particularly in Chile. The Cuban Revolution and its devastating effects on Cuba's ultraconservative Church gave Chilean Catholic leaders additional incentives for proceeding with both reform and social initiatives. Their urgency was heightened by the 1961 congressional elections, when the electoral coalition (FRAP) formed by the Communist and Socialist parties captured 30.6 percent of the votes, more than any other party or bloc.

One of the first things the bishops did was reorganize. New parishes were established in slums on Santiago's northern, western, and southern fronts, where migrants from rural areas were settling in massive numbers. Many were staffed by foreign missionary priests and sisters who had access to resources from abroad and further strengthened reformist tendencies within the Chilean Church. Priests and sisters were also given greater latitude, and rapidly devised new programs and materials tailored for the lifestyles and environments of the people with whom they worked. In 1963, evangelical missions were organized in Santiago and in other dioceses. These combined radio broadcasts with follow-up visits to local schools, factories, neighborhood centers, and social clubs by teams of priests, nuns, and lay workers. They provided those who rarely attended Sunday mass with instruction in basic beliefs and Catholic social teaching. In Santiago, more than 300,000 people, nearly 13 percent of the city's population, were reached during a three-month period.

Within the Episcopal Conference, the position of the social Christian faction was strengthened by the appointment, in 1961, of Raúl Silva Henríquez, director of Caritas Chile (the major relief agency of the Chilean Church) and a friend and ally of social Christian bishop Manuel Larraín of Talca, as archbishop of Santiago. Silva quickly became a leading force within the conference, and the principal spokesman for the Church as a whole. Under his leadership, its commitment to social reform grew stronger and more explicit. In pressing these issues, Silva and the other social Christian bishops invoked the recent encyclicals of John XXIII (*Mater et Magistra*, 1961, and *Pacem in Terris*, 1963), making it difficult for conference conservatives to oppose them. Conservatives often succeeded in injecting anti-Marxist statements in these letters but could not challenge their endorsement of agrarian and other reforms.

The popularity of Frei and the Christian Democrats further legitimated the spread of social Christian ideas. Frei was an attractive figure with a reputation for technical competence and unshakable personal integrity. He was the perfect "secular carrier" for social Christian ideas, a political centrist whose intense anticommunism was joined with an equally intense disdain for hidebound conservatism and laissez-faire capitalism. His notoriety enabled him to convey social Christian ideas to larger and more attentive audiences. He brought the concepts and positions of an enlightened minority into the social and political mainstream, enhancing their prestige and influence within the Church and in the political arena.

Close association with Christian Democracy (1962–67)

The Second Vatican Council (1962–65) helped to accelerate pastoral changes in Chile. Its endorsement of work for justice, decentralized Church structures, and a larger role for the laity in its ministries, reinforced initiatives of the Chilean bishops in these areas. During the early and mid–1960s, the bishops also began to call for structural reforms in Chilean society, backing up their words with social programs of their own.

In its words and actions, however, the hierarchy closely identified the Church with Chile's Christian Democratic Party. Social Christian bishops who had studied with the party's leaders in the 1920s and 1930s, or had known them from Catholic Action circles, began to shape the Church's pronouncements and programs in the early 1960s. These never mentioned the PDC explicitly but had a distinctively partisan air.

In a 1962 pastoral letter on the Chilean peasantry, for example, the bishops analyzed the need for agrarian reform in the same terms later articulated by the PDC.[29] Several months later, they took a similarly reformist position with respect to industrial policy.[30] The PDC's 1964 platform had not been finalized when these letters were issued. However, some of the "experts" who would help to draft it were advising the bishops at the time, and their analysis, tone, and policy recommendations prefigured Frei's platform in several areas.

While issuing letters and statements, the Church also developed organizations aimed at attacking the causes, not just the symptoms, of poverty. With financial support from abroad, housing cooperatives, peasant training programs, slum-dweller organizations, and trade union federations were set up under Church auspices. In fact, many of these were repositories from which Frei's government would draw programs and personnel once it took office in November 1964.

The Church's greater social involvement enhanced its vitality and influence. It helped project its message to wider audiences than could be reached through publications, pastoral pronouncements, and Sunday homilies. According to survey data collected in Santiago a month before the 1964 election, most Chileans (64.8 percent), and three-fourths (76.6 percent) of the regularly practicing Catholics, were aware that the Church favored social change, and roughly two-thirds of each group (67.6 percent and 67.3 percent respectively) approved of its stand.[31]

One might conclude from this that the Chilean Church had broken definitively with its earlier conservatism. Motivating many of the bishops, however, was the fear of a possible Marxist victory in the 1964 election. Christian Democrats and "Independents for Frei" labeled Salvador Allende a Soviet pawn who would set up a dictatorship of the proletariat if elected. The Catholic bishops, under the urging of ideological conservatives in their ranks, warned of "persecution, tears and bloodshed" for the Church in the event of a Marxist triumph.

However ambivalent some of the bishops were with respect to Frei's reforms, however, the Church's association with Christian Democracy hurt it during Frei's presidency. Many of its lay activists traded their apostolic commitments for positions in his government, robbing Catholic Action and other organizations of some of their most valued cadres. The number of seminary students fell to an average of 150 between 1967 and 1970, well below the annual rate of 176 between 1960 and 1963. In short, Catholics who had been working in and for the Church suddenly found the PDC an attractive alternative.

Impatience with the content and pace of reform under Frei arose early and spread quickly to the Church. Progressive intellectuals, dissident labor and peasant activists, and half the party's youth sector (many of whom were products of Catholic Action) thought that Frei's government should move faster to curtail the power of landed and industrial interests. Their sentiments were echoed by priests, sisters, and lay activists who were living and working in poor neighborhoods (*poblaciones*). The improvements that came in housing, education, and employment during the first several years fell short of what their people needed, and they too became frustrated.[32]

Rightist Catholics were vocal as well. Upper-income Catholics who had supported Frei only to stop Allende began to criticize the hierarchy's "partisan support" for the Christian Democratic agenda. Organizations such as Tradición, Familia y Propiedad (TFP) termed the bishops' support for agrarian reform a partisan position and "an abuse of episcopal authority," which, they argued, Catholics were free to ignore. Finally, the secretive Opus Dei organization attracted upper-income Catholic university students who favored authoritarian forms of governance for both church and society.[33]

The government's economic and political fortunes declined in 1967. As copper prices and government revenues fell, U.S. aid declined, and the rate of inflation began to rise again, Frei cut social spending and imposed wage controls to curb demand.[34] The 1967 municipal elections (in which PDC candidates polled 35.6 percent of the vote, as compared to 42 percent in the 1965 congressional elections) marked the beginning of the party's political decline. They led the bishops to conclude that the PDC was a merely mortal political force, and that a more neutral stance would be better for the Church in the now more likely event that a right- or left-wing candidate would win the next presidential election.

Moderating role amidst polarization (1968–73)

Beginning in late 1967, the bishops took steps to distance themselves from the PDC and to assume the role of nonpartisan moral tutors. In both their public pronouncements and private initiatives, they sought to mediate among opposing political tendencies. Against mounting ideological discord, they stressed the importance of democratic institutions and nonviolent methods of conflict resolution. Their efforts were not entirely successful. The political differences which they hoped to moderate grew stronger, further straining relations within their ranks and in the country at large.

Under Frei (1967–70). During the last two years of Frei's government, the bishops criticized its emphasis on the technical aspects of development. They spoke of the deficiencies and tensions attending all development strategies, including the PDC's. In the months preceding the 1970 election, they stressed the need for further initiatives in housing, employment, and education, and called on all Chileans to redouble their commitment to building a more just society, acknowledging, in effect, that Frei's government had failed to fulfill its mission.[35]

In distancing themselves from the PDC, the bishops endorsed structural changes favoring the poor but left Chilean Catholics free to decide how to accomplish this. However, it was hard for people to believe that the bishops had really abandoned the Christian Democrats. They were critical of Frei but their calls for moderation, and their criticism of "leftist and rightist excesses," implied a continuing preference for "third way" alternatives; in Chile, in the late 1960s, this meant the PDC. What the bishops may have intended as a return to nonpartisan social Christianity, therefore, still looked and sounded like Christian Democracy.

Dissent within the Church itself became more public and confrontational in the late 1960s. Twelve of the country's twenty-three dioceses held synods to discuss how best to implement the directives of the Vatican Council, which had ended several years earlier. At these meetings, priests, nuns, and laypeople alike complained that the Church was still too authoritarian, and focused unduly on middle- and upper-income groups. In 1968, Iglesia Joven (Young Church), a new, radical movement with some 200 members, occupied Santiago's main cathedral, claiming that the Church was not decentralizing rapidly enough or making a sufficiently serious commitment to the poor.

The Chilean bishops rejected these indirect attacks on their authority and pastoral priorities. "A Church separated from its legitimate pastors . . . [and] agitated by whirlwinds of doctrines," they argued, "is not the Church of Christ." And while the Church should be concerned with the poor, it should not allow "those who are not poor or young to become alienated," since its mission was to all people.[36]

Given the country's increasingly agitated political circumstances, however, the Chilean bishops were remarkably open and constructive in the months preceding the 1970 election. As the election approached, Cardinal Raúl Silva Henríquez and Bishop José Manuel Santos of Valdivia, president of the Episcopal Conference, issued statements stressing the Church's neutrality, urging Church personnel to avoid partisan involvements,[37] and underscoring the value of democratic procedures and the importance of avoiding civil war and military rule. Notably missing from their remarks on the other hand, were the dire warnings of "tears and bloodshed" of earlier elections. But by "allowing" Catholics to vote for Allende, the bishops may have hurt the Christian Democratic Party candidate, Radomiro Tómic. In fact, Tómic received fewer leftist votes than he had hoped to win and finished (with 27.8 percent of the vote) behind both Allende (36.2 percent) and Alessandri (34.9 percent).[38]

Under Allende (1970–73). The bishops maintained this nonpartisan, moral-tutorial role during Allende's presidency. They did so with timely statements urging respect for constitutional procedures, and by attempting to moderate conflict at critical junctures. Despite their efforts, polarization intensified, producing a political impasse, social chaos, and, in the end, military intervention. In fact, divisions in the Church worsened during the Allende years as radical Catholics attempted to throw the Church's weight behind Popular Unity parties and policies.

The heightened politicization had a draining effect on sacramental participation and on the vitality of many base communities.

The Church's relations with the political order evolved through four stages: the transition following the September election; early prosperity and success; economic decline and developing political crisis; and final crisis.

During the eight-week period between Allende's first-place finish in the election and his confirmation as president by Congress, forces within and beyond Chile tried to prevent him from taking office. The bishops refused to endorse intrigues by right-wing extremists (encouraged by the CIA) and rejected a request by national party leaders that they denounce Allende. They urged citizens not to become paralyzed by fear of radical change or to oppose it violently. They pledged their own cooperation "with changes in the country, especially those which favor the poorest."[39] While not endorsing Allende's confirmation, their refusal to participate in efforts to prevent it contributed to an atmosphere of dialogue and national unity, in which Christian Democratic and Popular Unity leaders were able to reach an agreement permitting Allende to assume the presidency.

Once in office, Allende began to implement the program on which he had campaigned. Using his prerogatives to take monetary initiatives, he imposed price freezes on basic commodities, raised wages by 66.6 percent, expanded public works projects, extended social services, and doubled the amount of money in circulation. These moves stimulated demand, increases in production, and led to palpable improvements in living standards for the poor.

Following Allende's inauguration, Cardinal Silva told an interviewer that the Popular Unity government's reforms were "supported by the Church," and that socialism contained "important Christian values" and in many ways was "superior to capitalism."[40] In May 1971, the bishops issued a declaration implying that some forms of socialism could be compatible with Christianity, and offering the Church's support for those government policies "that were liberating for the poor."[41] They did this following Pope Paul VI's release of an apostolic letter in which he discussed Marxism in more nuanced, less negative, terms than in previous social encyclicals.[42]

Among Catholics in poorer neighborhoods, there was considerable support for Allende's government and its programs. Thanks to increased state subsidies, Catholic schools and universities were able to offer more scholarship aid to the poor. Nuns, priests, and lay leaders helping to form base communities in poor and working-class neighborhoods urged their members to display "a generous constructive attitude" toward the transition to socialism and to participate in projects that the new government had initiated.

Many of these communities were engulfed by the tensions and politicization of their neighborhoods and work environments. As one priest put it, "the community was quite political, because the formation it received was intensely Christian. It brought faith and people's lives together, and people's lives were political."[43] Political divisions made it impossible for many such groups to function as religious communities. In fact, the resentments and antagonisms that divided Christian Democrats from radical Catholics, and that plagued the latter as well (they were usually divided into *miristas*, *mapucistas*, and Left Christians), tore many of them apart during the Allende years.

The government's economic policies ran into trouble by mid–1972. Production declined as private investors worried about possible nationalization; inflation rose with increased circulation of money, much of it in the hands of the poor. The cut-off of U.S. aid and trade and black-market profiteering to avoid price controls combined to produce a shortage of goods in middle- and upper-income areas. And, finally, the government's legally questionable expropriation of firms paralyzed by their striking workers simply exacerbated tensions. On the political front, the Christian Democrats joined with rightists to block the government's nationalization program and to hold up other policies as well. Street demonstrations and strikes by shopkeepers and professionals were orchestrated by both opposition parties, and these led to violent clashes with pro-government forces (consisting largely of working-class supporters).

The bishops did not take sides in these controversies, pressing instead for negotiated settlements. Three times during 1972, they warned against the consequences of violence, political sectarianism, and fratricidal war. They urged respect for law and compromise among contending groups, adding that most citizens favored changes benefiting the poor, and implying that they were needed. Their remarks were directed against violent extremists on both sides; those in government using extralegal means to achieve their goals, and hard-line oppositionists, including some Christian Democrats, bent on stopping the regime at any cost. In a word, the bishops used their moral authority to promote moderation and consensus.

They were unsuccessful. By late 1972, when a nationwide truckers strike paralyzed the country for weeks, there were few Chileans left who valued democratic principles more than the achievement of their own political and economic objectives. The bishops found no ready "secular ear" willing to heed their call for moderation. In fact, in a context in which everything was politicized, even their remarks seemed partisan. To many with commitments on the left, they sounded like crypto-Christian Democrats attempting to persuade Allende to abandon his socialist agenda. To those on the right, the hierarchy's commitment to a "constitutional" solution seemed designed to buy time for the left.

It did not help the bishops that as they addressed the country they were being challenged by radical priests and religious belonging to "Christians for Socialism" (CpS). CpS consisted of some 350 priests and nuns (roughly 5 percent of the total), most of whom worked in "popular" neighborhoods. They sought to align the Church with socialism generally and the Popular Unity government in particular, turning what the bishops had termed *one of several* legitimate options for Catholics into the *only* legitimate one.

CpS militants saw neutrality in the current political struggle as illusory. A cleric's responsibility, they contended, was to give the most effective witness he could to Christian values, which, in the present context, required support for Allende's Popular Unity government. Each time the bishops called for compromise and dialogue among political factions in society, CpS responded, urging Christians to support the left. During the congressional election campaign of early 1973, CpS activists campaigned on behalf of Popular Unity candidates. Some even demanded political support of the left as a criterion for membership in the base communities with which they worked.

Their statements and tactics amounted to a frontal challenge to three of the Church's core components described in chapter 1: the nonspecificity of its social teachings; the universal scope of Church membership; and episcopal authority. When Christians for Socialism ignored their repeated requests to cease and desist, the bishops formally condemned the movement and forbade priests and nuns from participating in it.[44]

The challenge of Christians for Socialism harkened back to the Falange-Church dispute of 1947. However, the "rebels" in that earlier conflict were publicly docile laypeople who were willing to retreat. Most CpS activists, on the other hand, were priests, and they were attacking core institutional concerns. The tensions between the bishops and CpS further complicated church-state relationships. Many UP leaders assumed that the bishops opposed their government, while most bishops apparently concluded that Marxists were bent on subverting the Church.

In the March 1973 congressional elections, Popular Unity candidates won 43.4 percent of the vote. Their Christian Democratic and National Party rivals divided the remaining 56.6 percent. Allende was then stronger than when first elected and had sufficient support to defeat attempts to impeach him, but he lacked the support he would need to complete the transition to socialism. Following the elections, he faced challenges from several quarters. One of these came from the bishops over a proposed Unified National Curriculum affecting both public and private (including Catholic) schools. For the first time since Allende assumed the presidency they strongly criticized his government. Not wishing a confrontation, Allende abandoned the proposal, only to face an abortive military uprising in late June and a renewal of the truckers' strike in July.

In these circumstances, Cardinal Silva attempted to bring opposing elements together, arranging face-to-face meetings between Allende and the PDC president, Patricio Aylwin. Two such encounters took place in the cardinal's office in August but accomplished nothing. Neither principal was in a position to speak for, or control, his supporters or coalition partners. At this point, confrontation of some sort appeared inevitable.

The bishops' efforts to avert this went for naught. Despite a less partisan moral-tutorial role than in the previous decade, they failed to arrest the polarization gripping the country. In subsequent years, their efforts during this period would be valued by many on the left. In the words of a Communist labor leader whom we interviewed in 1987:

> During the Popular Unity period, the Catholic Church was very important in influencing the policies of the popular government, particularly as a vigilant observer of what the government was trying to do in creating a new society. Its vigilance was not negative, but positive, because the Church gave the government advice on many issues. Here, in Santiago, Cardinal Silva was particularly active in doing this, and his perspective always took into account the needs of the whole country. The

> Church, then, engaged in dialogue with the government and helped it understand where things were not going well. This advice was well received in the government.[45]

But at the time, the Allende years were costly to the Church. The percentage of Catholics active in religious organizations fell from 4.1 in 1970 to 2.8 in 1973.[46] Similarly, the number of vocations to the priesthood, already declining in the late 1960s, reached new lows (averaging 112) during the Popular Unity period. Additionally, many base communities became highly politicized. Participants injected partisan political issues into group discussions and Bible study, and clashes erupted regularly between groups with different party preferences. Allende's government was careful to respect religious freedom, but the atmosphere of intense social struggle in which the Church had to operate made it difficult to escape or deflect destructive political tensions.

Surrogate for Resistance to Authoritarianism (1973–82)

During the first half of the military's years in power, the Church was a more effective surrogate for social organization and service than a moral tutor. During the first three years of military rule, the bishops avoided confronting the new government. Their public statements were cautious and somewhat inconsistent. Once it was clear that the military would not soon relinquish power, and when its repressive arm began to affect Church programs and people, the bishops became more critical. At no point, however, did the Church's efforts as moral tutor have an ameliorating effect on the policies of the military government.

In contrast, the hierarchy moved quickly after the coup to establish a network of social service organizations. Its facilities and services provided immediate relief to Chileans facing joblessness, malnutrition, arrest, imprisonment, despair, and torture. They also had the indirect but intended effect of challenging the military's authority when no other force could do so. Over time, Church programs developed and helped to sustain a critical mass of resistance that would later (in the mid 1980s) give rise to a united opposition movement.

Cautious tutor, active surrogate (1973–76)

During the Frei and Allende years, Church leaders spoke often and clearly as moral tutors. Many of their social and pastoral functions, however, were assumed by secular institutions. These tendencies were reversed during the early years of military rule. The hierarchy's moral-tutorial voice grew hesitant and equivocal, as the Church became an increasingly crucial dispenser of social services and arena of social participation.

The hierarchy was publicly supportive of civilian rule to the end, but many bishops actually welcomed the coup when it came in September 1973. Most believed that the polarization of the Allende years had been harmful to the Church, that things could not have gone on as they had for most of 1972 and 1973, and that a coup was unavoidable.[47]

The bishops' first public statement as a group came on September 13, 1973. It was cautious and conciliatory, and gave tacit moral legitimacy to the new regime. It decried the spilling of "the blood of civilians and of soldiers," expressed trust in the "patriotism and selflessness" of the military, and asked citizens to cooperate so that institutional normalcy could be restored.[48] In the ensuing months six of the thirty active bishops expressed their individual support for the coup.

Many of the bishops lived in the provinces where repression was either less severe or less visible. Most believed that conveying their concerns directly and personally to local military commanders would be more productive than "confronting" them. In some instances they were able to get clemency for specific individuals. Nevertheless, early estimates were that as many as 11,000 were killed in the first three months after the coup.[49]

Repression became more systematic and spread to small cities and rural areas in early 1974. The bishops' first criticism of regime abuse came in April, in their letter "Reconciliation in Chile." They lamented the "climate of insecurity and fear" in the country and expressed concern over "interrogations which include physical or moral constraints," but they hastened to affirm their confidence in the "good intentions" and "good will of our governmental authorities."[50]

During the ensuing year, rumors and evidence of human rights abuses abounded. Some bishops, like Cardinal Silva of Santiago, condemned these violations within their own dioceses. The Episcopal Conference, however, remained silent until September 1975, when it issued "Gospel and Peace." This document combined oblique criticisms (of human rights violations and the lot of the poor) with expressions of gratitude to the military for "freeing" the country "from a marxist dictatorship which appeared inevitable and would have been irreversible."[51]

Most bishops were privately critical of the government at this point but were willing to temper their public statements so as not to jeopardize the assistance they were providing victims of repression. In late September 1973, the Catholic, mainline Protestant, and Jewish communities had formed the National Committee to Aid Refugees (CONAR) to assist leftist emigrés who had come to Chile during the Allende years. In early October, Jewish and Lutheran leaders joined Cardinal Silva in forming the Cooperative Committee for Peace in Chile (COPACHI), whose mission was to provide legal and other assistance to people dismissed from their jobs, to those who had been detained or had "disappeared," and to their families.[52]

In fact, Silva was a thorn in the military's side from the beginning. His was one of the first voices to call publicly for the early restoration of political freedoms and normal judicial processes. He also strongly opposed the military's "takeover" of the Catholic University of Santiago, of which he was the titular head. It was his outrage at the government's extrajudicial persecution of its enemies, however, that prompted him to establish and expand COPACHI.

Actually, the politically astute Silva was one of the few people in Chile to realize the extent of the tragedy that was unfolding. As a leading leftist intellectual told us in 1987,

> In September 1973 we ourselves were not fully aware of what was happening, we militants of the left who were the principal victims of this long period of persecution that was beginning. . . . On the day of the coup, . . . we had telephone contact with Cardinal Silva Henríquez. From the start he sensed with an impressive clarity what was transpiring. Much better than any of us, much better than political leaders, he realized, with a kind of Chilean "popular and peasant-like" intuition, that unleashing the military was going to bring about a tragedy.[53]

Military authorities took an immediate dislike to the cardinal. They were suspicious of his ties to Frei and the Christian Democrats, and of his efforts to promote a negotiated settlement prior to the coup. They could not accept his willingness to give moral and material assistance to people they regarded as enemies of Chile. They viewed his Committee for Peace as disloyal and subversive, and blamed it for the international protests being directed at them at the time. During 1974, the

Junta sought repeatedly (without success) to persuade Vatican officials to "reassign" Silva to other duties.

In 1974 and 1975, COPACHI's operations were expanded to include cooperatives, credit programs, health clinics, feeding programs for the elderly and children, and other social service projects. With Congress and political parties in recess and the poor without jobs or recourse, people turned to the churches for assistance. Aided by COPACHI, base communities, parishes, and regional Church compounds (the *vicarías* into which the Santiago archdiocese had been subdivided) became surrogate social service centers supported by financial assistance from Europe, Canada, and the United States.

COPACHI programs provided people with safe havens in which they could share concerns, compare ideas and experiences, and coordinate future activities. Drawing on information from its provincial offices and from local churches, the organization prepared reports for the Church on arrests and disappearances, verified cases of torture, and was able to counter the lies and distortions of the controlled media. The Church paid for its efforts in these regards. Harassment and arrest of COPACHI personnel (especially in Santiago), police raids on its service projects, and the detention of volunteers working in them, were all routine by early 1975. In addition, a quarter of the foreign priests (over 300 of the more than 1,200 in the country) were obliged to leave during the first two years of military rule. Most worked in projects aiding those persecuted by the regime, and some had been members of Christians for Socialism.

The Church's educational institutions also suffered restrictions. Following the coup, the new government assigned military rectors to all universities, including the Catholic universities of Santiago, Valparaíso, and Antofagasta. Administrators and faculty sympathetic to the left or the PDC were fired, "controversial" centers and programs were eliminated, and governmental subsidies to low-income Catholic University students were cut back, as were grants to Catholic primary and elementary schools.[54]

When priests, nuns, and COPACHI workers helped four members of the Movement of the Revolutionary Left (MIR) gain asylum in foreign embassies after a shoot-out with the police in late 1975, the government pressured the sponsoring churches to disband the organization. They agreed, but almost immediately (January 1976) Cardinal Silva created the Vicariate of Solidarity (Vicaría de Solidaridad) in its place.

The new organization was made an official ministry of the Archdiocese of Santiago, and was less vulnerable to government pressure.

Because of the size and importance of his archdiocese, Silva lent the Church a more combative image than was warranted in 1974 and 1975.[55] By late 1975, however, the bishops as a whole were becoming more critical. Their shift in attitude resulted from several factors. One was the regime's heavy-handed efforts to discredit Silva and other Church people who opposed its policies. Another was the government's decision to go after Christian Democrats and other "respectable" political groups. Finally, by late 1975, it was increasingly clear that the military would not soon abandon power, and that its economic policies were imposing heavy burdens on the poor.

Forceful tutor and continuing surrogate (1976–82)

In 1976, the Episcopal Conference became more consistently critical of government policies. It did so following new attacks on labor unions and other dissidents, on public figures close to the Church, and on Church authorities themselves. These attacks coincided with the expansion of a new security force known as the DINA (The National Bureau of Investigations).

With the DINA's creation, antiregime activists became the objects of forceful interrogations and exemplary punishments. The mistreatment of union leaders who participated in workshops sponsored by the Church angered a number of the bishops, as did the expulsion from the country, in June 1976, of two officials of the National Human Rights Commission for "imperilling its national security."[56]

In addition, the bishops were attacked by right-wing Catholics who supported the military government. Tradición, Familia y Propiedad (TFP) had kept a low profile during the Allende years, but it reemerged following the coup. In February 1976, it published a book (apparently with government approval) accusing Church leaders of heresy, and calling on Catholics to denounce them and to refuse the sacraments from their hands.[57] TFP thus attacked episcopal authority in much the same manner as Christians for Socialism had done earlier. The bishops reacted in kind, condemning the attempt to establish a "parallel teaching office" and declaring that these people had "placed themselves outside the Catholic Church."[58]

A third and even more decisive factor was the public mistreatment of three bishops by government agents and supporters in mid-August. The victims were vocal regime critics returning from a meeting in Ecuador.[59] Led by undercover security officers, pro-government demonstrators met the bishops at the airport, throwing stones, shouting insults, and intimidating the bishops, their relatives, and their friends. The Episcopal Conference's Permanent Committee responded forcefully, reminding the country that canon law provided for the automatic excommunication of those involved in "violence and verbal aggression" against a prelate.[60]

These were the conference's first unequivocally critical statements in almost three years of military rule. Apparently, most of the bishops had harbored serious criticisms of the government for some time but only recently had concluded that private or subtle overtures were not going to affect government policies or practices. It seems that attacks on their authority, and on people whom they respected, persuaded them that more aggressive moral-tutorial efforts were required.

Their decision produced few positive results. Early in 1977, the military government announced that it would remain in power indefinitely and proceeded to dissolve all political parties (not merely those on the left), and to close down the Christian Democratic Party's radio station. The bishops countered with a statement criticizing the regime's "structural" weaknesses, demanding information on citizens who had "disappeared," charging the military's economic policies with benefiting only the privileged classes, and calling for an early return to constitutional government.[61]

In the months that followed, the bishops issued statements calling for amnesty for political exiles and denouncing the government's harassment of the clergy and continuing refusal to provide information on the disappeared. Then, unexpectedly, Pinochet became more conciliatory. In April, he sacked a justice minister who had given a speech critical of the Church. In subsequent months, dissident magazines were allowed to circulate, on a limited basis, and restrictions on political discussion and debate were relaxed. And finally, in July, he announced a timetable for phasing out military rule and restoring democratic institutions. Opposition groups were critical of his plan, but the Church responded more favorably, terming it an important step forward, and applauding the regime's apparent flexibility. Its statements in the last half of 1977 carried the same optimistic tone.

The bishops were less happy with the government's economic policies. GDP grew by more than 6 percent per year from 1977 to 1979, while inflation dropped from 374 percent in 1975, to 92 percent in 1977, and to 33 percent in 1979. But these feats came at enormously high social and human cost. Official unemployment remained at roughly 17 percent from 1977 to 1979, and the percentage of people actually out of work was much higher. And, although the economy itself was growing, the purchasing power of workers had fallen by 50 percent between 1974 and 1978. The bishops thus continued to criticize the economic model in 1978 and 1979, denouncing low wage and salary levels and the lack of credit and technical assistance for small farmers.[62]

Pinochet abandoned his limited political opening in mid-1978 by refusing to provide information on the cases of disappeared persons and by renewing his efforts to undermine Cardinal Silva's position in the Church. These moves short-circuited the thaw in church-state relations and led to further deterioration over the next several months.

A major factor in this process was the discovery, in late 1978, of mass graves containing the remains of regime opponents at Lonquén, Cuesta Barriga, and Laja. This placed the issue of the disappeared at center stage once again, and the bishops called for an immediate accounting. The government's response was to pass an amnesty "law" (covering all "noncriminal" actions for which its personnel may have been responsible since September 11, 1973), and to order the reburial of bodies without informing family members. Church-state relations improved briefly following successful papal mediation of a dispute with Argentina over the Beagle Channel, but the regime quickly resumed its attacks on Silva and other Church officials. The cardinal countered, decrying the government's failure to heed repeated episcopal calls for peace and the restoration of democracy.

At this point Pinochet announced a plebiscite for September 1980, to approve a new, authoritarian constitution extending his term as president for an additional nine years. The bishops responded skeptically, saying that if the results were to be valid, opponents would need equal access to the media and "the privacy of conscience" would have to be respected.[63] As things turned out, the opposition had virtually no opportunity to present their views, there were no provisions for ballot privacy, and in the end the government reported that 67 percent of those taking part had voted "yes." This figure was probably inflated. However, with Christian Democrats, Socialists, and Communists continuing to disagree on important issues, they failed to project themselves as a viable alternative to military rule. And as Cardinal Silva is said to have conceded privately, the opposition probably would have lost the plebiscite even if the balloting had been conducted fairly.

The results enhanced Pinochet's power and authority. They put the government in a triumphal mood, strengthening the hand of hardliners who would rather attack than accommodate their adversaries. The moral-tutorial role that the bishops sought to exercise from 1976 to 1980 thus failed to effect changes in government policy or orientation.

Thanks to the work of the Vicaría de Solidaridad, however, the Church had a more decisive impact as a provider of social services and as a staging ground and arena for political participation during this period. At its peak, the Vicaría employed between 350 and 400 administrators, lawyers, doctors, and social workers. Most of these worked in greater Santiago, with another 100 or so spread among the fifteen other dioceses in which the organization established a presence. The Vicaría's work was funded almost exclusively by outside sources, primarily the German Bishops' Fund (*Misereor*) and the governments of Sweden and other Scandinavian countries through the World Council of Churches. Although figures are not readily available, one high-ranking official told us that in some years the organization's total budget exceeded $3 million.

The Vicaría began in 1976 with labor and penal departments. The former assisted people who had lost their jobs for political reasons; the latter assisted those who were victims of government repression of various sorts. These activities evolved over the next several years. The labor department began to assist workers, including non-Catholics and Marxists, in rebuilding their labor unions. In 1977, a separate Vicaría de Pastoral Obrera (Vicariate of Worker Ministry) was formed, and under its auspices and influence Christian Democratic and leftist workers overcame the obstacles that previously prevented them from working together.[64]

In the penal or legal area, the Vicaría provided services to those accused of political offenses. Vicaría lawyers made inquiries, filed petitions, visited detainees, comforted their friends and families, and kept a thorough record of all proceedings. In the long run, the organization's ability to keep the country reliably informed of the extent of human rights violations, and to publicly challenge the government when no one else could do so, would be its most important contribution to the restoration of democracy.

During the late 1970s, the departments of communications and popular sectors were created. The former produced the monthly newspaper *Solidaridad*, which reached 30,000 people by the mid-1980s, and which helped to keep human rights violations at least minimally in the public eye. The popular sectors department oversaw a network of nutrition programs and employment organizations that assisted more than 700,000 people (primarily at the parish and base-community levels) between 1976 and 1980.

Vicaría services enabled many Chileans to survive physically and psychologically hard times. Their primary objective, however, was the development of popular organizations through which people could partially resolve their own problems. In the absence of other outlets and activities, the programs constituted public spaces in which people could come together and develop feelings of solidarity with one another. They became vehicles of organizational, social, and political development for the duration of the period of military rule.

Not surprisingly, the period of military rule was one of institutional revitalization for the Chilean Church. Vocations to the priesthood, which had fallen to an average of 112 per year during Allende's presidency, jumped to 208 in 1976, 374 in 1978, and 895 in 1982. Mass attendance increased slightly in relation to the previous period, with almost 43 percent attending either weekly or monthly, and more people considered themselves "religious" and "very religious" than in either 1970 or 1973. Most significantly, perhaps, involvement in Church organizations (prayer and reflection groups, base communities, catechetical groups, youth groups, etc.) rebounded dramatically from a low of 2.8 percent (of all Catholics) in 1973 to more than 10 percent in 1979, and low-income groups showed the most dramatic increases among organizational Catholics.[65]

Its social activism helped the Church to recapture the affection, and in some cases the faith and allegiance, of cultural Catholics with leftist political views. They were among the first Chileans to link up with COPACHI and the Vicaría, where they found refuge, sympathy, and the resources with which to help their comrades and others in need. The admiration that the Church gained among these lapsed Catholics was considerable. A Socialist Party (PPD) leader describing himself as a *creyente distante* ("distant believer") was typical of those whose view of the Church changed dramatically:

> But in these past fourteen years I have rediscovered a church which is completely involved with the problems of the people—especially the priests and nuns. . . . In two ways it has been most impressive: acting as a voice in defense of those suffering violations of their human rights, and (although with less vigor) making an option for the poor. I think these are practical things the Church is doing, and not just words. I believe these are also historically and qualitatively changing the role of the Church in Chilean society for the twenty-first century. I believe it will be a very different Church from the one in the twentieth century.[66]

As social and political space narrowed, practicing Catholics (especially Christian Democrats) joined cultural Catholic leftists and some secular leftists first in COPACHI and then in the Vicaría. Having confronted one another for most of the preceding forty years, they began working together. Their respective parties did not immediately or formally reconcile, but when political activity resumed in 1982, militants that had worked together in Church projects found it easier to collaborate politically.

Predictably, the Church's prominence as a social and political surrogate angered Catholics who supported the military government. Catholics in the military, and those from middle- or upper-class backgrounds who supported the regime, viewed Church activists in social programs as either subversives or naive moralists. As the bishops became increasingly critical of military policies and actions, some conservative Catholics began to challenge their pronouncements. This alarmed many bishops. Publicly, the hierarchy was critical of Catholic "dissenters," but privately it worried that one of the Church's perennial core concerns, its openness to Catholics of all political or ideological persuasions, was at risk.

When we asked a pro-government UDI activist in 1987 to rate the Church's performance during the years of military rule, he expressed ambivalent feelings. He conceded that the Church had a right and a duty to defend human rights, but he objected to the involvement and methods of many Church people:

> . . . the Church has been called upon to play a very unappreciated role. It has had to defend the innocent, and many times the not-so-innocent. It has had to perform a very difficult task because it has to defend human rights at the level of principle, from a biblical perspective. And yet, in my view, it has erred in some instances, . . . it has become too involved in the political and social aspects of the issue. . . . some priests by politicizing their pastoral role have caused Catholics to lose faith in the Church. . . .

> We have heard from residents in shanty-towns [*poblaciones*] that they meet in the Church to engage in politics. Many times arms have been hidden along with subversive pamphlets. . . . This has caused people to lose respect for their priests. It is, therefore, a very complex situation because the clergy can go to such extremes as to cause people to leave the Church.[67]

Following the 1980 plebiscite, the bishops sought to improve church-state relations and to reach out to pro-regime Catholics. In November, Cardinal Silva invited Pinochet to participate in the Eucharistic Conference being held at the Temple of Maipú, prompting criticism from base communities and from priests and nuns working in popular neighborhoods. Later, in another symbolic move, he agreed to celebrate a *Te Deum* mass marking the reopening of the presidential palace (*La Moneda*) and the inauguration of the new constitution. Finally, in April 1981, Silva announced cutbacks in the operations of the Vicaría de Solidaridad, the embodiment of the Church's opposition to the military regime.[68]

In these decisions, Silva resisted the advice of his progressive colleagues. He did not believe that reconciliation with the regime was either possible or desirable, but he thought it was wrong for the Church to engage in prolonged struggle with a government that had just won a national referendum by a substantial margin. His efforts to forge a middle-of-the-road policy failed to impress the radical priests, sisters, and lay activists who felt he was lending legitimacy to the regime, or the government, which continued to attack him mercilessly.[69]

Silva spent the next two years preparing for his retirement. His projects included overseeing the decentralization of programs and decision-making (along zonal lines), developing mission programs for young people, putting Vicaría programs on a more stable financial footing, and grappling with the question of Church authority over the Catholic University. In 1981 and 1982, he continued to stress the same themes on which he had spoken for years: the need for genuine reconciliation, the importance of determining fates and responsibilities in cases of human rights violations, and the urgency of restoring democracy and adopting policies that would alleviate the suffering of the poor.

Silva's moderation toward the end of his term was disappointing to some Catholic activists but was accepted and applauded by others. The views of Chileans outside the Church were even more favorable. A non-Catholic Christian Democrat with whom we spoke in 1986, for example, praised Silva's astuteness in guiding the Church:

> When it came to playing the game with the military regime, the cardinal showed great moderation and balance. He presided over a *Te Deum* on September 18, 1973, one week after the military coup. However, a month and a half later he created the Committee of Peace, the forerunner of the present Vicariate of Solidarity. The cardinal, therefore, knows that the Catholic Church must be very wise in negotiating this world of politics. This confounds anyone who wants to judge polemically. . . . Is the real cardinal the one who gave a *Te Deum* seven days after the death of Salvador Allende, or the one who created the Committee of Peace a month and a half after Allende's demise? Such balance by the cardinal provides an element of stability in a world where politics truly has many extremes and where at times even the most perceptive political leaders lose a sense of equilibrium and of reality.[70]

Conclusions

By way of conclusion, we return to the hypotheses advanced at the end of chapter 1.

The first hypothesis predicted that in a country suffering from substantial poverty, the more amicably and completely the Church distanced itself from the state and conservative elites, the more likely it would be to develop new pastoral strategies along social Christian lines. The process of Church-state separation in Chile in the early 1920s partially confirms this prediction.

Formal separation was the final step in an erosion of the Chilean Church's legal privileges that had begun in the middle of the nineteenth century. It reflected the growth and strength of secular political forces. However, the care and accommodation with which it was negotiated left both sides relatively satisfied. With anticlericalism having subsided, the Church no longer had to oppose liberals or identify with their political adversaries. Separation also freed the Church from state control over important internal matters. Henceforth, it could create dioceses, select bishops, and define pastoral ministries and strategies without first obtaining permission or resources from state authorities.

If a relatively conflict-free separation from the state was thus *necessary* for the development of new, socially progressive pastoral strategies, however, it was not a *sufficient* condition. In Chile, important groundwork for helping the poor had been laid earlier by progressive clerics, and the Vatican was instrumental in easing separation and encouraging socially oriented programs in ensuing years.

The Constitution of 1925 did not entirely dissolve the strong Conservative sympathies of all of Chile's bishops and priests. Old allegiances lingered where class, ideological, and personal ties were strong, but Rome's repeated insistence that Church people avoid political entanglements prevented many of them from publicly reaffirming those ties in the 1930s and 1940s.

Finally, the Church's separation from the state in 1925 was not complete. The new constitution left open the possibility of future government support for Church programs. This provided a tacit foundation for future state endorsement of Catholic positions on divorce, abortion, and subsidization of Catholic schools.

The second hypothesis affirmed that the successful incubation of social Catholicism in a national Church would depend on Vatican support, tolerance by national church leaders, and the absence of serious threats to hierarchical authority. The Chilean experience confirms the influence of these factors.

The Vatican was crucial for inaugurating Catholic Action programs committed to serving the poor and for their continued development in ensuing years. Similarly, the leadership of priests like Alberto Hurtado made Catholic Action very appealing to young Chileans during the 1940s, and the influx of foreign priests to both urban and rural areas in the 1950s provided a similar push for social Catholicism generally. In contrast, however, the Falange Nacional's conflict with the bishops in 1947 underscored the vulnerability of these new pastoral strategies.

Hypothesis three suggested that to be legitimate and effective the Church's social pronouncements should be sufficiently concrete to be relevant but without tying the Church to partisan causes or movements. It also stressed the importance of a relevant audience or "ready ear" in secular society (i.e., policy makers and/or political forces) that would take its advice seriously. The Chilean experience indicates how difficult it is for these conditions to be fulfilled simultaneously for any length of time.

In functioning as moral tutors on behalf of social justice in the early 1960s, the Chilean bishops stopped decrying the symptoms of poverty and began calling for the transformation of the social structures underlying it. As they did, they found a "ready ear" among broad sectors of the politically active population, i.e., large numbers of people willing to accept reformist policies on their own merits or as a means of undercutting the appeal of Marxist groups. In the process, however, Church leaders became identified with the Christian Democratic Party. Their language, style of analysis, and policy recommendations were those of the PDC, and the Church's neutrality was thus again compromised, as it had been earlier by its ties to the Conservative Party.

During the late 1960s and early 1970s, the bishops sought to distance themselves from the PDC and to restore the neutral base on which their moral and political credibility rested. They were clear and forceful in their support for both social justice and democratic politics but without identifying with any particular political group. In the 1970 election and the tense weeks that followed, they lent important moral legitimacy to the Allende government's objectives, although not to Allende himself, and in 1972 and early 1973 the bishops were forceful advocates of both civility and constitutionalism. Their calls for negotiations and compromise among contending factions were aimed at preserving the country's democratic tradition, which they saw as benefiting all. In the climate of polarization that prevailed from mid-1970 on, however, their efforts were rejected by both Popular Unity and opposition forces.[71]

Their fears of Marxism affected the bishops' moral-tutorial role late in the Allende period and in the early years of military rule. In the charged social and political conditions of early- and mid-1973, some bishops blamed Marxists for corrupting Christians for Socialism, and feared that both the Church and the country would suffer irreparable damage unless Allende were overthrown. Initially, under the military, on the other hand, their pronouncements appeared to favor the imposition of order and depoliticization. Instead of rising to the defense of justice and human dignity, they lent tacit moral legitimacy to the ruling junta. At a time when more prophetic denunciations would have found a "ready ear" in both national and international arenas, their statements were ambiguous and on occasion obsequious.

Church leaders began to speak clearly and critically in 1976, largely moved by attacks on their priests, their friends, and some of their fellow bishops. But by this point, their window of opportunity had closed. Foreign governments and international banks had given the Chilean government the resources it needed to survive, and it was well ensconced in power. The moment of the "ready ear" was gone.

Ecclesial, not societal, interests were the determining factors in the bishops' decisions in these instances. At this point, at least, the hierarchy's regard for such core Catholic concerns as unity, hierarchical authority, and institutional survival was greater than its commitment to human rights, social justice, and democratic politics.

The fourth hypothesis stressed the importance of "secular carriers" through which social Christian ideals and perspectives could be projected and applied in practice. It also speculated that divisions within such carriers can weaken and neutralize the impact of these principles and perspectives.

The Chilean case clearly underscores the importance of local dynamics in making social Catholicism a national political force. The rise of Eduardo Frei and his Christian Democratic Party in the late 1950s greatly enhanced social Catholicism's visibility and prestige. Frei was an ideal alternative for those voters who feared the left but were willing to support progressive reforms. His emergence as a national political figure popularized social Catholic principles, and his party quickly filled a significant political void.

But the Chilean case also highlights the problems of maintaining unity in such reform parties as well. Once in power, the Christian Democrats had difficulty agreeing on a single course of action. Social Catholicism, it seems, was understood and applied quite differently by the party's various factions. Its principles could not produce a coherent ideology, program, or strategy around which progressive Catholics could unite. And as the Frei, and later the Allende, governments were forced to choose in these regards, Catholics split along class and ideological lines.

This is not surprising. One cannot leave lay Catholics free to apply principles in complex situations *and* expect a unified response in politics and policy-making. One *could* hope that Catholics belonging to different parties or factions would value their shared values more deeply, and respect one another enough to work together. But neither of these hopes were fulfilled under either the Frei or Allende governments.

Hypothesis five pointed to the increased capacity of post-Vatican II Catholicism to act as political protagonist and provider of social services in times of crisis. It speculated, however, that the longer the Church performed these roles the greater a challenge they would pose to its perennial religious mission.

The Chilean case supports both aspects of this hypothesis, suggesting limits to the Church's effectiveness as an opponent of authoritarian regimes. Networks of assistance, solidarity, and communication were set up at local- and intermediate-Church levels, and foreign financial resources were mobilized to support them. At the height of repression (1973–80), the Church served as an umbrella under which many Chileans found refuge. With other institutions outlawed or severely restricted, its priests, nuns, and lay leaders saved countless lives, provided hope to many on the brink of despair, and won the Church new credibility in the eyes of cultural Catholics and non-Catholics alike.

But even massive efforts by courageous and resourceful Church people failed to thwart the agenda of an authoritarian regime that had the support of international financial institutions and was willing and able to repress its citizens on a massive scale. Under such circumstances, the Church can be a second line of defense for such citizens but cannot change or prevent state policies merely by denouncing them. Only a unified and resourceful opposition movement, to which the Church might contribute, can do this.

In fact, the Chilean Church was quite vulnerable in its dealings with state authorities. Its schools were dependent on public subsidies, and half of its priests were foreigners serving in Chile at the good graces of the military government. Pinochet's government exploited these dependencies and periodically intimidated Church personnel in an effort, often successful, to limit the scope of their activities. Such restrictions further weakened the Church's capacity as an opposition force.

Finally, the longer the Church performed its surrogate role, the more adversely was its religious mission affected. Its willingness to respond to all in need opened the Church to penetration by social and political forces of all sorts. It was no longer setting its own agenda but reacting to the largely social and political demands of others. In doing so, it touched the lives of many people it would not otherwise have reached, although it was often at the expense of its preaching and sacramental functions.

The years of military rule were years of great vitality for the Chilean Church (e.g., increasing numbers of seminarians, more low-income people participating in the base communities, and increased credibility for the Church among cultural Catholics with leftist sympathies). However, there were losses on other fronts. Upper-income Catholics became disillusioned with a Church that was associated with resistance to a government that they supported. Military officers distanced themselves from Church leaders and activists who seemed hostile to them and their policies. These phenomena were sources of concern for many Chilean bishops and for the Vatican as well. In the interests of Church unity and remaining open to all social and political forces, they were determined to redirect the Chilean Church along more conventional pastoral lines.

3

The Peruvian Church: A Historical Overview

Peru is a more deeply Catholic country than Chile. Its people are strongly attached to their Catholic priests, saints, festivals, and devotional practices. Protestantism, while growing of late, has not made the inroads into Peru that it has in Chile.[1] Institutionally, however, the Peruvian Church is rather weak. As of 1984, it owned or administered 32 hospitals and 585 clinics and dispensaries, and operated 15 major seminaries, 5 universities, roughly 50 technical or vocational schools, 536 primary and secondary schools, 35 homes for the elderly, and more than 300 social service facilities. These institutions and facilities exert a significant impact, but for a country of 22 million people they are woefully inadequate.[2]

So too are the Church's financial and human resources. The Peruvian Church was not the large landowner it was in other countries,[3] nor was it as closely aligned with the wealthiest families.[4] More important, its Catholics-per-priest and Catholics-per-sister ratios are among the continent's highest,[5] and would be higher still were it not for more than 1,100 foreign-born priests and 2,000 foreign-born sisters (more than half the national total in each case) currently assigned to it.[6]

Actually, the absence of strong challenges from either Protestant or secular movements helps to explain the Peruvian Church's institutional weakness. The strong anticlerical and Marxist parties that prompted Chilean Church leaders to adopt new pastoral programs and strategies were absent in Peru. The Church's extrication from its traditional social and political alliances came later and has been less thoroughgoing. Nor has its social Christian movement developed as fully or coherently.

During the 1960s and 1970s, the Peruvian Church exhibited many of the progressive tendencies seen in the Chilean Church, i.e., the assumption of a moral tutorial role on behalf of social justice, decentralization of functions and responsibilities, and an expanding network of Church-sponsored social services. However, the authoritarian impulses of many Church authorities have made them wary of democracy, and the absence of effective secular carriers for social Christian values has diluted the Church's impact generally.

In what follows we discuss the evolution of Peruvian Catholicism from the mid-nineteenth century until 1975. As in chapter 2, we stress changes in the Church's social composition, pastoral orientation, and internal structures and relationships, relating each to the interplay of societal and ecclesial forces. We identify three distinct phases: a period of *continuing union with traditional allies* (1850–1955); a period of *partial extrication and pastoral renovation* (1955–68); and an era of *moral tutorial and institutional support for reform* (1968–76).

Continuing Union with Old Allies (1850–1955)

During this period, the Church maintained its traditional pastoral orientation and political alliances. It ministered sacramentally to a largely upper-class faithful, and ceremonially to a religious but unchurched mass population. It looked to congenial or sympathetic political elites for help in defending its interests and institutional prerogatives. From independence until the 1930s, it faced moderate or token pressure from anticlerical forces, making only minor concessions to them. Beginning in the 1920s, anticlerical elements were strengthened by the emergence of new social and political forces. In Chile similar forces were powerful enough to force the issue of church-state separation, but Peruvian Church leaders were able to fend them off with the help of both military and civilian allies. Lay Catholic organizations and programs were also part of this effort. These programs, however, were spiritual and their primary function was the defense of Church interests and privileges, unlike in Chile, where a more social thrust led to a new, competitive reform movement. None of the political movements spawned by Peruvian lay Catholics became serious contenders for power or effective carriers of Catholic social teaching to the broader population. The organizations did have a transforming effect on their own members, however, and, in the longer run, on Church structures as well.

Traditional Church privileges and strategies (1850–1930)

The Peruvian constitution of 1834 provided for union of church and state. It committed the government to the defense of the Roman Catholic faith, and explicitly forbade the practice of other religions. For much of the nineteenth century, liberal political elites tried but failed to amend the constitution to allow for religious toleration. Military leaders, supported by landed interests with ties to the Church, controlled the presidency for nearly fifty years following independence (1824–72). Fifteen different constitutions were produced during this period, all of them protecting the established position of Roman Catholicism.

During the last half of the nineteenth century, Church officials made several concessions to anticlerical forces, but resisted calls for separation of church and state. In 1856, they agreed that clerics accused of crimes could be tried by civil courts, and later they relinquished their control of cemeteries and the practice of obligatory tithing. In 1897, under the presidency of the devoutly Catholic but politically pragmatic Nicolás de Pierola, the Church accepted a reform that allowed non-Catholics, virtually all of them foreign immigrants, to be married civilly. Finally, under the presidency of another practicing Catholic, José Pardo, Church leaders acquiesced in the passage of the Toleration Act of 1915, which authorized non-Catholic acts of worship for the first time.[7]

Despite these concessions, the Church continued to defend its status as the one "true" church. Catholic newspapers, which were the dominant print media in Cusco and Arequipa, and competed with leading secular publications in Lima and elsewhere, were an important line of defense and counterattack. In addition, the bishops promoted and encouraged lay organizations like the Peruvian Catholic Society, the Catholic Union for Ladies, and the Catholic Union for Gentlemen, whose purpose was to defend the Church and its privileges. According to a leading historian of the Peruvian Church, the membership of these organizations was drawn predominantly from the middle and upper classes, and from "older, aristocratic families," although not from the country's oligarchy as such.[8]

Church leaders worked closely with conservative allies in Congress to limit these concessions, minimize their impact, and preserve the Church's established status.[9] When Congress debated the Act of Toleration, these leaders encouraged devout Catholic women to fill the galleries and to shout down the bill's proponents. When it passed anyway, the bishops persuaded President Pardo to delay its promulgation. Later, in response to requests from Church authorities, dictator Augusto Leguía refused to enforce the civil marriage or toleration laws, and in 1923 formally dedicated the country to the Sacred Heart of Jesus.[10]

It is difficult to speak of the Peruvian bishops as a single actor during this period, however. Most were ideological and ecclesial conservatives who identified with the country's socioeconomic and political elites and were comfortable dealing with them. But they seldom spoke with a single voice or acted in conjunction with one another. Each was a veritable feudal lord in his own diocese, reporting only to Rome. Each worried about problems within his own jurisdiction, and dealt directly with its social and political elites.

Moreover, the character and interests of these elites varied from region to region. And because there was neither a Catholic nor a "conservative" party operating at the national level,[11] "the Church" was not identified with a particular class or party, as in Chile. Some bishops had good ties with the military, others were close to the legislators of their areas or the civilian president of the day. Finally, some were close to conservative circles, while others were on good terms with more liberal elements. All were comfortable making personal appeals to those in power, but few were inclined to define themselves in politically specific terms.

Mounting anticlericalism and Catholic Action (1930–55)

As the 1930s began, the Church appeared strong.[12] But as Lima's Archbishop Emilio Lissón y Chaves lamented in a 1928 pastoral letter, "Some Peruvian churchmen, owing to their conviction that they could not minister to the spiritual needs of the faithful unless they controlled the political structure had failed both on spiritual and political fronts."[13] According to Lissón y Chaves, the Church had neglected its own followers and structures, relying instead on the power of socioeconomic or political patrons. The strategy was adequate as long as these forces remained unchallenged. But during the 1930s, the country's middle and lower classes began to stir. Economic decline in the highland valleys pushed small holders off the land and into coastal cities in search of work. Others were drawn by the hope of opportunities for their children if not for themselves. Between 1920 and 1930, the population of Lima jumped 60 percent from 176,000 to 281,000.[14]

These immigrants were potential recruits for the Alianza Popular Revolucionaria Americana (APRA), the country's first mass-based reform party, which Víctor Raúl Haya de la Torre founded in 1924. Those who were literate registered to vote, while others joined APRA trade unions and neighborhood groupings, or attended the party's public rallies. APRA drew support from labor, students, and marginalized sectors of the middle class as well.

These forces soon challenged the Church and its elite patrons. In 1931, Haya and military strongman Luis Sánchez Cerro ran for president on platforms calling for separation of church and state. The bishops and the Catholic press attacked both. *El Amigo del Clero*, the official organ of the Lima Archdiocese, portrayed Haya as "incessantly against . . . all Catholic action" and charged his party with espousing "part of the program of Moscow, especially the antireligious part." Some Catholic workers (in Arequipa, for example) supported Sánchez Cerro, but the bishops rejected him and attempted to rally support for the candidate of the Popular Union, an alliance of conservative and reform Catholics that they had tried to put together. During the campaign, priests and sisters were enlisted to carry the bishops' message to their parishioners and students.[15] In the end, Haya and the victorious Sánchez Cerro shared 85 percent of the vote, while the Popular Union candidate finished a distant third. Clearly, most Catholics ignored their bishops, and voted as they saw fit. Apparently they did not view Haya and Sánchez Cerro as "dangerous anticlericals" or Church interests as the central or sole important issue. Their support for candidates against whom the Church had campaigned was a blow to the image of the Church as a moral and political force.

A second successful challenge to the Church was the legalization of divorce in 1933. It had been another of Sánchez Cerro's campaign promises, and he pushed it through the Congress despite the objections of the Catholic bishops. But not even Sánchez Cerro insisted on formal church-state separation. He and other military leaders were more concerned with keeping APRA from power, and did not want to alienate such potential allies as the Church.[16] The furthest that Sánchez Cerro (who was assassinated in April 1933) and his successor, General Benavides, were willing to go was to approve a constitution that assured other religions "freedom in the exercise of their respective beliefs," and yet, "respecting the sentiments of the national majority," affirmed special protection for "the Roman Catholic Apostolic Religion."[17]

The Church was thus cast in a defensive and unfavorable light for much of this period. It proved to be a top-heavy institution with limited influence and an equally limited following. Never very good, mass attendance had fallen off in recent years, and in terms of their political loyalties lay Catholics were difficult to distinguish from the population as a whole. When they had the chance to vote for APRA candidates (in 1931 and again in 1945), most middle- and working-class Catholics did so, while wealthier Catholics supported conservative parties or anti-APRA military leaders like Benavides (1933–39). Church leaders tried to shore up their defenses against anticlerical forces on a number of fronts. Following the 1931 election, a number of bishops lent their support to conservative movements in Cusco, Arequipa, and other parts of the south. In other areas, they strengthened their ties with local military and civilian elites. Some, like Lima's Archbishop Guevara (1948–55), were ideological conservatives and viewed wealthy and politically powerful Peruvians as natural allies. Others, like Pedro Pascual Farfán, were willing to align themselves with anyone capable of defending Church interests.[18]

Msgr. Farfán became archbishop of Lima in 1933 and, along with Msgr. Mariano Holguín of Arequipa, actively promoted the development of national lay Catholic organizations. By the late 1920s, the staid Catholic Unions (for Ladies and for Gentlemen) had all but ceased operations. They were replaced by groups appealing to younger Catholics. Some of these groups were more interested in sports and social activities, while others developed into study groups concerned with religious and social issues.[19] With the hierarchy's encouragement and the guidance of priests familiar with lay organizations in Europe, they developed into a national Catholic Action movement.

For the bishops, however, these people were surrogate clerics and instruments of institutional defense. A 1935 document stated their functions as:

> The sustaining and dissemination of "good press"; the teaching of Religion and Catechetics in primary and secondary schools and outside school settings as well; the crusade against indecent entertainment, particularly films, and the defense of our most holy Religion against Protestant propaganda, utilizing the press, the public school system, and health care facilities . . . to counteract the damage that it can do.[20]

Most Catholic Action activists came together in groups around their priest advisors. At first they used parish structures and networks, but these only attracted members who were already regularly practicing Catholics. In time, they found one another's homes, university campuses, schools, and offices more effective in reaching and attracting new people. Catholic Action groups had invigorating and reenforcing effects on their members. Some were attracted to the ideas of Maritain, Mounier, and other social Christian thinkers to whom their advisors or fellow activists had been exposed in Europe. They helped to form lay and clerical leaders who played important roles within and beyond the Church during the 1950s and 1960s. Fathers Gerardo Alarco, Gustavo Gutiérrez, José Dammert, and Emilio Vallejos were among the early Catholic Action activists who later became influential priests and bishops.[21] Others were Catholic laymen who later helped to launch the Christian Democratic Party.[22]

Catholic Action groups were not particularly effective in defending Church interests or influencing public opinion, however. Unlike their Chilean counterparts, they did little to assist workers or the poor, nor did they have anything comparable to the Juventud Conservadora or Falange Nacional to serve as political outlets. Moreover, they had no access to the media, and they were generally regarded as mouthpieces of the Church and of conservative interests.[23] In the absence of a national conservative or Christian reform party, activists were divided in their political preferences. Those in outlying provinces tended to support local conservative forces. Those from Lima avoided partisan politics, although many supported José Bustamante y Rivero, an independent social Christian who was elected president in 1945, only to be overthrown by military strongman Manuel Odría three years later.

The Church's most visible response to the challenges of this period were the several Eucharistic Congresses held between 1935 and 1954. Archbishop Farfán and other bishops used them to strengthen the resolve of the Catholic faithful, and to intimidate and discourage potential adversaries. The congresses consisted of several days of workshops, public liturgies, and processions. The large numbers of people attending, and the decline in anticlerical fervor in the late 1940s and early 1950s, gave the impression of a Church that was in relatively good health.[24] In addition, the Odría dictatorship's populist policies appeared to be attending "successfully" to the social demands of at least some lower-class groups, to be keeping the dangerous APRA at bay, and to be preserving the Church's institutional interests.

The congresses did little to influence society or the political process, however. In fact, both the congresses and Catholic Action in Peru were intended primarily to defend the Church and not, as in Chile, to promote social justice. Moreover, Peruvian parish structures were inadequate for reaching people or enlisting their support, and the congresses were a means of circumventing rather than addressing this problem. Finally, concern for the quality of people's material and spiritual lives was secondary, and only some Catholic Action advisors exposed their members to the social encyclicals of the popes and social Christian thinking generally.

During the early 1950s, Church authorities were confronted with startling evidence of the low level of religious development of most Peruvian Catholics. A study of religious belief and practice published in 1953 had the same jolting effect on the Peruvian hierarchy that Father Hurtado's study had in Chile during the 1940s.[25] Most Peruvians, the study indicated, were Catholics in name only, with little interest in either doctrine or religious practice. Others were Quechua- or Aymara-speakers who retained their pre-Christian beliefs and practices at least partially for lack of pastoral attention. Finally, most practicing Peruvian Catholics were "standard" Catholics whose sacramental or ritualistic faith centered around personal, rather than social, concerns and was unconnected to their family, work, or social relationships. Catholicism in Peru, the study concluded, was stagnant, and Peruvian Catholics were woefully "under-churched."[26]

These conditions were one consequence of the chronic shortage of priests in rural areas and in the slums and poorer neighborhoods of larger cities. Single priests had to cover large areas and minister to tens of thousands of people. In these circumstances, they had no time to get to know people, give them a firm grounding in faith, or help them to apply their faith to the problems and issues they faced on a daily basis.[27] But the Church's problems were not simply a matter of numbers. Their traditional reliance on the patronage of wealthy families and civil authorities, and the absence of overt crisis, prevented the Peruvian bishops from embracing new pastoral ideas and strategies. Younger, more progressive clerics and lay leaders managed to touch a bishop or two, but until the extent of the Church's underdevelopment became clearer, the ideas and proposals of these leaders lacked the hierarchical support needed for implementation at the national level.

Partial Extrication and Pastoral Renovation (1955–68)

The development of a new conception of mission was the work of Msgr. Juan Landázuri Ricketts, a Franciscan priest and canon lawyer who was appointed archbishop of Lima in 1955. During the 1950s, Vatican authorities began to prod Latin American Churches to reorganize and to pay more attention to the social conditions in which people lived and worked. Landázuri's appointment as head of the country's largest single diocese was a key move in this regard. Under his leadership, the Peruvian Church gave important support to the movements for reform and democracy. Unlike the situation in the Chilean hierarchy, where a strong nucleus of bishops exercised progressive leadership roles in the the late 1950s and 1960s, in Peru Landázuri and his auxiliary, Bambarén, were virtually alone in embracing the social Christian perspective, and there was no competitive Christian reform party capable of channelling progressive Catholicism politically. In the end, the only group to undertake and carry out significant social and economic reforms was the Peruvian military, an institution that was not known for its democratic sentiments or respect for the political rights of its critics.

Expanding pastoral commitments to the poor

Under Landázuri, the Peruvian Church rediscovered the country's urban and rural poor. As it did, it grew increasingly critical of the country's social and political elites. Sociopolitical and ecclesial factors combined to facilitate these moves.

In 1956, conservative president Manuel Prado launched a program of import substitution industrialization. It called for subsidies to new industries while fixing prices on agricultural products. These policies drove peasants and small landowners off the land and into coastal cities. On the political front, in order to attract urban working-class support, Prado offered to legalize APRA and to appoint *apristas* to his cabinet in exchange for the party's support. A new party, Acción Popular (AP), emerged around the architect Fernando Belaunde Terry, who appealed to peasants, workers, and middle-class voters interested in "reform" but afraid of APRA.

These developments suggested possible changes in the country's social and political relationships. They strengthened the determination of Landázuri and other Church leaders to distance themselves from socioeconomic and political elites. The new archbishop had the support of lay and clerical collaborators anxious to put social Christian ideas into practice, of newly arrived foreign priests and nuns willing to work in rural and urban slum areas, and of Vatican officials as well. But he had to overcome the resistance of his fellow bishops, most of whom were theologically and ideologically quite conservative. His remarkable blend of astuteness and congeniality helped him enormously in this regard. So too did new ecclesial structures that enhanced his stature and administrative authority.

An ecclesial and theological moderate concerned with maintaining Church structures and unity, Landázuri was open to the world and was deeply touched by the plight of his country's urban and rural poor.[28] Although he did not always share their views, he was very supportive of the pastoral efforts of his progressive priests. One of his first initiatives was the citywide Lima Mission launched in May 1957. As part of the mission, teams of priests and nuns spent weeks in shanty towns, assessing people's circumstances, distributing food and clothing, administering the sacraments, and teaching catechism. The program succeeded in winning back some lapsed or fallen-away Catholics. It also had a formative influence on the priests, nuns, and laypeople involved, many of whom ended up living and working permanently in the areas they had visited. A related initiative was the holding of a *Semana Social* two years later. Modeled on the French and Belgian Catholic experiences, it brought priests, sisters, lay intellectuals, university students, workers, and peasants together to discuss social problems and the role of the Church.[29] It anticipated many of the concerns of the Second Vatican Council (which would begin in 1962), and had a energizing effect on all who attended.

These developments would have influenced other parts of the country in any event, given Lima's prominence in national life. But the emergence, during the late 1950s, of a National Bishops' Conference further enhanced their impact. Previously, individual dioceses and archdioceses were virtually autonomous units of Church organization. Archbishops Farfán and Guevara (both of Lima) introduced the practice of informal consultations with other bishops, but "ordinary" bishops remained sovereign in their respective jurisdictions, reporting directly to Rome.[30] In 1957, however, the practice of consultations was formalized with the establishment of a national Episcopal Conference. It was hoped that a national ecclesial authority would assure uniformity of practice, more efficient use of resources, and more orderly relations between the Vatican and the local Church. In Peru, as in Chile, it enhanced the authority of the conference president and/or head of the country's largest diocese at the expense of bishops from smaller cities and less-populated areas. They could say and do what they felt appropriate in their respective dioceses, but with the conference speaking regularly for the Church of Peru, their stature and impact diminished. In fact, Archbishop Landázuri used conference machinery to create a Peruvian Church, and to redirect its energies and resources to lower-income Catholics in popular neighborhoods and shantytowns.[31]

Another key figure in this shift was Msgr. Luís Bambarén, a Jesuit priest who was named Landázuri's auxiliary bishop in 1968. Bambarén was assigned to oversee the squatter settlements that were mushrooming on Lima's outskirts. He coined the term *pueblos jóvenes*,[32] emphasizing the positive potential of the new settlements, and encouraged priests, sisters, and lay people to come to the shanty towns to work pastorally and in community development projects. By early 1968 his methods had been adopted by Church agencies in other cities, and Bambarén formed a national organization, Pueblos Jóvenes del Peru (PUJOP), to coordinate their efforts.

As in Chile, the Church's interest in ministering to the popular sectors in Peru coincided with arrival of foreign missionaries. More than 200 priests and 1,400 religious women came in the late 1950s, and by 1964 foreign priests constituted more than 53 percent of all priests, and foreign sisters only a slightly lower percentage of all women religious.[33] They came originally to parishes and schools in middle- and upper-class areas, but ended up in shantytowns and isolated rural areas.[34] Between 1955 and 1964, fifty-three new parishes were established in the slums and shantytowns of Lima. Virtually all of them were staffed by missionary priests and sisters, and provided social services (e.g., libraries, meeting rooms, medical clinics, and recreational facilities) that were not otherwise available to the poor.

Many of these missionaries developed close ties with progressive Peruvian priests who had been working in popular-sector communities for some time. Two such priests were Carlos Alvarez Calderón, who worked with the Young Christian Workers, or JOC, and his brother Jorge, who was pastor of San Juan Bautista parish in the Lurigancho section of Lima and advisor to the Christian Workers' Movement, or MTC. A third was Catholic University professor Gustavo Gutiérrez, who guided the national union of Catholic university students (UNEC). Gutiérrez believed that Christian values should offer people hope in the context of their social circumstances, and that religious ministry required social and political commitment on behalf of the poor.

Gutiérrez and the Alvarez Calderón brothers helped many foreign priests and sisters to change their conceptions and strategies of pastoral ministry. They left the comfort and isolation of their congregations' houses for apartments and rooms in the areas where they were working. They began to meet with their people in the more intimate settings of *hermandades*,[35] catechetical groups, reflection groups, and Christian communities, in which many lower-class Peruvians felt more comfortable.[36] They began to encourage, advise, and assist people belonging to block organizations, district committees, soup kitchens, and other social organizations. A desire to "accompany" the people in their faith and their daily lives led them to experience social reality from the standpoint of the poor, and to feel pastorally responsible for supporting them in their struggles.[37]

These priests and nuns developed organizations that helped them to play more active and effective roles in both Church and society. One such organization was ONIS, the National Office of Social Information, which was established in March 1968 by priests, women religious, and lay Catholics hoping to prepare lay Catholics to deal more effectively with the country's "chronic injustice, oppression, and public immorality."[38] It evolved, under the influence of Gutiérrez and the Alvarez Calderón brothers, into a clerical pressure group with substantial influence on the bishops and on public opinion generally. It also served as a support/reflection group for priests active in social ministry, both Peruvian and foreign-born.[39]

Some observers have argued that the bulk of ONIS's members were foreign priests (who were more difficult for local bishops to control), and that their increasingly radical positions and tactics were unrepresentative of the Peruvian clergy.[40] In fact, most of the organization's permanent members, as opposed to those who would come and go, were Peruvian-born (and largely diocesan) priests, its relations with most bishops remained good until it ceased functioning in 1979, and its public statements were remarkably consistent over time, reflecting the progressive, but hardly radical, tenor of Latin American social science (e.g., dependency theory) at the time.[41]

Other groups active during this period included: the National Conference of Religious, an organization representing priests, sisters, and brothers,[42] Father Gutiérrez's Centro de Educación Popular, which prepared religious education materials, the Bartolomé de las Casas Institute, a research and educational center directed by Father Gutiérrez; and the Episcopal Conference's Social Action Commission (CEAS), headed by Msgr. Bambarén. These groups developed their own identities within the Church, although some were more, others less, "official."

CEAS's statements and documents bore Bambarén's signature and formally represented the country's bishops, although the statements seldom had their unanimous endorsement or support. Initially concerned with assisting priests and lay catechists who were working in rural areas, CEAS quickly became a platform for moderate and progressive bishops and their advisors. Its executive secretary, Jesuit theologian Ricardo Antoncich, wrote most of its statements and had a hand in drafting other conference documents.[43] Assisting him was a small staff of lay graduates of the Social Science faculty of the Catholic University, most of whom were close to Father Gutiérrez.

The other groups "spoke for themselves" and not for the Church as a whole, although they had close ties to Archbishop Landázuri and other members of the Episcopal Conference's Permanent Committee. Their public statements and remarks were often taken as representing "the Catholic community" if not "the Church" as such. Conservative bishops viewed them with alarm and saw them as challenging their traditional monopoly over Church representation. Their net effect was to represent "popular sector" Catholics in both Church and society.

With greater attention being given to social issues, differences of opinion in Peruvian Catholicism became more apparent. While Landázuri and the Episcopal Conference were dominant at the national level, clerical and lay Catholics at lower levels grew more independent of their ecclesial superiors. Most progressive priests looked to Landázuri for help in their dealings with conservative bishops and parishioners but seldom consulted him before acting.[44] Their greater independence and latitude during the 1960s and 1970s was partially a function of the shortage of both priests and nuns. In this context, bishops thought twice about suspending or dismissing a priest or a congregation of nuns.[45] Pastoral agents also benefited from their knowledge of pastoral methods and procedures (to which many bishops were willing to defer) and from the difficulty that Church authorities had in monitoring their actions and initiatives.[46]

The Church as moral tutor: support for reform and democracy

Despite changes in internal dynamics and in its relationship with social and political elites, the Peruvian Church had little impact on politics during this period. It lacked a secular "carrier" like the Chilean PDC to bear its message of social reform in the political arena. For one reason or another, the political figures and movements to which most bishops and priests looked for leadership of the reform process (e.g., the Peruvian PDC, and Belaúnde's Acción Popular) were not equal to the task.

Church authorities sought to infuse a sense of social responsibility in political leaders and the population generally. Landázuri believed that the Church should play a "guiding" role in the country's political and social development and persuaded his fellow bishops to endorse structural changes, a more just distribution of wealth, and worker participation in company profits and decision-making. In 1961, they denounced "those who were standing in the way of economic and social reform" and spoke of the government's special obligation to aid "less favored" citizens. In 1963, they called for an "orderly" agrarian reform, and warned that the current state of "social decomposition" could lead to violence unless there were "profound changes," and unless those in power overcame their "fear" of the masses.[47]

Such statements and positions must have been troublesome for socially and theologically conservative members of the conference.[48] Thanks to his own imposing figure, to the documents of the Second Vatican Council (1962–65), and to the recent encyclicals of Popes John XXIII (*Mater et Magistra* and *Pacem in Terris*) and Paul VI (*Populorum Progressio*), however, Landázuri convinced them that the Church should endorse social reform and that it should distance itself from established social structures and forces.

These new positions closely paralleled those of Peru's small Christian Democratic Party (PDC), which had been founded in 1956. Unlike its Chilean counterpart, however, the Peruvian party never got off the ground. Its bland, and mostly provincial, middle-class leaders were overshadowed by APRA's Haya de la Torre and the elegant "reformer" Fernando Belaúnde. The Christian Democrats had no one (like the Chilean PDC's Eduardo Frei, for example) to compete with either, and ended up splitting in two, with progressives retaining control of a smaller PDC, and more conservative elements breaking away to form the Popular Christian Party (PPC), which won 10 to 15 percent of the national vote in several ensuing elections.[49]

There were elements within the Peruvian military, however, who shared the Church's concern for social change. In 1962, following an inconclusive election,[50] and with APRA likely to support the former dictator Odría if matters were left to the Congress, the high command of the Armed Forces set aside the results and assumed executive power until new elections could be held. Actually, the military had been worried about the country's economic and political conditions for some time. Senior officers had established the Centro de Altos Estudios Militares (CAEM) in 1950 to train candidates for high command. CAEM's directors undertook to redefine the concept of national security as requiring "national control over the extraction and production of petroleum" (currently monopolized by foreign companies), more systematic pursuit of industrialization, agrarian and tax reforms "to assure internal stability and neutralize revolutionary potential," and the stripping of power from the "traditional oligarchy for its responsibility for Peru's underdevelopment and dependent ties with foreign capital."[51]

The provisional military government moved on several of these fronts. Tax reforms were introduced forcing exporters (mostly agricultural) to pay more. Unused and poorly managed estates were expropriated and social projects were instituted in regions where guerrilla insurgencies had arisen. Projects that helped to integrate indigenous peoples into national life, that subsidized middle- and low-income housing, and that involved the military in public works and community development, were introduced. And a new election law was decreed to reduce fraud and intimidation during voting.

Church officials were impressed with these programs and with the thinking that underlay them. They came to see the military as a potentially reformist social and political force. It may have been easier for them to do so since the democratic credentials of civilian forces (in terms of both policy positions and political conduct) were highly suspect. Neither the bishops nor the military took sides in the 1963 election, although both were sympathetic to Belaúnde. His program of agrarian reform, national planning, expanded social services, and national integration was the closest to the thinking of reformist elements in both institutions. In the end, Belaúnde captured 39 percent of the vote, more than the one-third necessary to become president; Haya polled 34.3 percent and Odría 25.5 percent.

Once in office, Belaúnde moved to fulfill his campaign commitments, making headway on several fronts. An agrarian reform law was passed. Programs expanding health, education, and housing services were launched. Roads linking the coast with the interior were built, programs in which the military worked with local indigenous communities were begun, and municipal elections were restored after a forty-five year lapse. The new president had only minority support in the Congress, however, and APRA and UNO (Odría's party) combined to block his initiatives. In some instances, they refused to consider his proposals at all; in others, they watered them down or approved them but then rejected the tax increases needed to finance them.[52]

Further complicating Belaúnde's position was the suspension of U.S. Alliance for Progress funds because of the nationalization of properties of the International Petroleum Company. Belaúnde was forced to finance his reforms by Central Bank emissions and loans from private foreign banks, which extended the country's foreign debt and hampered efforts at containing inflation.[53] Guerrilla insurgencies also arose during Belaúnde's term. They were routed by the military, but the officers involved came away convinced that radical reforms were needed if further upheaval was to be averted.[54]

A final obstacle to Belaúnde's fulfillment of campaign commitments were his government's own divisions and defects. Moderate, centrist, and radical populist advisors and supporters vied for control of his agenda from the outset. Their divisions delayed the initiation of programs, diluted the government's reformist image, and undermined its support in both military and ecclesial circles. In the months preceding the October 1968 coup, ONIS's public statements stressed the need for swifter and more radical reforms. The bishops, on the other hand, were troubled by the evidence of scandals and improprieties involving government officials. These had been a hallmark of all recent "democratic" governments but were particularly blatant and extensive under Belaúnde.[55]

In this context, Church officials came to view the prospect of a reformist military government more favorably. The Catholic backgrounds of leading military progressives,[56] and the Catholic roots of military thinking on social and economic matters, also influenced some bishops.[57] A certain affinity thus existed between the Church and the military prior to the October 1968 coup. By mid-1968, many churchmen appeared to regard the promotion of social justice and reform as having greater urgency than either constitutional or liberal democratic principles.

The coup itself was the result of a generalized loss of confidence in the ability of "democratic" institutions to carry out necessary reforms. It was precipitated by the discovery of the excessively generous terms of the government's settlement with International Petroleum Company. When the terms were made public in late September,[58] Belaúnde was denounced from virtually all political quarters. Archbishop Landázuri issued a strongly worded statement, saying that "the matter directly affected the country's sovereignty and well being," and that "the principles of patriotism, justice, and public and private honesty (had to) be upheld, and their violation be sanctioned in prompt and exemplary fashion."[59] While not intended to promote or justify a coup, his remarks lent the military's rationale additional credibility. In the wake of the coup itself, which came a week later, Church leaders neither defended nor condemned it. But the impression of many, which they did nothing to counter, that somehow they approved.

Moral Tutorial and Institutional Support for Reform (1968–75)

The first seven years of military rule drew the Peruvian Church into unprecedented political involvement. With normal political channels and institutions suspended, Catholics played an unusually active role in policy discussions and general political debate. The Peruvian Episcopal Conference's leaders, most of them moderate progressives, supported the military government's general direction, although they were critical of certain provisions and methods of implementation. Ecclesially and ideoleogically conservative bishops, priests, and laymen, on the other hand, opposed Velasco and his policies across the board. As Velasco's reform programs and economic policies turned sour, and as his government lost both its patience and its reformist zeal, the Church was less able to speak or act effectively in its defense.

Initial period of social reform (1968–73)

The new government's character was not immediately clear when the military ousted Belaúnde. Velasco favored the reformist agenda of the CAEM people, but other officers did not. Discussions continued well into 1969 before Velasco marshalled sufficient support for the agrarian (June 1969) and industrial (July 1970) reforms.[60] Archbishop Landázuri and a few other bishops[61] were anxious to help progressive elements bring the others along. They spoke of the need for reform generally, and publicly endorsed specific initiatives as they were announced. They created the impression of a Church identified with the goals and policies of the new government.

This was not really the case. Landázuri and the other moderate and progressive bishops who made up the conference's Permanent Committee set the agenda for conference assemblies and determined the content of its documents. Outspoken ecclesial and theological conservatives like Jurgens of Trujillo and de Orbegozo of Chiclayo[62] remained an isolated minority, and most other bishops were pulled along by the reformist spirit that the Latin American bishops had embraced at Medellín in August 1968. The fact that Velasco's reforms were things that the Church had been urging for decades, that Velasco appointed Landázuri's brother-in-law, General Montagne, his minister of interior, and that several other members of his inner circle were products of the Church's *cursillo* movement, made it easier for some conservatives to view Velasco and his reform agenda favorably.

Moderate and progressive bishops were thus able to speak for a much more conservative conference during the first several years of military rule. Six days after the coup, Cardinal Landázuri publicly congratulated the new government for nationalizing the IPC. Although he requested an early return to constitutional rule, he did not press the point.[63] In contrast with other Latin American military governments of the period, there were no arbitrary detentions, instances of torture, or summary executions with which to be concerned. Indeed, many bishops seemed relieved that the political stalemate of recent years had been broken, and that the new government, although not conventionally democratic, was moving ahead with the reforms.

A statement issued by the conference following its annual assembly in January 1969 was particularly pointed. It attributed the country's poverty and misery to the "concentration of economic and political power in hands of the few and of the international imperialism of money which operates as an accomplice to the oligarchy." When the agrarian reform was announced in June, several bishops endorsed it,[64] but were critical of some of its provisions and of the manner in which it was later implemented in some areas.[65] Catholic landowners and religious orders whose lands were affected, on the other hand, opposed the law and sought "exemptions" from local officials. People who worked in Church organizations during this period note (critically) that a number of the bishops opposed the government's land reform program because it affected the holdings of wealthy friends or religious congregations with which they had ties.[66]

In July 1970 the government announced its industrial reform law (the *ley de industrias*), under which workers could purchase stock in the firms in which they worked.[67] It was challenged by businessmen and others concerned with property rights, but Landázuri and other bishops gave it a ringing endorsement. They saw it as assuring meaningful worker participation in the ownership and management of their firms, something that Catholic social teaching had been advocating for years, and as "another positive and important test in the transition process we are living."[68] While praising the government's effort to address a serious social problem systematically, however, the bishops made a point of leaving evaluation of specific features and mechanisms to "qualified experts."

A month later, the government introduced its education reform proposals. Initial reaction was mixed, with some bishops praising efforts to increase the number and quality of schools serving lower-income children, while others questioned provisions dealing with religious instruction. The matter was referred to the Episcopal Conference's Educational Commission, which was caught between Catholic parents determined to defend "quality" (private) schools for their children, and more progressive priests and educators who wanted higher-quality education for all. The commission's head acknowledged that difficult issues were involved, but thought that these were resolvable, and that many Catholic parents were simply interested in defending their privileges.[69]

In general, the Velasco government viewed the Church as a potentially valuable ally. Within days of assuming power, the new president called Bishop Bambarén to ask for suggestions on how the government could deal effectively with the urban poor. Bambarén offered advice from his PUJOP experience, and two months later the government established a National Office for the Development of Young Towns (ONDEPJOV), applying several of Bambarén's ideas at the national level. In 1971, ONDEPJOV became SINAMOS, a program of state-chartered neighborhood organizations to organize the poor and provide them with social services.[70]

Bambarén had close ties to the military government, and was very supportive of its policies and objectives. When a priest reported to him that there was something wrong in his parish, Bambarén would call one of his friends and have the matter "resolved." He relished his notoriety and his influence with Velasco and other officials. He used them effectively to convey the Church's concerns and to promote the cause of squatters, local-level social organizations, and popular-sector groups generally.

Bambarén quickly became a leading Church figure and spokesperson. Velasco himself confirmed this during the "Pamplona" affair in May 1971. Bambarén was arrested by the minister of interior, General Artola, for presiding over a liturgy honoring a squatter who had died of injuries suffered in an attempted land invasion on the outskirts of Lima. Velasco had the bishop released, and later dismissed Artola. Bambarén responded by accepting the government's apologies and later urging "the entire Peruvian family . . . (to) . . . be united around the revolutionary process that we are living."[71]

A 1971 document ("Justice in the World") prepared for a synod of Bishops in Rome reinforced the impression of a hierarchy solidly behind the Velasco government. In it, the bishops advocated:

> that the Church commit itself to supporting governments that arise seeking to implant in their countries more just and more humane societies, helping to breakdown prejudices, recognizing their aspirations, and encouraging them in the search for their own path toward a socialist society with humanistic and Christian content.[72]

The Peruvian bishops issued this document just as their Chilean counterparts were acknowledging that socialism could be compatible with the social teachings of the Church.[73] But the Peruvians had not insisted, as the Chileans had, that socialism be pursued within a liberal democratic framework. Their strong position was a function of Landázuri's leadership, and of the language and other skills of his advisors. Klaiber explained the matter in these terms:

> At that time the Episcopal Commission for Social Action (CEAS) was very active and . . . (its) president was Luís Bambarén. But more than that, the Secretary was Ricardo Antoncich. Antoncich was a key influence. . . . Landázuri's role was always passive really. He would just say 'go and write something and I will look at it'. . . . So . . . Ricardo and others would write these documents. . . . Then, they would present the document to the Episcopal Assembly . . . and at the time there was a lot of pressure to approve social documents. So, in other words, . . . the bishops were under the impact of changes in the Church, and the Pope. . . . These

> are the years of Pope Paul VI. So I do see a very strong influence of CEAS and the bishops being persuaded by the impact of the moment.[74]

Apparently, it also helped that the conservatives were willing to let others write the statement on social justice as long as they could dictate the terms of a parallel document on the priesthood.

During the early 1970s, Catholic groups sought to influence public policy from within and beyond the Velasco government. The most prominent cases of insider influence were the Christian Democratic and independent Catholic intellectuals. Two Jesuit priests, Ricardo Morales and Romeo Luna, and the independent Catholic intellectual Augusto Salazar Bondy, were responsible for drafting Velasco's educational reforms. Christian Democratic (PDC) leader Hector Cornejo Chávez helped to fashion the industrial reform bill and later became editor-in-chief of one of the newspapers that the government expropriated in 1974. Fellow Christian Democrats Francisco Guerra, Carlos Franco, and Carlos Delgado helped to direct the efforts of SINAMOS and other government agencies promoting social and political participation.

Most observers think that the Christian Democrats filled an ideological vacuum in the Velasco government and had considerable influence over it. Others insist that the military remained in control and used the the PDC (and other Catholics) to put a Christian veneer on policies to which they were already committed. As a conservative political leader with whom we spoke put it,

> General Velasco headed up a military government. . . . It was the armed forces, and only the armed forces . . . that governed Peru. It was a chemically pure military dictatorship. But confronting Belaúnde's movement and APRA, Velasco needed to simulate . . . popular participation. . . . And into this situation comes a political expert, Carlos Delgado, who says there's no problem with participation, and . . . quickly develops a movement and even an office of participation. It was thus a government of absolutely no participation that tried to appear as participatory, for which it needed small groups, just about anybody, to embrace it. And the anybody that appeared was Christian Democracy.[75]

Intermediate-level Church organizations also took part in both public and private policy debates. ONIS and a lay counterpart, Faith and Solidary Action, held press conferences, took out newspaper ads, and circulated materials in which they criticized government policies. They were particularly concerned with the government's repression of strikes and demonstrations, and, in general, they pressured the regime to broaden and accelerate its reforms.[76]

This was also the time that Father Gustavo Gutiérrez published his path-breaking *Teología de Liberación.*[77] The book was a synthesis of Gutiérrez' biblical reflections, his sociological studies at Louvain, and his years of pastoral experience among Lima's poor. In it, he argued that religious salvation was neither an otherworldly phenomenon nor an experience mediated exclusively through prayer, sacramental practice, or personal acts of faith. It rather involved the liberation of the full human person, body and spirit. Conversely, sin was both individual and collective, and to liberate the poor was to reject sin and to begin a process of personal and collective salvation. In fact, if the Church were to truly give witness to the gospel, it had to join the poor in their political struggle for liberation from poverty and injustice.

Gutiérrez's ideas helped priests and nuns to integrate the religious and social dimensions of their ministries. His Bartolomé de las Casas Institute and the Centro de Educación Popular, which he directed, provided them with pastoral and socio-analytical materials that helped them advise and strengthen popular organizations in their areas. Among these, the ecclesial base communities (*comunidades eclesiales de base*), which had been endorsed by the Medellín Conference of Latin American bishops (in August 1968), grew significantly in the early and mid-1970s.[78]

Social and political involvement by nuns and priests accelerated as the government turned its attention to the *pueblos jóvenes.* Velasco showered resources on neighborhoods and organizations that would cooperate, thereby confronting priests and other pastoral agents with difficult political decisions. Many were critical of certain policies and methods but welcomed the government's interest in assisting the poor and in providing a context in which popular social organizations and movements might develop. In virtually all instances, they ended up working with local-level government officials. Parish facilities were used as meeting places, and priests served as mediators between the government officials and neighborhood residents. The criticisms and suggestions of local Catholic leaders and activists helped to improve specific policies and programs and often had the effect of broadening public support for them.

Many of the grass-roots organizations were advised by Catholic intellectuals and activists working out of "alternative" centers and institutes.[79] Funded by international Church sources, they designed community development projects, directed popular education programs, and provided forums for the discussion of political and ideological issues. They were havens and laboratories for Catholic activists who were skeptical of both the military government and the "vanguard" leftist parties that began to operate in 1972 and 1973. They provided valuable formation and organizational support for pastoral agents and their lay collaborators. They tended to encourage people's expectations and to pressure the Velasco government to do more to alleviate the social and economic problems of poor people.

The Catholic organization most consistently critical of the Velasco government was ONIS. It was particularly insistent that more be done in terms of popular participation and autonomy. It wanted the government not only to work for people, but to enable them to control their own organizations and to carry out their own agendas.

ONIS priests loaned their churches to popular organizations for meetings, hunger strikes, and other purposes. They consistently supported workers in their conflicts with their public- and private-sector employers, and attacked the government for refusing to recognize worker and peasant organizations and their concerns.[80] Unlike other Catholic groups, ONIS was willing to take positions on specific issues. With respect to industrial reform, for example, its spokesmen endorsed the idea of worker-owned and worker-run firms, but argued that unless workers were given a voice in the structure of the sector and market in which they competed, and unless restrictions were placed on what they could produce, their firms would not survive.[81]

Structural changes introduced in the Church in the late 1960s made it easier for priests, sisters, and laypeople to develop their own positions with respect to the military government. In the archdiocese of Lima, for example, the emergence of deaneries (which included from six to ten parishes) and vicariates (that included as many deaneries) provided horizontal contacts among the priests and pastoral agents of a given area. They made it possible for them to develop common positions and to press superiors into taking positions as well. Such structures enabled bishops like Bambarén to coordinate their priests more effectively, and to assume responsibilities traditionally reserved for the archbishop.

In Bambarén's case, they reenforced the impression of a Church solidly behind the regime. His public statements invariably defended the "legitimacy" of its policies and initiatives from the standpoint of Catholic social teaching, thereby countering the arguments of conservative Catholics. In July 1973, in a sermon given at an Independence Day mass, he challenged them frontally:

> To be a brother is to effectively overcome the source of inequality, (but) without a false and hypocritical reconciliation of interests, and without defending private property when it is precisely private property that prevents most men (and women) from enjoying the fruit of their labor in a dignified manner. . . . Anyone who finds themselves in a situation of extreme necessity has the right to take what he needs from the wealth of another.[82]

Also providing support for the Velasco government was the document on evangelization and liberation, which the bishops issued the preceding January. It was written by one of Landázuri's auxiliary bishops, Germán Schmitz. It warned radical and progressive pastoral agents against getting caught up in politics, confusing their religious missions with temporal political concerns, and being used by partisan political forces. But it defended the Church's "political mission" and its preferential option for the poor and the oppressed. It further denounced the government's human rights violations and the obstacles that others, including "Catholics who were more concerned with external practices than with social commitment," were placing in the path of necessary reforms. And, finally, it stressed that "much more remained to be done" in this regard.[83]

The insistence that Velasco's reforms were legitimate and in line with Church teaching left an impression of political support that not all bishops liked. The distinction between the legitimacy of general objectives and the provisions of specific policies allowed Bambarén and other progressives to think that they were avoiding partisan or contingent politics. But such niceties were usually lost on the general public. A generally favorable reference to a policy created the impression of support for that policy, whatever the reservations with which it was expressed.

It is not clear, however, that the Church actually "legitimated" Velasco's government in the eyes of rank-and-file Catholics. Its pronouncements may have helped some middle- and upper-class Catholics to tolerate initiatives they otherwise might have opposed, but there is no evidence that pastoral agents "converted" significant numbers of people to the government's side. They may have encouraged those already supporting the government's policies. They may have helped them to defend or pursue their interests more effectively. But they do not appear to have sustained the "populist" programs of an authoritarian government that faced little opposition in this initial phase.[84] Nor, with the exception of the educational reform proposals, perhaps, did the criticisms or pressures of Catholic groups appear to affect policy debates or relations among the government's various factions. As in the case of Chile, when some bishops attempted to save the Allende government in 1973, the Peruvian bishops' public authority by itself was insufficient to shore up a government that had lost its ability to rule effectively.

Regime retrenchment (1973–75)

Between 1973 and 1975, the Peruvian economy, and with it Velasco's government, began to falter. As conditions worsened, so too did his health and equanimity, and he resorted increasingly to force in dealing with critics and adversaries. His nationalization of the country's newspapers in 1974 was a final blow aimed at elite interests; but it created as many problems as it solved, and left him isolated in military and broader political circles. In late August 1975, Church authorities and activists watched helplessly as a government with whose general objectives they identified was pushed aside by more conservative military leaders.

The economic difficulties arose from several sources. Domestic and foreign investment declined in response to the government's nationalization policies. Inflationary pressures were building as a result of wage increases granted since 1968. In addition, revenues from exports fell short of anticipated levels, as changing ocean currents led to a sharp decreases in anchovy exports (1973), and copper prices plummeted (1974). Finally, oil deposits discovered in 1971 and 1972, on whose promise the government borrowed more, spent more (on social welfare programs), and worried less, proved more modest than initially estimated.

In these circumstances, Velasco found himself pressured from all sides: by banks for debt service, by manufacturers for lower labor costs, and by military colleagues for new equipment and the containment of labor unrest. These measures amounted to the abandonment of his reformist project. They alienated unions and popular organizations,[85] and yet failed to stabilize the economy.[86] Velasco's personal health problems further complicated matters, forcing him to curtail his activities and depriving his government of strong leadership at a critical juncture.

Elements of the Catholic community responded differently to the tensions and confrontations that ensued. The hierarchy refrained from speaking out on specific labor conflicts, and actually mediated a number of them. The bishops expressed concern for the problems faced by the popular sectors, but with the exception of statements by individual prelates and regional groupings they did not challenge the government directly.

Sensing its political isolation, the government sought official Church support. Following a meeting in August 1974 with the conference's Permanent Committee, the head of SINAMOS, General Zavaleta, told reporters that "the bishops had reaffirmed their support for the revolutionary process and were in complete agreement with the newpaper expropriations and the New Press Statute." The bishops protested, insisting that they had asked for the meeting to convey their "grave reservations and concerns with respect to certain aspects of the revolutionary process," that many of their comments had been critical, and that they had outlined the principles that ought to govern the operation of the expropriated newspapers and had not passed judgment on the expropriations themselves.[87]

Landázuri and others who had supported Velasco through 1973 went along with this more critical public stance. They were unhappy with the government's increasingly repressive methods, and they were troubled by their own divisions with respect to specific policies (e.g., the government's educational reform proposals) and the propriety of Church involvement in politics generally. The conference's composition had not changed markedly since the late 1960s, but with the government under attack skeptical bishops were willing to speak up and to resist the arguments of Landázuri and other progovernment colleagues.

On the other hand, intermediate organizations like ONIS, CEAS, some religious orders, and lay groups, were critical of the Velasco government's retreat. They endorsed union demands, and accused the government of failing to bargain in good faith, and of repressing dissident union and political leaders. Many Christian communities, neighborhood groups, and parish teams tried not to take sides in these matters, but others joined with leftist parties and unions in attacking the government, particularly where lay leaders were active in secular movements and political organizations.

Such politicization led to increased tensions between Church authorities and local parishes and communities. As social pressures and forces enveloped the Church, priests and pastoral agents were forced to set aside their own agendas and to accommodate the personal and social demands of their people. Tensions never reached the point of open conflict, nor did they precipitate challenges to core Catholic concerns, as in Chile during the Frei and Allende years. Most priests and activists at the local level avoided partisan political commitments, and their political judgments and commitments were easier to accept as part of the Church's broader commitment to assist the poor in their struggle for liberation and social justice.

During this period, only the Commission on Social Action (CEAS) came close to proposing concrete solutions to the Velasco government. It continued to argue for the legitimacy of noncapitalist institutions and experiments, although it conceded that the government's efforts could be undercut by legal or technical defects or by problems of economic viability. It also acknowledged that the Church had no particular expertise in these areas. But in May 1975, it spoke of a crisis in the revolutionary process, and of the need for "new arrangements and understandings regarding popular political participation." It argued that the military should view popular organizations as channels of participation rather than social control, and actually concluded that in the present circumstances it would be best for it to "transfer power to the people."[88]

Had the bishops and other Church groups united around such a position, they might have helped Velasco to shore up his position within the military. But with Church people holding a variety of positions, and with many either saying nothing or speaking in highly general, normative terms, the impact of progressive, pro-reform elements was diluted. In fact, Church authorities did not know how to help to sustain the reform process at this point. They knew that conditions were bad and getting worse; they knew what needed to be done in the longer term but were not sure what to do at the moment or how to advise others. Their message at the end of 1974 was a classic expression of politically impotent moral indignation. It lamented that many Peruvians did not have enough to eat and were being denied their socioeconomic rights. It implied that further reforms, rather than retreat or consolidation, were needed. It looked forward to "peace," but not "one that accepts injustice or returns to a sinful past or to a paralyzing and conservative stagnation."[89] But having said all this, the bishops fell silent. They had nothing to say, for example, about strategies or objectives around which Christians could unite in the short term, settling for less but heading off potentially greater disasters. And they had nothing to suggest in the way of appropriate compromises, adjustments, or bases of consensus because they themselves were unsure or badly divided on such matters.

During the Velasco years, the Church grew steadily more fragmented. Its public pronouncements between 1973 and 1975 were much less consistent and less effective than earlier. The bishops' Commission on Social Action held conferences and issued declarations but spoke less for the Episcopal Conference than for Bishop Bambarén and his advisors. Catholic intellectuals and social activists, and intermediate and local-level groups such as ONIS, promoted and defended popular organizations, pressed the regime to fulfill and extend its reform commitments, and occasionally upstaged or outflanked the bishops. But Catholics with children in elite academies, or whose farms or factories were under siege by their workers, joined others in opposing the government's "attack" on religious education and private property. In these circumstances, Church authorities found it difficult to address issues effectively or to control the initiatives and commitments of their faithful, even when these "represented" the Church in some capacity or another.

Conclusions

We draw several conclusions from this review of Peruvian Catholicism through 1975. First, we believe that the changes in social composition, ecclesial structure, and pastoral orientation that enabled the Church to play a progressive role in Peruvian politics beginning in the late 1950s were fundamental and lasting.[90] From that time through the mid-1970s, a once arch-conservative Church grew more moderate, if not more progressive, theologically, ecclesiologically, and politically. In the process, it distanced itself from the country's socioeconomic and political elites, and won back to its fold workers and others who had drifted away during the nineteenth and early twentieth centuries.[91]

Second, these changes were the result of an interplay of societal and ecclesial forces. Social developments produced challenges to which new pastoral strategies and conceptions of Church arose as responses. These changes would not have been adopted, however, had they not been consistent with core features or concerns of the Catholic tradition. Once adopted, they helped to accommodate and channel additional pressures. They conditioned the Church's social and ecclesial experience (at all levels), and in turn had an impact on the social and political processes.

During the 1960s and 1970s, however, the Church's capacity for influence was weakened by both societal forces and changes in its own structure and dynamics (which were also affected by these forces). During these years, the complexity of socioeconomic and political issues, the ability of others to take on political responsibilities, and the presence or absence of political crises, affected the Church's social and political impact more directly than the orientations of individual bishops, priests, or lay Catholics. Their attitudes affected the extent to which the Church's potential for influence was fulfilled but could not create it in the first place. Moreover, the Church's social and political involvement brought to light tensions and conflicts in its own ranks and helped to generate new ones. These forces produced a pluralism among the bishops, clergy, and laity at various levels. As the Church extended the scope of its moral concerns and authority, its cohesiveness and impact declined.

These conclusions permit us to consider the hypotheses articulated in chapter 1. The Peruvian Church's experience exhibits distinctive features, but, like the Chilean case, it underscores the importance of societal contexts in shaping ecclesiastical choices and impact.

The first hypothesis predicted that the more amicably and completely a national church extricates itself from the state and from traditional conservative allies, the sooner it will develop new strategies for influence in society. The Peruvian Church's extrication from traditional alliances began much later (than in Chile) and was not yet complete by 1975. The Peruvian clergy (especially the bishops) continued to exercise political influence through direct, personal contact with elites through the late 1950s. In Chile, episcopal and political elites helped to shepherd the negotiations that produced separation in the 1920s and helped to facilitate the Church's subsequent distancing from the Conservative Party. Their country's more diversified economy, social structure, and party system both enabled and forced them to do so. The process was handled well by enlightened leaders on both sides, but the need to face extrication was a consequence of changes in the broader society, whose timing could not be controlled by religious or political leaders.

Peru's economy, in contrast, remained predominantly agricultural and its class structure much less diversified. Landed and commercial elites dominated political life until the 1950s, and the absence of an expanding middle class and/or sizable industrial workforce reinforced these trends, thereby limiting the potential constituency for anticlericalism. In the resulting context, churchmen continued to depend on elites pledged to preserve their status and prerogatives. In fact, in Peru, the *push* of its social and political context was more important than any ecclesial *pull* from Rome, and may well have determined it. The Chilean Church developed the resources to handle and survive extrication, and did so in defense of its core religious interests. But the Vatican felt no compulsion to encourage the Peruvian Church in this direction until much later, since not only were its bishops willing to continue operating as they had been, but local conditions made this appear feasible.

According to the second hypothesis, successful incubation of social Catholic or social Christian tendencies depends on the extent of Vatican support, commitment from the national clergy, and the absence of threats to hierarchical authority from lay groups. The Peruvian experience again underscores the importance of social context, however, and calls for qualification of this hypothesis.

Catholic lay programs in Peru were inaugurated with less tension between the laity and hierarchy than in Chile. At no point since their establishment in the 1920s did any of Peru's lay movements challenge the hierarchy. They began and continued to grow under the watchful tutelage of bishops for whom they were instruments or extensions of authority. For most of their early existence, however, Catholic Action and other lay programs in Peru lacked a strong social thrust. They were used by the bishops to defend Church interests and prerogatives. Only a few of the groups became training grounds for laypeople committed to social change, and only a few linked up with secular reform movements.

The Peruvian case thus shows that at least in the short term such programs can exert a conservative impact on church and society, and that their direction depends at least as much on cultural and political contexts as on their endorsement by the bishops, their approval by the Vatican, and/or the presence or absence of threats to episcopal authority. In fact, local cultural and political contexts help to determine the goals or objectives that Church leaders will define for these new organizations.

The third hypothesis suggested that for clerics to play an effective moral tutorial role, their pronouncements on behalf of social justice would have to be sufficiently contextualized to be relevant to issues of the day, but not so specific as to tie the Church to partisan ideologies or movements. It added that there must be a willing and ready "ear" in secular society to heed such pronouncements. The Chilean Church's experience in the late 1950s and 1960s confirmed this hypothesis but underscored the difficulties of maintaining conditions favorable to such a role for an extended period of time. The Peruvian case suggests, however, that the clergy's efforts are unlikely to bear fruit where progressive institutions and forces are not firmly established.

The episcopal conference (under Landázuri), its Social Action Commission, and the priests' group ONIS, attempted to play moral tutorial roles on behalf of social reform and economic justice in Peru from the late 1950s through the 1970s. None of them crossed the delicate line by supporting specific parties, as the Chilean bishops did (on behalf of the PDC) in the early 1960s, and as some Chilean clerics did (on behalf of the Popular Unity government) in the early 1970s. But the Peruvian Church still failed to exert effective moral tutorial influence. Many of its bishops and priests found it difficult to give strong and consistent support for either democratic principles or social justice. The Peruvian hierarchy was much more divided than the Chilean on whether the Church should support social reform, and this diluted its political impact greatly. Moreover, most civilian policy-makers (Belaúnde, APRA, etc.) resisted the call of prelates who did lend their support, unlike the Chilean case, where both the Christian Democratic and Popular Unity governments attempted to implement a number of the reforms that many bishops and priests had been urging. And even the country's progressive military leaders were unable to sustain their commitment to reform or accommodate dissent beyond the early years of the Velasco period.

The fourth hypothesis stated that a national Church must have viable "secular carriers" through which its laypeople can act and exert influence if it is to have a progressive impact on society. In Chile, the Christian Democratic Party provided a multiplier effect for social Christian values at the national political level. In Peru, social Christianity was a weaker force, whose possibilities for political outreach or expression were much more limited.

On the other hand, the character of Peruvian Catholic Action, and the lack of strong episcopal support for constitutional rule, hampered social Catholicism's development and limited its impact. Peruvian Catholic Action produced far fewer progressive lay Catholics from which a future Christian Democratic Party might later draw. Catholic Action groups in Peru were instruments of institutional defense more than schools of social Catholicism, and the Peruvian bishops were anything but enthusiastic supporters of either democracy or social reform during the 1940s or 1950s.[92]

In the external political arena, on the other hand, the small size and factionalism of the Christian Democratic Party, the early dominance of the political center by the anticlerical APRA, and the inorganic nature of Acción Popular limited the possibilities for effective influence on the part of progressive lay Catholics. Under Archbishop Landázuri, social Christianity became the dominant tendency in both the episcopacy and Catholic Action circles during the late 1950s, but by then the political space in which a Chilean-style Christian Democratic Party might have operated successfully had been filled by forces not particularly attractive to progressive Catholic voters. APRA enjoyed substantial support among industrial and agricultural workers and progressive sectors of the middle class, but its anticlericalism had little appeal for middle- and working-class Catholics. Accion Popular's Belaunde, on the other hand, had some appeal to Catholic voters, but his "party" was simply an electoral movement controlled by political brokers and elites whose influence would be difficult to challenge.[93]

Hypothesis five dealt with the capacity of post–Vatican II Catholicism to be a surrogate political actor and provider of social services in contexts of political authoritarianism and/or economic austerity. It speculated that such roles could be sources of tension and conflict with respect to the religious mission of a national Church. During the early Velasco period, the Church had easy access to the military government, respectfully criticizing specific measures within a posture of general support for the government and its program. Its PUJOP program, its expansion of local base communities, its social service network, and its priests and sisters in countless neighborhoods and districts, provided the government with ideas, resources, organizational conduits, and mediation with the civilian population. In these ways, it served as a surrogate partner and collaborator of a regime that lacked its own institutional or organizational network.

But these networks and services were not so essential that the regime's legitimacy depended on them, nor could they pick up the slack when the government began to cut back on its initial commitments. The Church was thus a useful "vanguard herald" for the Velasco administration, "showing it the way" at the outset. But once the government's projects were in place, and the military's enthusiasm for reform abated, it retreated, and Church programs and personnel became "rear guard surrogates" trying to meet the needs of people formerly provided for by government programs.

As the Peruvian Church moved from a "vanguard" to a "rear-guard" surrogate role, it experienced the same tensions that the Chilean Church faced under Pinochet. As Velasco's reforms sputtered and economic problems arose, the bishops were unable to speak with the same coherence and force that they did earlier. And as economic and political conditions deteriorated, the time and resources of local pastoral agents were taken up in meeting the material needs of their parishioners and of members of the wider community. These tensions never produced serious conflict between higher and lower clergy, but they gave more conservative forces additional reasons for backing away from social and political issues. They would intensify between 1975 and 1980, and with the resumption of civilian rule as well. As they did, they attracted the attention of ecclesially and theologically more conservative Vatican authorities, as we shall see in chapters 6 and 7.

4

The Chilean Church and the Transition to Democracy

The preceding two chapters help us to better understand the Church's capacity for supporting democratic values and practices. The Chilean and Peruvian cases offer ample evidence of its willingness to offer its moral and institutional support on behalf of social justice, human rights, and popular participation in social and political life. We also have seen Vatican and national Church authorities hesitate, however, when the defense of such ideals appeared to threaten core religious interests. Additionally, we found that some Church leaders were less interested in elections and civilian rule than in social justice, and that others were hesitant to condemn human rights violations openly if this jeopardized institutional interests and prerogatives. Finally, we learned that churchmen providing moral support for democratic and constitutional values were not always successful in winning over key political actors.

In fact, given their historical experiences, one might expect the Chilean and Peruvian Churches to be cautious, inconsistent, and even superfluous in their respective countries' transitions to democratic (civilian) rule. Transitions involve complex political issues on which civilian forces must speak and act for themselves. Moreover, as the transitions to democracy were getting underway in Chile (in the early and mid-1980s) and Peru (in the late 1970s), Vatican authorities were pressing Latin American Churches to limit their political involvement. They wanted clerics and lay pastoral workers to steer clear of politics, and were appointing and promoting bishops on whom they could rely in this regard.

These factors pointed to a reduced role for the Church in most countries, and the bulk of the literature on redemocratization in Latin America leaves the impression that this is what happened.[1] Yet Church people played active roles in several transition processes during the 1980s. Bishops in El Salvador, Nicaragua, and Guatemala were appointed to commissions to mediate between government and civilian leaders at various junctures. In Brazil, in the early 1980s, Christian community activists supported strikes and demonstrations aimed at restoring civilian rule, as they did in Haiti several years later.[2] Local Church activists also took part in popular movements that brought an end to authoritarian rule in the Philippines in 1986, and in East Germany (where they were mostly, though not exclusively, Lutherans) and Hungary in 1989. In Poland, Catholic bishops and intellectuals alike were instrumental in resisting communist rule and in discussions (between 1985 and 1989) leading to the relegalization of Solidarity, the development of political pluralism, and the holding of free elections.[3]

The issues and arrangements involved in restoring democracy are clearly beyond the Church's normal competence. The strengthening or reform of political parties, making the state more accountable to people, defining the military's future political role, and brokering consensus around economic goals and strategies for the post-military period are technical and contingent matters in which Church leaders had neither experience nor special credibility. However, successful management of these challenges requires an atmosphere of trust between regime and opposition leaders. And since such trust is usually one of the first and most enduring casualties of military rule, impartial mediators are needed to help the two sides explore areas of potential convergence without appearing to compromise their respective core interests. The Church's commitment to all persons places it in a unique position to arrange and/or host such talks, and to help social and political forces to shape new policy and procedural consensuses.

In addition, successful democratization requires the development of organizations free from state and party control. These are crucial for pressuring the military to relinquish power and in holding accountable the civilian elites that are negotiating with government representatives. But their impact depends largely on the values and beliefs of their members. If they shun compromise, and instead resort to violence, or make inflexible demands, they could short-circuit the process of accommodation, and actually prolong the life of the military regime. But here, too, the Church could assist the cause of democratization. Under military rule its networks of social organizations and activities expanded greatly in both Chile and Peru. It became a holding space for activists and intellectuals under seige from military authorities; it was a source of value formation and organizational training for a new generation of community and labor leaders. In both countries, the Church enabled people of differing ideological persuasions to work together in defense of universal human values, an experience that would help them to resist the pull of radicalization and to better manage the anxieties and tensions of the transition process.

In this light, the Church emerges as a potentially important influence in transitions from military to democratic rule. Bishops and pastoral agents could provide an *overarching moral framework* on which a consensus might be built. Church leaders might act as *surrogate political brokers* bringing estranged military and civilian elites together. Finally, through the *leavening action* of its lay following (organizational, sacramental, and cultural Catholics), the Church could serve to moderate or counter potentially obstructive radical elements.

If its efforts on behalf of redemocratization were to bear fruit, however, Church leaders would have to remain or become credible with the political forces negotiating its terms.[4] This would not be easy. To begin with, the issues were complex, and political forces were sharply divided and distrustful of each other. Secondly, the ecclesial and theological conservatives appointed as bishops in the 1980s were likely to press for greater attention to spiritual or personal moral matters rather than socioeconomic or political issues. And, finally, the efforts of Church leaders would be affected by both practicing and cultural Catholics, whose attitudes and initiatives they might not be able to control. Leftist activists and sympathizers who "returned" to the Church through contact with its social-service or human rights programs, for example, might or might not have changed their thinking as a result. They might retain their antiliberal views and these could spread to previously or otherwise moderate organizational and sacramental Catholics. If this were the case, people would be more likely to seek confrontation with military authorities, resort to violence, make demands that the government and its supporters would find unacceptable, and undermine prospects for a peaceful transition.[5]

But the opposite might also be true. Scholars (e.g., Mutchler, Langton, Gismondi, and Grigulevich) skeptical of the Church's seemingly progressive character credit the clergy with tactical maneuvers to ward off the growing appeal of leftist movements in the 1970s. But they argue that most rank-and-file Catholics remained staunchly conservative. If they are correct, sacramental and organizational Catholics would probably steer clear of social and political involvement during the transition to democracy. In fact, they might sympathize with rightist and authoritarian movements, focus their energies on church-type activities, and completely avoid emerging social and political movements. From this perspective, the Church is more a refuge from a cruel and hopeless world than an incubator of democratic or radical politics. Its lay activists are unlikely to support democratic institutions or processes, and might even back resurgent rightist forces.

It thus clearly matters what lay Catholics in each country actually think. The next several chapters look at the Church's impact on the transition to and consolidation of democracy in Chile and Peru. They examine the bishops' effort at mediation and the effect of their public statements and initiatives. They also analyze selective data on Catholic lay attitudes, the nature and level of Catholic social and political involvement, and the extent to which either can be attributed to Church programs and initiatives.

This chapter examines the Church's role in Chile's transition to democracy. It argues that the institutional Church had more of an impact than most observers have conceded,[6] but that Catholic activists and Christian communities were less of a factor than many believe.[7] The Chilean transition began formally with Pinochet's defeat in the October 1988 plebiscite.[8] But it really dates from the early 1980s, when deteriorating economic conditions stripped the military government of its once substantial support, fueled massive protests, and revitalized a previously divided and demoralized opposition.

Neither the protests nor the emergence of a broad opposition front succeeded in dislodging Pinochet. But they weakened his hold on power and set the stage for opposition victories in the 1988 plebiscite and the 1989 elections. The Catholic Church played an important role in both the protests and the emergence of a united opposition movement. It successfully challenged the government's legitimacy long before the opposition emerged as a viable alternative. Its pastoral agents and Catholic activists alike were mainstays in local social and political organizations, and played important roles in the protests in Santiago and other cities. More important, the bishops and other elements of the Catholic community helped the country's remergent social and political forces to develop an alternative to the regime by uniting around moderate positions that most Chileans and key political and military elites could, and later would, support. Other forces were pushing them in the same direction, but Church initiatives and efforts were an important part of the overall process.[9]

In what follows, we first consider the impact of the bishops and then the involvement of the larger Catholic community.

The Bishops

Between 1982 and 1989, the Catholic bishops of Chile continued to retreat from their active opposition role of the late 1970s. They sought to mediate between the government and its critics and to reconcile Chileans of all political persuasions. As in previous years, however, they stressed issues of human rights, social justice, and the need for a dialogue that would permit a return to democratic rule. Their public statements and positions were phrased in terms with which most Chileans could identify, and were more difficult to discredit or dismiss than those of other groups. Finally, the bishops were one of several forces pushing the parties in the direction of greater unity and moderation, helping them to become an alternative to military rule, and contributing to a climate of compromise and reconciliation in which electoral consultations could take place and their results be honored.

The years in question encompassed three phases or stages of episcopal strategy: an initial period of attempted retreat (1982–83), a subsequent period of attempted formal mediation (1983–86), and a third period of retreat and informal mediation (1986–89).

Attempted retreat (1982–83)

Although opposition sentiment and activity developed apace in 1982, the bishops maintained the cautious stance they had adopted following the 1980 plebiscite, when they began to seek a less confrontational relationship with the regime. They were pushed in this direction by pressure from Rome, by their own concerns for Church unity, and by the reemergence of political parties after almost a decade of forced inactivity. Undermining their pursuit of better church-state relations, however, were their faithfulness to Catholic social teaching, Pinochet's unwillingness to reciprocate, and the massive political protests that began in May 1983.

Pressure from Rome came through the papal nuncio, Msgr. Angelo Sodano, and from new episcopal appointments. Sodano firmly supported Pope John Paul II's efforts to pull the Church back from political involvement[10] and continually pressed the Chilean bishops in this regard. He also showed a preference for theological and ecclesial conservatives in the nominees for bishop that he proposed to the Holy See. Seven of the ten episcopal appointments made by Pope John Paul II between 1980 and 1983 were of theological and ecclesial conservatives.[11] Among them was Juan Francisco Fresno, who replaced Raúl Silva as archbishop of Santiago, in June 1983.

The appointment of conservative bishops encouraged right-wing Catholic "dissidents." In Fresno, they felt, they had finally "gotten one of their own" after twenty-one years of liberal leadership under Silva.[12] The appointment of conservatives did not lead to dramatic changes in Church policy or practice, but did strengthen the moderating turn that Silva himself had taken earlier.[13]

Concern for internal Church unity made the Chilean bishops more willing to accede to the Vatican's wishes. In a 1982 pastoral letter, the bishops acknowledged that some Catholics were leaving the Church, that the number of "dissident" Catholics was growing, that a "popular" Church was emerging, and that the faith lives of some Catholics had become completely politicized.[14] They viewed these phenomena as reactions to the Church's heightened political involvement in recent years and were determined to retreat, to contain politically active priests, nuns, and communities, and to begin stressing the themes of dialogue and reconciliation.[15]

Somewhat surprisingly, Fresno retained most of Silva's advisors and collaborators,[16] and affirmed his support of programs (such as the Vicaría de Solidaridad, the Vicaría de Pastoral Obrera, and the Academia de Humanismo Cristiano) that were frequent objects of government surveillance, harassment, and vilification. Ecclesial progressives retained their majority and continued to occupy the principal leadership positions in the Episcopal Conference. Its letters and statements continued to stress the themes and concerns of previous years, albeit with greater emphasis on traditional religious and ecclesial themes and in more conciliatory terms when dealing with social and political matters.

Several things prevented the bishops from leaving the political arena entirely. For one, neither conservative nor progressive bishops could retreat from the principles of Catholic social doctrine. Torture and other violations of human rights were always immoral; the poor should never be disproportionately burdened with the social costs of economic policies; workers had the right to belong to labor unions; and democracy was the most appropriate form of political organization in today's world. Progressive and conservative bishops alike affirmed these principles and the need to defend them publicly.

In doing so, they ran afoul of increasingly defensive military government officials, in whose eyes the bishops were guilty of meddling in matters in which they had no competence. From the regime's perspective, the bishops' defense of such values was naive and one-sided; it failed to recognize the need for certain extraordinary measures, and had the effect, whatever its intent, of hurting the government's image while legitimating that of its enemies.

Confrontation between the custodians of these universal moral values and an increasingly besieged and defensive regime was probably unavoidable. Pinochet's personal characteristics further complicated matters. He had difficulty thinking in relative terms and in appreciating the value of a less conflictive relationship with the Church. He remained a military man who saw things in either-or terms, reveled in the climate of war, and resisted concession and compromise almost instinctively. The reciprocity that the bishops needed for improvement in relations was thus missing. Some of Pinochet's advisors urged him to offer a temporizing or accommodating gesture, such as discouraging proregime shock troops from attacking Church personnel and activities. But most viewed the new Episcopal Conference as they had previous ones. Public criticisms of torture, human rights violations, or unemployment, however measured or circumspect, were acts of hostility to be answered in kind.

The government's responses included intimidation of Church personnel and harassment of Vicaría de Solidaridad programs and officials. In March 1983, it expelled three Irish priests who had participated in masses and hunger marches organized by their parishioners in defense of human rights. Pinochet refused to allow Church officials to address the issue on television, not even on the station of the Catholic University (Channel 13), of which the archbishop of Santiago was the titular head. The expulsions were particularly ominous in that more than half the priests working in Chile at the time were foreign-born, and were being warned that they could be next.

These and other hostile acts stifled the development of a more conciliatory spirit within the conference. Even conservative bishops, it seems, resigned themselves to an adversarial relationship with the regime. In their December 1982 letter, "The Rebirth of Chile," the bishops spoke of "having exhausted [the potential of] private overtures," and vowed "to speak out publicly on the country's economic, institutional, and above all moral crisis," and to continue calling for the restoration of democracy, "thanks to which we have lived for years in peace and been admired throughout the world," lest "the [current] level of tensions leads to a possible tragedy."[17]

The bishops' efforts to develop better relations with the regime also were undermined by mass protests that began in May 1983 and continued on a monthly basis through October 1983, and sporadically through August 1986. The protests were fueled by an economic recession that produced a sharp drop in GDP, an upsurge in unemployment, and the collapse of a number of banks and savings and loans associations.[18]

These developments punctured the government's aura of invincibility and weakened its political position generally. The protests attracted tens, and at times hundreds, of thousands of workers, slum-dwellers, students, and former political activists. They brought together middle- and lower-class elements whose methods and intensity of involvement often differed greatly.[19] In time, leadership and initiative passed from the hands of trade unionists to the resurgent political parties, most notably the Christian Democrats and moderate socialists. In the process, these parties recovered some of their previous vigor and appeal, and forged relationships with a new generation of Chilean citizens.

The bishops were cautious in their dealings with the protests. Calling them "legitimate expressions of popular frustration and dissent" for which there were no other outlets, they nonetheless sought to discourage them, fearing that they would make the government more rather than less intransigent. Although charging the government with responding in a "disproportionately repressive" manner, they also denounced "the provocations, the threats, the violence, the intransigence, and the excessive repression of both sides."[20] Their conciliatory efforts were undercut, however, by the prominence of priests and Christian community activists in protest activities, and by the dynamic of the protests themselves.

In fact, the protests afforded Pinochet a context in which he could remain the embattled commander and a regrettable but necessary autocrat. Although most protestors were peaceful and remarkably disciplined, a few who were not sought confrontation with security forces. The government was happy to oblige. In the June, July, and August protests, a total of 35 people were killed, 170 injured, and 3,415 were arrested. The government used these "incidents" to discredit the opposition, to depict the country's alternatives as order (the government) and chaos (the opposition), and as a pretext for repressing real and imagined adversaries.

Pinochet's temperament made concessions of any sort difficult. But as moral authorities and leaders of Chile's most respected national institution, the bishops were not as easily dismissed as other critics. Moreover, they could challenge the government's legitimacy in ways that others could not.[21] Their public statements and positions offered an alternative to the "order" of dictatorship and the "chaos" of a still divided opposition movement whose radical elements gave many Chileans pause. In effect, the bishops made it possible for Chileans to reject the government before there was a viable alternative to it.

The protests also made it difficult for Church leaders to persuade some of their own rank-and-file activists to negotiate with the government. For some opposition groups, including Catholic radicals, the protests were proof that the regime was about to collapse, and there was no need to talk. With just one or two more protests Pinochet would fall or be sacrificed by military colleagues for whom he was a growing liability.[22] The popular support that the protests enjoyed during 1983 caused many Catholic activists to view the bishops' call for moderation as unduly and unnecessarily concessionary, and to persist in their efforts to force the government from power.

They underestimated the uneasiness that the protests inspired in many middle- and lower-class Chileans. They also failed to appreciate how difficult it would be for military leaders to relinquish power to forces unwilling to talk with them. Their continued active participation in the protests undercut the bishops' efforts to bridge the gap between the government and its critics. It gave the protests a certain Catholic flavor, prompting Pinochet to dismiss Church leaders as either duplicitous or unable to control their activists and rank-and-file followers. In either case, there was no reason to take their calls for moderation seriously.

Non-neutral mediation (1983–86)

Failing to develop a nonadversarial relationship with the regime, the bishops spent the next several years attempting to mediate between "reasonable" elements on both sides of the political divide. Their early initiatives included sponsorship of a dialogue with interior minister Jarpa in 1983 and the brokering of the National Accord in 1985. They helped to overcome divisions and antagonisms among regime opponents and to encourage them to adopt moderate positions more acceptable to forces that had once backed Pinochet.

The Dialogue with Jarpa. The first initiative was Archbishop Fresno's sponsorship of a dialogue between government officials and moderate opposition leaders in August and September 1983. In early August, Pinochet appointed Sergio Onofre Jarpa as minister of the Interior, authorizing him to establish contacts with "responsible" political forces. Pinochet hoped to diffuse the protests, win back right-wing supporters, and persuade the Christian Democrats and other "moderate" opponents to press their concerns through official channels, thereby isolating the left. He also signaled a willingness to make concessions to financial and industrial interests that had been critical of his neoliberal advisors and policies,[23] although not to abandon neoliberalism as such. Moderate and progressive bishops were not impressed with the relatively minor economic adjustments but joined with conservatives in welcoming the political opening.

One of the first people with whom Jarpa made contact was Archbishop Fresno. When Fresno suggested getting together with representatives of the newly formed Democratic Alliance,[24] Jarpa agreed, and met them twice at Fresno's residence and once at the papal nuncio's.[25] The archbishop or one of his auxiliaries hosted and facilitated the sessions. The participants discussed political normalization and possible changes of aspects of the 1980 Constitution dealing with the transition. The Church's auspices and presence made it easier for people to meet and to hear one another's proposals. No one was giving in to anyone else; all were deferring to a widely respected third party. Unfortunately, negotiations were complicated by the glare of publicity, and by the emphasis that some participants gave to some of the divisive issues.

In the end, Jarpa was unable to make the concessions demanded by the Alliance parties. In turn, they were unwilling to suspend their protests or to accept institutions or mechanisms in which their influence would be marginal. Left to themselves, "moderate" critics and Jarpa-like "reformers" might have agreed to a number of things. But each had to persuade and control more intransigent partners, i.e., opposition "radicals" and "hard-line" elements respectively,[26] and this was not possible in either case.

Jarpa's credibility, in particular, was undermined by his uncertain stature within the regime and by Pinochet's unwillingness to restrain security and paramilitary forces. The latter attacked the leaders of the Democratic Alliance during the September protests, and continued to target Church personnel and facilities in the ensuing weeks and months.[27] Were it not for the attacks, the bishops might have played the more neutral, mediating role that many of them wished to. But with their priests, lay workers, and church buildings under attack, and with their Christian Democratic friends targeted by security forces, the bishops closed ranks and persisted in calling for changes (e.g., an end to repression, the negotiation of differences, and a return to democratic rule) that in the regime's mind linked them to the opposition. In effect, Jarpa was unable to elicit from Pinochet the gestures of good faith necessary for people even to consider the possibility of a negotiated settlement.

During 1984, the bishops continued to urge both sides to make concessions in the interests of justice and peace. Unfortunately, the regime would not talk to anyone who did not accept the 1980 Constitution, and this meant virtually all opposition groups. Church-state relations were further aggravated by the government's refusal to grant safe passage to *miristas* who had sought and been given asylum in the papal nuncio's residence in early 1984.[28] In March, Archbishop Fresno asked for a "gesture" that might contribute to the overcoming of differences, but Pinochet refused, apparently believing that he was being asked to resign. The Episcopal Conference repeated Fresno's call in early April, urged opposition groups to renounce violence and to refrain from harsh language when referring to the regime, but also pressed it to clarify the circumstances in which more than a hundred opposition activists had died over the last three years.[29]

These appeals were ignored, and church-state relations continued to deteriorate. The regime refused to talk with its opponents or to accept criticism from any quarter. It dismissed the bishops as "naive" and "excessively concerned with political matters," and further tightened its grip on society. Additional restrictions were placed on the media. Antiterrorist legislation that permitted the arrest and detention of anyone perceived to be a threat to public order was imposed. And Church people and programs were hit with a new round of attacks by security forces and paramilitary groups.

In September, a French-born priest was shot to death by army troops while reading the Bible in his residence in Santiago's La Victoria neighborhood. Later that month, a church in Punta Arenas was firebombed by unidentified assailants, bringing to seven the number of churches attacked that year. In early November, the government condemned a meeting that several bishops had held with Chilean exiles (among them Marxists) in Rome. And later that month, having declared a state of siege,[30] it refused to allow the publication of one of Archbishop's Fresno's official statements, to permit the holding of the Church's annual *semana social*, or to let Father Ignacio Gutiérrez, the head of the Vicaría de Solidaridad, back into the country following a visit to his native Spain.

During this period, Pinochet's position began to erode as supporters and former supporters grew increasingly disillusioned with his economic policies. But while broader, the opposition remained sharply divided over the acceptability of the 1980 Constitution, proscription of Marxist parties, and the propriety of protests, demonstrations, and other forms of social mobilization. In fact, Pinochet frequently defended his unwillingness to dialogue with his opponents by asking: "With whom should I be talking? No one represents anyone."

The National Accord. In this context, Archbishop Fresno launched his effort to develop a National Accord.[31] It was an effort to bring opposition groups together, and thus to provide Pinochet with a coherent interlocutor. Meetings were held at Fresno's residence beginning in March 1985. In issuing the invitations and structuring the meetings themselves, the archbishop was advised by Christian Democrat Sergio Molina, who had been treasury minister in the Frei government; Catholic businessman José Zabala; and Fernando Leniz, an agnostic who had been treasury minister in one of Pinochet's early cabinets.[32] At each meeting, the invitee discussed his views with Fresno, while Zabala took notes. Zabala later drafted a summary of the leader's position, identifying common ground with other leaders, bases for possible accommodation, and issues for further discussion.

In the midst of these sessions, Fresno was named a cardinal, and spent most of May and June in Rome. His elevation in status further enhanced his prestige and influence. In late July, Zabala produced a text that identified the areas of agreement and contention among the participants. All were invited for "further conversations" at a Jesuit retreat house outside Santiago. Meeting as a group for the first time, they learned that they agreed on more things than not, and were persuaded to authorize the drafting of a document that formally endorsed a peaceful transition and the principles that would govern a postmilitary government. Prepared with the help of Christian Democratic and Socialist party "experts," it was presented at a second meeting, to which Christian Left leader Luis Maira was also invited.

The document called for the immediate normalization of political life (i.e., an end to all states of exception, to internal and external exile, and to restrictions on party activity), the holding of direct presidential elections (as opposed to the single-candidate plebiscite called for by the 1980 Constitution), and changes in other aspects of the Constitution, e.g., the procedures for amending it and the powers of the Congress and the Council of State, as well. It also defined principles that would guide social and economic policies in the postmilitary period. The only stumbling block was whether to proscribe "antidemocratic" ideologies or "antidemocratic" behavior. Léniz got Maira and Andrés Allamand of the rightist MUN to agree to the proscription of parties with "antidemocratic objectives," by which something more than abstract principles but less than outright actions was implied.

The National Accord on the Transition to Full Democracy was signed in late August by representatives of the National Union movement, the Liberal, Republican, and National parties, the Christian Democratic, Radical, and Social Democratic parties, the Briones Socialists,[33] the Christian Left, and two smaller socialist parties. The Almeyda Socialists and Communists spoke favorably of some of the document's provisions but were unhappy that it did not call for Pinochet's immediate ouster, and did not sign on.[34]

The text was released on 26 August, but Fresno was unable to present it directly to the government. The minister of the interior refused to return his phone calls, and Pinochet ignored a letter that Fresno wrote in early September.[35] Rumors at the time indicated favorable interest on the part of some high-ranking officers (Army general Benavides, Air Force general and junta member Matthei, and others). Pinochet himself, Admiral Merino, and other "hard-liners," on the other hand, flatly rejected the initiative, and effectively condemned it to failure. In their view, the bishops were not credible mediators deserving of respect, but rather naive meddlers being used by "political elites" (*los señores políticos*).[36]

In the ensuing months, tensions and divisions developed among the signatories, with the MUN and Christian Left abandoning the accord and the MAPU signing on. Differences over how to deal with the regime (dialogue versus social mobilization) resurfaced, as did debate and discussion with respect to violence, confrontation, and repression.

Although rejected by the government, the National Accord was more substantial and successful than the dialogue with Jarpa. It made important inroads in political circles and in public opinion generally. It reopened the question of the government's continuation in power, and helped in developing longer-term support for a united opposition front. It was the broadest coalition of forces to date to call for the restoration of democracy, and the first of a series of agreements, each more extensive and specific, that would culminate in Pinochet's defeat in the October 1988 plebiscite.[37]

Difficult decisions still lay ahead for the opposition: the Christian Democratic Party's selection of the more conservative Patricio Aylwin as its presidential candidate, the Almeyda Socialists' willingness to participate first in the plebiscite and then in the Concertation, and the compromise with Renovación Nacional and Interior Minister Cáceres on amendments to the 1980 Constitution (prior to the 1989 elections), were all steps that remained to be taken, and none was assured by the accord. But it was a crucial initial advance in the direction of accommodation and convergence. Finally, the accord's provisions for transferring power to civilian hands, and the principles it established with respect to policies in the postmilitary period, helped the opposition to project an image acceptable to the moderate right and to elements within the armed forces as well.[38]

Publicly, Church officials insisted that they had merely brought the parties together, and that the parties themselves were responsible for the agreement. But the mediating influence of Fresno and his advisors was substantial, if not decisive. Fresno was able to convene people who otherwise would not have met with one another. Additionally, the Church's auspices, Fresno's presence, and the sequence of bilateral sessions followed by group meetings, all made it easier for the participants to speak their minds, acknowledge common ground, and make concessions without appearing to do so. Thirdly, the moderate, centrist formula that the archbishop and his advisors had in mind from the outset excluded "problematic" forces, but facilitated an agreement with which virtually all others could live.[39]

Most important, Church sponsorship gave the accord a stature and appeal that went beyond the negotiation of particular interests. It represented the reconciliation of a divided people and the recovery of a national identity and unity. The Church generally, and leaders like Fresno in particular, embodied these notions and spoke for them in ways that ordinary social and political forces could not. In doing so, they were able to lift such forces out of constraints from which they might not have escaped on their own.[40]

Fresno's role in these matters was not widely understood. For many Catholic activists, he was insufficiently forceful with the regime. They wanted a champion to stand up to Pinochet, as Cardinal Silva had done. But Fresno had little of Silva's political instincts or inclinations, and was frequently made to look beholden to the regime and easily manipulated by it.

A painful example of this lack was the visit he paid Pinochet in December 1985 to ask him to reconsider his rejection of the National Accord. When Fresno broached the subject, Pinochet cut him off abruptly. "Look, Cardinal," he said, "I know you want to talk to me about this accord you have made. No. No. Let's turn the page instead."[41] Fresno was embarrassed and depressed by the experience. Opposition Catholics were devastated, as were some Church workers. They were looking for short-term impact, and were worried that his conciliatory manner was being exploited by the government in ways that were helping to sustain it. A lay activist in the Santiago archdiocese's southern zone (Vicaría Sur) saw the Cardinal's efforts at mediation as a retreat from the Church's earlier commitment to the people:

> As we say here, I believe Don Pancho Fresno is stepping back to a more cautious position, being more careful about what he says. I think he is afraid that the idea of real reconciliation costs too much and that (practical) reconciliation means not being on bad terms with the dictatorship. I believe that . . . this is a distortion of what Puebla and Medellín have said about the Church's role of service.[42]

A Catholic activist belonging to the Izquierda Cristiana expressed the same thought somewhat differently.

> Fresno clearly constitutes a step backward for the Church, and . . . I believe his behavior has been insidious [*nefasta*]. Far from being a true reconciliator, he has moved further away from it because under him the Church has lost its valid role of interlocutor. The good thing that the Church had previously in playing the role of mediator . . . was that it chose sides. It was different from its present position because logic indicates that what one should do as mediator is what Cardinal Fresno is doing: distancing himself from both parties in the conflict. Aristotelian

> logic advises this same strategy: "distance oneself from both parties, take the middle position, and there one finds an effective solution." However, the logic of politics in Chile, in Marxist terms, is different because there are exploiters and exploited. To (really) mediate in such a context, one must take a position and it must be with the exploited. From that stance one becomes a true interlocutor with power. This is the only mediating position that has a chance of success, and that is what Cardinal Silva well understood.[43]

A former seminarian and now Socialist Party militant (having belonged to both the PDC and MAPU previously) thought the problem was simply Fresno's weakness as a leader:

> the Church . . . has lost leadership; it does not have direction, it is leaderless. . . . A good friend of mine, a Polish priest who is now working in Chile, and who served for a time in Poland, told me that at the beginning of Fresno's term he thought the same thing would happen [in Chile] as occurred in Poland when Archbishop Glemp became primate. In reality, the appointment of Glemp reflected a commitment to negotiate with the government. But it was an option for negotiation taken by a leader with a firm position. This is not the case with Cardinal Fresno. In his case, it is simply an absence of leadership.[44]

These assessments were offered shortly after the accord was completed. In the eyes of these people, all close to the Church, Fresno had failed. They could not see beyond Pinochet's rejection of the proposal. But in the longer run, Fresno's initiatives would help the opposition to hammer out an acceptable plan for transition. Although rejected, the accord gave the opposition an aura of reasonableness and moderation that the country's parties had not enjoyed since Allende's downfall. In this respect, Fresno was even more effective than Silva in getting people who had once supported the government to take a step away from it and to agree on a timetable for restoring civilian rule.[45]

Fresno's contribution was not as a public critic but as convener and interlocutor *among* critics. During the late 1970s and early 1980s, the Church had been a public critic and the opposition had operated under the protective umbrella of its agencies and institutions. Fresno's strategy was different. With the reopening (however partially) of the political arena, the parties needed to emerge on their own and begin to speak for themselves and the Church was again free to mediate between opposing sides. Fresno was effective in doing this. In the wake of someone like Silva, he appeared to be weak, inconsistent, and/or inept. But precisely because of this he was a more effective interlocutor among the opposition parties, and between them and former or now ambivalent government supporters.

During the first eight months of 1986, the bishops (actually the Episcopal Conference's Permanent Committee) continued to decry the upsurge in violence on both sides, and to press the government for a more flexible transition to civilian rule.[46] At the same time, however, many of them, including Cardinal Fresno, ordered priests and other pastoral agents to stop collaborating with opposition groups in their *poblaciones* and popular neighborhoods. Priests in Santiago, for example, were told to deny the use of parish facilities for political or human rights work, on grounds that it was not an "evangelical" activity. Some officials, such as recently appointed auxiliary bishop Antonio Moreno in Santiago's northern zone, enforced this policy with zeal; others ignored the directives or allowed their communities to decide how facilities should be used.

These moves reflected the desire of many bishops to define a less committed, if not entirely "neutral," position. Their interest intensified with the September 1986 attempt on Pinochet's life, in which five of his armed escorts were killed. That incident, and a related discovery of cached arms, led to retaliatory killings and to the jailing and internal exile of thousands of regime opponents. It underscored the divisions plaguing the opposition, damaged its claims to be a viable alternative to Pinochet, and gave him renewed vitality and leverage at a politically crucial juncture. It also made it more difficult for liberal bishops to oppose the retreat that moderates and conservatives were urging.

Catholic rightists thought that Church leaders should be concentrating on personal moral issues such as contraception, and should confine themselves to enunciating or reaffirming the principles governing the social and political orders, leaving individual Catholics free to decide how to apply these principles in practice. In the words of one conservative Catholic intellectual,

> "I believe . . . in terms of what one can hope for from the hierarchy in the clarification of principles on the moral and political order and the relation between the two, that the discourse of our pastors has been extraordinarily poor. They have created tremendous unresolved doubt in people's minds by addressing questions of a contingent nature, e.g., whether or not democracy is good, whether the 1980 Constitution is legitimate or not. These are issues about which Catholics have the

freedom to decide one way or another, whatever the opinions of their pastors. But those points that are essential to doctrine, I do not see the bishops addressing."[47]

While failing to impress Catholic rightists, however, the Church's pullback angered or disappointed activists within its own ranks. One whom we interviewed in mid-1987 yearned for the days, under Cardinal Silva, when Church agencies and parishes opened their doors to secular groups.

> At one point the Church played a role of gathering together especially those who needed to organize themselves, which was the only defense at the time against their various problems. I had the experience, for example, of serving in a committee of *allegados,*[48] in a debtors' organization, and in cooperative meal programs [*ollas comunes*]. There would have been no possibility whatsoever [of solving problems like these] if there had not been a Church of the sort I remember under the archdiocese of Silva Henríquez. Now I sense that there has been a change in the Church under Fresno. I'm talking about a change that is taking place at the base level, at the community level, which has obliged the priests to be much more cautious [*precavidos*], in other words, to be more careful in their commitments to the popular sectors. There are more limitations on their activities. I do not want to pass judgment, you know, but I have noticed a change, I am feeling it.[49]

A secular, but somewhat religious, community worker in the Vicaría Oeste saw the same thing happening but viewed it in primarily ideological and political terms. Under Silva, she said,

> The Church began to accompany the poor, and to link up with their organizations. There came a moment, however, when [it] distanced itself and retreated. It began to vacillate and to walk on the opposite side. In so doing, [it] appears to be falling into partisan political territory....
>
> At the time when the Church saw new organizations emerging in Chile [in the early 1970s] and these were trying to act autonomously and included Marxist leaders and even those with extreme left leanings, it was not afraid.... But now other leaders have appeared to head the political system. Therefore, the Church says: "Ah, we have crossed the line. Let us retreat. We have become too involved with this side. Let us shift to the other side." This is the case, no? Now the vast majority of the Chilean hierarchy has tilted to the other side of the political spectrum.[50]

Catholic activists took issue with the bishops' shift because they saw it moving the country's center of political gravity rightwards. In their view, the only way to be done with Pinochet was to force him out, and anything that diminished the intensity of opposition to him could only strengthen his position. To more sophisticated political observers, however, the important thing was to build a united opposition movement. In their view, the Church's conservative shift might facilitate this by appealing to groups in the center and on the right. And, in any case, it was better for the country in the long run if their Church were not directly involved politically. As one leftist intellectual put it,

> I do not think it appropriate . . . that the Church identify closely with the political opposition. . . . From this point of view, I see the changes brought about by Fresno as healthy, [although] not because it was planned that way [I have no doubt that he is a political conservative], but . . . [because] I believe we have reached a point where uniting all dissident activity is important, [and it is important that we do it] not looking for protection behind the clerical cassock.
>
> The [best] moment to do something politically different is precisely the moment in which the Church is less disposed to play a public role of opposition to the regime. This requires that parties and other political organizations accomplish this task.
>
> The other [problem] is [that] the Church which today is playing the role of the democratic opposition could play a very conservative role tomorrow if linked to, or too closely associated with, a partisan political option. I prefer a more independent civil society and . . . given the conservative attitude which now characterizes the Church's leadership, at least Cardinal Fresno, it is best that the Church retreat a bit from political involvements. Surely this may weaken some opposition against Pinochet now, but it could be better from another perspective.[51]

Retreat and informal mediation (1986–89)

The period encompassing the attempt on Pinochet's life, the visit of Pope John Paul II in April 1987, and the plebiscite and elections of 1988 and 1989 was one of retreat and informal mediation on the part of the bishops. The appointment of additional conservative bishops (seven of the nine named between 1983 and 1987) swelled the ranks of those eager to further deemphasize "politics" in favor of traditional pastoral concerns.[52] In the wake of the assassination attempt, moderates and some progressives agreed to the shift as well. They wanted to avoid further deterioration of church-state relations, and any suggestion of support for the recourse to political violence.

From late 1986 on, the bishops tried to position themselves between the two sides rather than against the government. They continued to urge dialogue, reconciliation, respect for human rights, justice for the poor, and a return to democratic rule, but in more balanced and less forceful terms than before. Their shift in tone and emphasis was noted and applauded by some regime supporters, although it failed to win back alienated upper-class Catholics. It also heightened rather than eased tensions within the Church, bringing critical responses from progressive bishops and from local-level priests, nuns, and lay workers, many of whom had become better organized and more difficult to control than in previous years. In the end, however, the bishops contributed significantly to the transition to civilian rule by helping to make it more acceptable to conservatives and to erstwhile regime supporters.

The period did not begin well. Progressive bishops Tomás Gonzáles of Punta Arenas and Carlos Camus of Linares appeared in the media frequently, denouncing the government and contributing to the image of a fragmented and dispirited Church. In October 1986, a meeting was held in Chillan, at which other bishops expressed their concern for people who had "lost their faith" in a Church whose "leaders were afraid to speak out" and "had capitulated to the regime." Apparently, many Catholic activists thought that the bishops should have protested more forcefully the kidnapping and abuse of a Dutch priest who was forced to leave the country earlier in the year, and the subsequent expulsion of the three French priests who had been working in slum neighborhoods. In fact, no Church official went to the airport to see the French priests off, and the papal nuncio, Msgr. Sodano (later named Vatican secretary of state by the pope), actually appeared on television, as they were leaving, at a reception for Pinochet.

The failed attempt on Pinochet, coupled with the discovery of cached arms that the communists may have intended for use against future civilian governments,[53] persuaded reluctant opposition parties to abandon the idea of forcing the regime from power and to move toward confronting it electorally within procedures established by the 1980 Constitution. The assassination attempt enhanced Pinochet's political status and support. It also made it clear to progressive Christian Democrats and moderate (Nuñez) socialists that the strategy of social mobilization and pressure was uniting the military in support of Pinochet and alienating middle-class Chileans to boot. The arms discovery made it easier for renovated Almeyda socialists to begin persuading their Leninist colleagues to set aside their revolutionary expectations, to distance themselves from their communist allies, and to embrace the *vía electoral* in a broad front with centrist and right-wing regime opponents.

The PDC's decision to challenge Pinochet on his terms led it to elect Patricio Aylwin as its new president. Since 1984, Aylwin had been calling for acceptance of the 1980 Constitution, arguing that its more objectionable provisions could be amended. He also belonged to the party's conservative wing, and was thought to be more acceptable to potential antiregime allies on the right than his progressive rival Gabriel Valdés. In fact, however, Aylwin was determined to reach out to both the moderate and not so moderate left, and his long-standing friendship with Clodomiro Almeyda, who spent most of 1987 in jail charged with "subversion," helped him to draw Almeyda's socialist fraction into a anti-Pinochet front that spanned the political spectrum.

Church leaders were pleased with these developments. They were what Fresno and others had been hoping and working for since 1983: a united opposition movement willing to negotiate a peaceful return to democratic rule. Apart from helping to set the process in motion several years earlier and giving it crucial moral legitimacy in its initial stages, however, they had less of an impact on the partisan political dynamics and decisions that carried them forward.

The Papal Visit. The papal visit of April 1987 helped the Church to defuse the growing polarization within its ranks and to strengthen its social and political influence in the months and years to come. The last half of 1986 and the early months of 1987 were devoted to preparing for the visit. With forcefulness and aplomb, Church authorities controlled the structure and schedule of the visit, minimizing its political utilization by either the government or the opposition, and enabling the pope to say what he wanted in the settings he deemed appropriate.[54]

The five days that the pope spent in Chile were replete with ceremonies and appearances in which the theme of reconciliation (with God and one's fellow Chileans) was stressed. There were moments of tension, and situations, such as the incidents at the outdoor mass and the pope's impromptu appearance with Pinochet on a balcony of the Moneda palace, that the government managed to exploit.[55] But for the vast majority of Chileans who followed or attended the pope's activities, the atmosphere that emerged from them was one in which the disposition to forgive one's adversaries, settle the differences that still divided people, restore democratic institutions, and build an economy and culture in genuine solidarity with the poor, were very substantially strengthened.[56]

In the short term, the papal visit enabled people to take to the streets for the first time since the assassination attempt, making them aware of their numbers and common concerns. Partially or fully-televised events, at which the pope and other speakers were critical of existing conditions and practices, legitimated and encouraged those who had been raising these issues for some time. The pope made a point of meeting with Carmen Gloria Quintana, who had survived a brutal attack by an army unit that had taken the life of her companion during a July 1986 work stoppage (*paro*). He also met with opposition representatives, including one from the Communist Party, warning them against the use of violence (it would be counterproductive), but lending moral support to their efforts at restoring democracy. These symbolic gestures did not go unnoticed, and helped to counter the regime's depiction of its opponents as unworthy of attention or respect.

For the Chilean Church, the pope's visit came at a fortuitous point in time. It helped the Chilean bishops to defuse the polarization that had been developing in their own ranks and to unite the Church's various factions and tendencies, thereby strengthening its moral authority and its social and political influence.

In June 1987 the bishops began to encourage voter registration with an eye to the plebiscite to be held sometime before 1989. They did so before it was clear that the plebiscite would be fair and its results honored, and despite the fact that many Chileans still believed that Pinochet could be overthrown, or, in any case, did not want to recognize the 1980 Constitution.[57] The bishops felt that registration would be useful whether a plebiscite or direct elections were held; with the attempt on Pinochet's life, they apparently concluded that changes in either the format or timetable of the transition process were unlikely, and that an up-or-down plebiscite was the most that could be hoped for.[58]

They defended these views over the next fifteen months. Most priests and sisters encouraged their local communities to register as well; a smaller number, fearing that Pinochet would find a pretext for remaining in power, argued against registration. By early 1988, however, it was clear that for all its risks and limitations, the plebiscite was as good an opportunity for challenging Pinochet as there would be. In fact, Catholic activists in Santiago were instrumental in the decision of several still skeptical parties to finally endorse registration and establish an alliance to promote a No vote (Concertación por el No).

As the plebiscite became a foregone conclusion, the number of people and parties willing to participate expanded, and the bishops began to question the conditions under which the campaigning was taking place and the plebiscite would be held (e.g., the continued imposition of states of exception limiting opposition activity, inadequate opposition access to media, partisan intervention by the military, the use of pressure tactics by government officials, etc.).

The Social Pact. A final episcopal initiative contributing to the transition to democratic rule was the brokering of a social pact between entrepreneurs and trade unionists in mid-1988. Building upon the meetings that the pope had held with both groups the previous year, the Permanent Committee met with them in April and then designated a committee of prominent laymen to see if they could reach an agreement on wages, employment practices, and related issues for the post-military period.[59] The powerful Confederación de Producción y Comercio (CPC), which had long supported the military government, declined to participate, citing the proximity of the plebiscite. Discussions with other entrepreneurs, union leaders, and Catholic economists affiliated with CIEPLAN proceeded, however, and a consensus was quickly reached.[60]

Shortly thereafter, the president of the CPC, Manuel Feliú, endorsed the agreement and showed it to the organization's member federations. All six of the federations agreed to sign, although on the morning of the scheduled signing two of them reneged and prevented the confederation as such from becoming a party to the accord. Fear of alienating a sitting president who might remain in his post was apparently the major stumbling block,[61] and following the plebiscite those involved expressed regret at not having signed. In their absence, the less prestigious Social Union of Catholic Entrepreneurs (USEC) signed on behalf of management groups, and in the ensuing months the agreement was subsequently ratified by unions and business groups throughout the country.

While they failed to produce a formal agreement between the country's principal trade unions and "peak" entrepreneurial associations, the negotiations did ease entrepreneurial fears concerning management-labor relations in the post-Pinochet period. They made clear that workers were not bent on revenge, that prospects for harmony were much better than most observers believed, and that Pinochet's alternative of "me or chaos" was not an accurate representation of the options available to Chileans at the time.

The pact may have seemed an enormous expenditure of energy for a relatively modest advance, but it helped to reweave another piece of the country's social fabric. This fabric had been torn apart by conflict between left and right and between workers and their employers both before and since 1973. By the late 1980s the military regime had lost most of its initial support. But these divisions persisted. They would have to be overcome if democracy was to be restored. Things like the social pact made significant contributions in this regard.

The Plebiscite. As president of the country's largest and most strategically positioned party, the PDC's Patricio Aylwin emerged as the opposition's leading figure during the year preceding the October 1988 plebiscite. Although left- and right-wing opponents of the regime might have preferred someone else, both groups saw him as a potentially appealing alternative to Pinochet and were happy to have him speak for them within the Concertation for a No Vote which the opposition parties had established in early 1988.

In the campaigning that preceded the plebiscite, the bishops remained neutral, although a majority of them personally favored the No position.[62] In Santiago, Cardinal Fresno forbade the use of Church facilities for anything except voter registration. This distinction was often difficult to enforce, as the priests and sisters active in the registration drive were known to favor a No vote. Some regional Church officials, like the vicar (auxiliary bishop) of Santiago's western zone, Mario Gárfias, came very close to urging a No vote in their pastoral communiqués, without formally doing so.[63]

The efforts of Fresno and other bishops to keep the Church from becoming identified with a No vote caused a certain amount of consternation among local church activists who were working in the No campaign. Leftist party leaders, although disappointed at first, came to see the benefits of such a policy. A local leader of the Socialist Party told us in 1988, for example,

> The meetings of the committees planning street rallies used to take place in parish facilities. Then came a circular by Cardinal Fresno in early

> May 1988 in which he forbade the use of the parish buildings for political activities during the plebiscite. . . . This caused problems in some *comunas*. I am a district leader for the Socialist Party in the *comuna* of La Florida, . . . where the best place for a meeting was the parish. There simply was no other alternative. It is evident, therefore, that the conflicts which were created were quite complicated.
>
> Well, one reaction was to blame the Church for our lack of capacity. The Cardinal's directive could thus be seen negatively in terms of its impact on the political game being played at the moment. However, it [also] had a positive aspect, namely, to force the parties to locate themselves in society by their own means, i.e., to return to a role in society distinct from that played by the Church.[64]

Assessments of the Church's policy thus depended in part on one's view of the transition process generally. Those, including many Church activists, who believed that a massive No vote would force Pinochet from power tended to be critical; others, including a good many leftists, saw the transition as a longer process, were concerned with strengthening their own organizations and believed that a more "neutral" Church might be better for them and more credible to the government as well.

Voters rejected Pinochet's bid for an additional eight-year term by a 54.7 percent to 43 percent margin. Most polls had projected an opposition victory, but Pinochet was nonetheless stunned by his defeat. During the evening of October 5 and the early hours of October 6, his military colleagues pressured him into announcing and accepting the results. The most widely believed account of events that night indicates that three of the four service heads (all except army general Sinclair) rebuffed Pinochet's attempts to set aside the results.[65] Confronted with his defeat, Pinochet charged that he had been betrayed, that the opposition's victory would lead to unacceptable attacks on the presidency and military unity, and that the junta should immediately grant him extraordinary powers. But the other commanders insisted on honoring their commitment to the provisions of the 1980 Constitution, arguing that this was in the best interests of the country, the armed forces, and even Pinochet himself.[66] While they thus had reasons of their own for doing so, it also helped that the Church and other opposition forces had cultivated better relations with the military in recent years.

Regime hard-liners were no more graceful in defeat than Pinochet. Some, like Sergio Gaete, the ambassador to Argentina, charged the Church with having "intervened" and caused Pinochet's defeat. Most analysts believed that a larger turnout would benefit the opposition, and Gaete was implying that in urging voter registration the Church was promoting the No vote, although he may have been thinking of Msgr. Garfias's remarks as well.

Pinochet's defeat set the stage for full-scale presidential and legislative elections the following year. The major Concertation parties each selected a candidate for the nomination. The PDC's selection was the most closely watched since its candidate was believed to have the best chance of winning the nomination. With assumption of real power a distinct possibility, however, differences between the party's conservative and progressive factions reemerged with the fierceness of earlier times. Deliberations continued for over two months, casting a pall on the Concertation as a whole. In the end, however, the progressive wing's Valdés embraced Aylwin's candidacy and persuaded his allies and supporters to do likewise. A month later, Aylwin was proclaimed the Concertation's candidate as well, with full and apparently enthusiastic support of its right- and left-wing components.

The opposition's victory in the 1988 plebiscite gave additional impetus to its developing consensus. It showed the parties themselves that unity worked. It also put additional pressure on the government to agree to amend the 1980 Constitution. In fact, in early 1989, a series of reforms were hammered out among the Concertación, the right-wing Renovación Nacional, and Pinochet's minister of the interior, Carlos Cáceres.[67] These changes greatly improved the climate for the 1989 elections. Aylwin's victory was almost a foregone conclusion given the 1988 plebiscite results and the likelihood that former regime supporters would divide their votes between two candidates (Büchi and Errázuriz).[68] But with the reforms, the opposition had more reason to believe that the powers it would inherit would be worth its efforts, while forces that had previously supported Pinochet could view their likely defeat as less than a complete disaster.

Aylwin's margin of victory (55.2 percent to 29.4 percent for Büchi and 15.4 percent for Errázuriz) was roughly that of the No vote in October 1988. It suggested that much of the Si vote may have been real support for Pinochet and not simply fear of reprisal. Opposition candidates did about as well in the congressional elections, at least in terms of their percentage of the vote. Restrictive electoral laws (particularly the 66 percent clause) denied them additional seats in both the Senate and the Chamber, but on balance did not treat them too badly.[69]

The pronouncements, initiatives, and activities of the bishops, while less assertive or contentious than in earlier years, helped to move the transition along at key junctures. Their criticism kept the regime's legitimacy at minimal levels during the period (1982 to 1987) even though the opposition was divided and unable to offer an alternative to military rule. Msgr. Fresno's mediation of the National Accord was an important factor in the emergence of the broad, democratic consensus that finally overcame Pinochet in the political arena that he had fashioned. Finally, and most importantly, the bishops helped to create an atmosphere of moderation and compromise that enveloped virtually all major players, i.e., politicians, trade unionists, entrepreneurs, popular-sector groups, and even, although apparently very few, military officers.[70]

In not pressing the regime more forcefully from 1986 on, the bishops disappointed many Church activists. The bishops' reticence probably made it easier for the government to insist on its original format and timetable for transition, although the failed assassination attempt was much more decisive in this regard. Over the longer run, however, the Church's encouragement of dialogue, reconciliation, and compromise helped to moderate and thereby enhance the opposition's image, and to persuade the political right and at least some military leaders that they could live with an opposition government.[71]

Chile's transition to democracy was not a categorical victory for the opposition. Despite economic crisis, protests, and the development of a strong and increasingly united opposition movement, Pinochet held out for his format and timetable, and made it difficult for any government that followed him to break decisively with his economic policies.[72] The dictator was denied the primary end for which his transition had been designed, however: his own continuance in power. Faced with a choice between eight more years of Pinochet and a government headed by opposition parties united around a moderate social and economic program, a majority of Chileans chose the latter.

The Catholic Community.

Members of the broader Catholic community also played a part in Chile's transition to democracy. In what follows, we look at how groups with very different relationships to the Church translated their Catholic values into political involvement and commitment. We distinguish among practicing and "cultural" Catholics at national and local levels.

Each of these groups played a role in Chile's transition to democracy. Practicing and cultural Catholic elites were important players in virtually all social movements and political parties at the national level. At the local level, protests and other antiregime activities were led or sustained by practicing Catholic activists and by nonpracticing cultural Catholics whom the Church had reached through its social service programs. The practicing activists were a small minority of both the larger Catholic population and the local Christian communities, but they often "represented" the Church in their dealings with local secular groups. The relationships of these groups to the Church varied considerably. Some were willing extensions or instruments of its authorities; others were allies or useful contacts whose activities were conditioned by these ties; yet others were assertive and occasionally rebellious subjects whose initiatives affected and committed Church authorities irrespective of their wishes.

In what follows, we examine the contributions of these groups, place them in relation to other Catholics, and attempt to determine the Church's responsibility for their impact on the transition.

Practicing Catholic elites

Many of the intellectual and political leaders of the Christian Democratic Party (PDC) and, though to a lesser extent, of MAPU, the Christian Left, the various socialist factions, and even the Communist Party, were practicing lay Catholics. Their Catholic identities and their close ties to bishops, priests, and other Church functionaries enabled them to serve as conduits and spokespersons for Catholic values. In many of their political dealings, they "represented" the Church; in others they spoke for, or had the support of, its bishops. Conversely, their ties to highly placed Church people enabled them to influence clerical understanding of social and political conditions and to shape episcopal positions on issues. Their advice and encouragement helped the bishops to speak out on behalf of a return to democratic rule before the parties began to function openly. They also helped the bishops to mediate among opposition parties and among antagonistic social forces during critical dialogues throughout the 1980s.

Two Christian Democrats, Sergio Molina, treasury minister in the Frei government, and businessman José Zabala, helped Msgr. Fresno to create and carry through the dialogues that produced the National Accord. Former *mapucistas* Enrique Correa and José Antonio Viera-Gallo, both of whom were close to a number of bishops, were crucial bridges between them and figures on the left at several points. Christian Democratic economists Alejandro Foxley and Rene Cortázar, the latter a long-time participant in Christian Life and other reflection-group activity, facilitated the discussions between business and labor groups that resulted in the Social Pact in 1988, and later served, along with Molina, in President Patricio Aylwin's cabinet.

Cultural Catholic elites

Most Latin American countries abound in nominal Catholics who have ceased practicing their faith but retain attitudinal vestiges of their formation. This was the case in Chile with many of the intellectuals and leaders of leftist parties during the Allende years. A significant number of them either sought or were given refuge in Catholic agencies or Church-sponsored organizations in the months following the coup. Some ended up working in one of the Church's educational or social service programs. A number of left-wing intellectuals, for example, were taken in by the archdiocese of Santiago's Academia de Humanismo Cristiano, where they taught, did research, and wrote on social and political subjects under the Church's protection.[73] In addition, hundreds of trade unionists and social activists were absorbed by Church-sponsored educational, cultural, and humanitarian organizations, where they enjoyed protective space and access to resources without which they would not have survived psychologically or organizationally.[74]

Many of these cultural Catholic elites were deeply affected by their reencounters with the Church. They developed personal ties to the bishops, priests, nuns, and Catholic laypeople. They resonated with their values, and were touched by the experiences of joy and sorrow they shared in the post-coup years. Some rediscovered their faith in the process; others developed a new appreciation for Christian values and principles.[75] Virtually all were affected by their new Church ties. Some enjoyed influential relationships with authorities (clerical and lay) in the Episcopal Conference. Others provided the bishops with access to leftist political parties and to the opposition movement generally.

Local practicing Catholics

Practicing Catholics at the local (parish or community) level included "sacramentals" and "organizationals." Together they made up roughly 30 percent of the Catholic population, far outnumbering those in the previous two categories.[76]

Sacramentals constituted the vast majority of these local practicing Catholics. They attended mass or other rituals at least weekly but were not involved in other Church-sponsored activities, spiritual or social. Given Chile's class structure, many sacramental Catholics lived in areas (*poblaciones* and lower-middle-class neighborhoods) where opposition to the military government was strongest during the transition. Their regular mass attendance and the strictly ritualistic character of their religious practice presumably made them passive carriers of Catholic values and the objects of Vatican efforts to depoliticize the Church and restore its waning discipline and unity.[77] What they thought about military rule, democratic government, and the strategies being employed by both Church and by secular leaders to reestablish democracy, would have a bearing on the Church's role in the transition process.

Most organizational Catholics were regular or occasional mass attenders living in popular (*poblaciones* and lower-middle-class) neighborhoods. Unlike sacramentals, they were also active in Church-sponsored pastoral organizations at the local level. These centered around spiritual (i.e., prayer and reflection) activities (*grupos carismáticos, catechumenados, sodalitium,* Mes de María), sacramental preparation groups (*mamás catequistas, grupos matrimoniales*), or social service projects such as meal programs (*comedores infantiles, comedores de ancianos*), human rights defense committees, employment pools (*bolsas de trabajo*), and rehabilitation programs (*alcohólicos anónimos*), etc. Some organizations, especially the base communities (CEBs) spanned both spiritual and social activities since they combined rituals, study, prayer, and involvement in social outreach activities.[78] Although relatively few in number (1 percent of all Catholics), organizational Catholics had the greatest exposure to Church personnel, pastoral guidelines, official teachings, and the statements of Church authorities. They were also regarded as the most socially and politically committed Catholics in most other Latin American countries as well. But organizational Catholics were not simply faithful extensions or instruments of Church authorities in politics. As we have seen above, the Church's social teaching is subject to divergent interpretation and application, and under the pressures and influences of Chilean society many of them acted frequently on their own. And this is not surprising, given the spirit of democratic participation with which many Chilean Catholics were formed religiously and politically since 1973.

From 1983 to 1986, the popular neighborhoods in which most organizational Catholics lived abounded in antiregime sentiment and activity. Many took part individually and as groups in protests, strikes, and demonstrations. Two organizational expressions of such involvement were the Sebastian Acevedo Movement against Torture, a group of clerics and laypeople that demonstrated regularly against the use of torture by intelligence and security forces, and the Coordinadora de Comunidades Cristianas Populares, a coalition of Christian communities of greater Santiago that was active in the protests, and each year organized a Way of the Cross during Passion Week that would visit "stations" at which acts of repression against popular groups had been committed.

Independent Christian communities and groups also played important roles in the protests, particularly in the western and southern zones of the archdiocese of Santiago,[79] and in other opposition activities. In early 1988, a group of organizational Catholics helped to persuade the leaders of several still skeptical opposition parties to encourage participation in the plebiscite later that year, effectively paving the way for the emergence of the Concertation for the No. They did so by taking up the cause of two young men, one a Christian community activist and the other an evangelical Protestant, who had gone on a hunger strike in order to shame opposition parties into endorsing the plebiscite. They formed a solidarity committee in Santiago, and brought party leaders together in January at a meeting that began with a Bible reading and reflection, and produced an agreement four days later (only the communists resisted), to work together to defeat the regime's candidate.

Many young Catholics were politically radicalized while participating in programs preparing them for First Communion and Confirmation. Some remained religiously active. Most continued to "believe," but dropped out of their religious communities because of the demands that the parties to which they were drawn (the Communist party, the Christian Left, MAPU, MAPU-Lautaro, and the MIR) made on their time.[80] One such person was the *mirista* Rafael Vergara, whose parish priest and confidant considered him "the young Christian most favored by the grace of God that I know, that I have ever known."[81] Others ended up abandoning their faith because politics offered "more meaningful" solutions to their country's problems and made their religious beliefs seem naive or superfluous. A nun who had spent thirteen years helping to prepare young men and women for Confirmation in Santiago's Pudahuel Norte section told us,

> Look, young people have no alternatives. They have no way out. What alternatives do they have? On the streets, drugs, the corner, having a good time, spending what little money they have, or taking a chance. And these parties give them the opportunity to take a chance. More than the communities! It was the communities that taught they must become committed, but we have not enabled them to achieve a synthesis [of their religious and political beliefs]. In fact, they see me in the street, and they tell me: 'Oh, thank you, sister, it was the community that opened my eyes.' [And I ask them] 'Well, where are you now?' [And they answer] 'In the *Jota* [the Young Communists], sister.' [And I ask them] 'Well, have you left your community?' [And they answer] 'Yes, I have. It does not do anything for me anymore.'"[82]

The impact of local practicing Catholics on the struggle to restore democracy in Chile is difficult to determine, as is the role played by Church authorities. It is difficult to weigh the countervailing effects of moderate and radicalized organizationals,[83] on the one hand, and those of apolitical or disengaged sacramentals, on the other. And, while Catholic bishops, priests, nuns, and pastoral programs almost certainly influenced the attitudes and levels of involvement of organizational Catholics, secular factors and forces may also have had decisive, and in some instances crosscutting, impacts.

Local cultural Catholics

These were baptized Catholics who were religiously formed (to some extent) as children but who long since had ceased to practice their faith, and had little or no contact with the Church in the years since. Some of them began to work as volunteers in Church-sponsored organizations providing social services to local communities following the coup. Others were recipients of these services. The Church opened its doors and resources to both groups without regard to their political affiliation or level of religious practice.

Volunteers and recipients alike were touched in new ways by Catholicism, which they found more compassionate and caring than ever before. Their contact with socially committed priests, nuns, and lay Catholics led some to rediscover their religious faith and Catholic values generally. Some became useful intermediaries between Church authorities and local activists. In fact, these local cultural Catholics were active participants in terms of participation in unions, community organizations, political parties, and social protest movements. Finally, local cultural Catholics were a vital force in reintroducing into the Church dimensions of working-class popular culture called for (officially) by the Medellín and Puebla conferences. Thus they represented a new, more secular Catholicism permeating the Church's outer layers in the years preceding the democratic transition.

The attitudes and activities of these Catholics were important for both politics and the Church during the 1980s. If they were moving in a radical direction (along with the movements and parties to which they belonged), they would undermine the consensus-building and reconciliation needed for successful transition, and would present a threat from below to the authority of newly appointed conservative hierarchs, thereby exacerbating ecclesial tensions as well.

Attitudes of Local-level Catholics

Sacramental and organizational Catholic elites figured prominently in the movement to restore liberal democracy in Chile. Their activities brought them into contact with the Church, but in no way threatened its traditional core interests. Practicing and nonpracticing Catholics, on the other hand, were less visible, and somewhat more divided, but nonetheless important participants in the opposition movement, in center and left parties, and in the plebiscite itself. Some favored radical strategies, were critical of the Church's efforts at mediation, and clearly made its central authorities nervous; others supported and applauded the bishops' efforts on behalf of a negotiated settlement.

Surveys taken during the mid-1980s indicate limited political interest and involvement among Catholics generally,[84] and among those active in local Christian organizations or communities.[85] Less than 1 percent of the Catholic population of Greater Santiago belonged to any type of Christian organization, and very few (7 percent) of them belonged to political organizations; more (10 to 20 percent) said they occasionally took part in social and political activities, and looked to their Catholic groups for help in integrating their social and spiritual concerns, but almost none saw their Christian communities (or other organizations) as political vehicles. Unfortunately, none of these surveys examines the religious or political views of individual members, of Catholics generally, or of people's political ideas or involvement at previous junctures, making it impossible to detect the impact of the communities themselves.

A survey we conducted in 1987 fills several of these voids, but can only hint at possible causes and influences. It covered all of Santiago, probed attitudes as well as involvement, and included Catholics with varying degrees of involvement in the Church. But it is based on a purposive, not a representative, sample.[86] Our respondents, 518 in all and 312 living in lower- and lower-middle-class neighborhoods, were identified by local priests and pastoral agents whom we asked to connect us with organizational, sacramental, and cultural Catholics, in a rough ratio of 3 to 2 to 1, in each of twenty-four Santiago parishes. Our Chilean colleagues were unanimous in advising against a representative sample. They felt that in the prevailing political climate (mid-1987) few respondents would be willing to candidly discuss delicate matters with someone suddenly appearing at the door, and that a modestly sized and priced representative sample (of all Catholics) would not yield enough organizational activists to provide a basis for generalization. Accordingly, our survey assured us of reliable responses to probing questions from a substantial number of local activists but does not provide a basis on which to generalize about either Chilean or Santiago Catholics, or Catholic activists as such.

We classified Catholics in terms of the degree of their involvement with the Church, i.e., as organizational, sacramental, or cultural Catholics. Organizational Catholics were those who participated in one or more Church-sponsored religious or social organizations (Christian communities, reflection groups, prayer groups, etc.) at the parish or subparish level. Most of them were also regular or occasional mass attenders. Sacramental Catholics attended mass regularly or occasionally, but were not involved organizationally.[87] Cultural Catholics were neither sacramentally nor organizationally active, but still considered themselves Catholics, and had some (albeit infrequent) contact with the Church.[88] To see how these Catholics would compare with those with no contact with the Church we added a category, *no religion*, which included fallen-away Catholics and Chileans from completely secular traditions.

We asked these Chilean Catholics, 312 of whom lived in *poblaciones* and lower-middle-class areas, a series of political and religious questions. The questions dealt with political tendency (self-defined), the country's "best" government in recent years, whether or not violence was ever justifiable as a political strategy or tactic, the propriety of collaboration between Christians and Marxists, and whether or not the respondent belonged to a social organization or political party.[89] On the attitudinal questions, Catholics as a whole were quite progressive, as reflected in the data presented in table 4.1, below. Over four-fifths of all Catholics identified themselves as having a leftist political tendency (82.4 percent), chose either the Allende (45.2 percent) or Frei (38.8 percent) government as the country's best, and favored collaborative activities between Christians and Marxists (80 percent). Moreover, well over half (55.2 percent) rejected violence categorically as a legitimate tactic. However, on each of these issues, those with no religious affiliation were further to the left than any of the Catholic groups, including the culturals.

Table 4.1 differentiates among the lower-income Catholics in our sample in terms of their relationship with the Church (organizationals, sacramentals, and culturals). Clearly, intensity of Church association did affect the respondents' views, with those most exposed to the institution holding the most moderate views on every question.

The low-income Catholics active in grass-roots religious organizations that we surveyed (including base communities) exhibited little of the radicalism that some scholars and journalists have attributed to members of Christian communities. But the views of these organizational and sacramental Catholics were hardly "conservative" in any

TABLE 4.1

Political Attitudes of Popular-Sector Catholics, Santiago, 1987.

	Left Political Tendency	Best Government, Allende	Best Government, Frei	No Violence	Christian-Marxist Collaboration	Participation in Social Organization	Participation in Political Organization	Range of N
All Catholics	82.4%	45.2%	38.8%	55.2%	79.9%	26.6%	31.2%	199–262
Organizational	74.7%	32.4%	46.8%	62.1%	75.8%	21.8%	23.1%	95–132
Sacramental	86.1%	33.3%	47.6%	67.3%	74.5%	15.4%	26.7%	36–52
Cultural (Lapsed)	76.5%	52.9%	29.4%	47.6%	73.7%	27.3%	28.6%	17–22
Fallen-away	96.1%	81.6%	16.3%	30.9%	96.3%	47.3%	56.9%	49–55

Source: Fleet, 1987a.

plausible sense of the term. More favored Frei's over Allende's government, but none chose Pinochet's, and very few opted for either Alessandri or Ibáñez. Moreover, between 74 and 86 percent of all organizational and sacramental Catholics described themselves as having leftist tendencies, and almost three-fourths of them favored Marxist-Christian collaboration, although both were generally opposed to the use of violence in pursuit of political goals. In fact, of the Catholics we surveyed, those with the greatest exposure to the Church were moderately progressive, and most likely to support a peaceful transition to civilian rule.[90] Those with little or no contact with the Church (cultural and fallen-away Catholics respectively) were even more progressive, but the former[91] less than the latter.

In terms of impact on the transition process, the openness of these low-income Catholics to collaboration with Marxists and to a political regime in which Christians and Marxists could work together permanently is worth noting. Four-fifths of all low-income Catholics in the sample favored "collaborative activities" (*actividades conjuntas*) between Catholics and Marxists. When asked to indicate what this might entail, three-fifths of these most involved with the Church (organizationals) mentioned some form of permanent alliance or single party composed of Catholics and Marxists. In fact, organizationals were significantly more positive (60.2 percent) than sacramentals (44.0 percent) and much closer to culturals (73.9 percent) in this regard.

These attitudes may reflect the day-to-day contact that many organizational Catholics had with Marxists in local Church-sponsored programs and in organizations under military rule. These exchanges could have made them more aware of the values that Catholics and Marxists held in common. Sacramental Catholics, on the other hand, had much less contact with Marxists and leftists and were not as positive about future political alliances, although the still substantial number of them (40 percent) who were positive in this regard suggests that non-Church-related influences may have been pushing people in this direction.

In view of their more extensive (and presumably left-wing) political involvement, it is not surprising that the sample's nonreligious respondents favored both overall collaboration and a permanent relationship between Christians and Marxists by much wider margins (54.9 percent and 70.6 percent respectively) than all three Catholic subgroups. Their attitudes no doubt reflect wholly political assessments and judgments, although the public pronouncements of the Chilean bishops during the 1970s and early 1980s, and the moral and political support provided by local Church-sponsored programs, probably made it easier for them to view Catholics as potential allies.

Of the Catholics we surveyed in Chile, the attitudes of those from relatively well-to-do areas differed from those of people living in lower-income neighborhoods.[92] Among middle- and upper-class neighborhood respondents, for example, 70.7 percent of the organizationals and 64.7 percent of the sacramentals opposed the use of violence to achieve political ends. In addition, wealthier organizational Catholics were more supportive (84.5 percent), but sacramentals less (70.6 percent), of Christian-Marxist collaboration. But they were both less enthusiastic (only 37.1 percent of the organizational Catholics, and 28.1 percent of the sacramental Catholics) about permanent political alliance between Christians and Marxists. These differences suggest the possibility that extra-ecclesial (i.e., societal) forces were shaping Catholic attitudes and behavior in these contexts.

To explore the impact of broader, secular influences on Catholic attitudes and behavior, we controlled our respondents for class, residential area, and parish types. Class appears to have affected attitudes only marginally.[93] White-collar workers held the most progressive views, while independent and blue-collar workers were the most favorably disposed toward relations with Marxists. But none of the "more progressive" groups (occasional sacramental and cultural Catholics) had large concentrations of lower- or lower-middle class respondents, and among organizationals, independent and blue-collar workers were slightly more progressive than white-collar workers and petit bourgeois elements in their assessments of previous governments and the possible use of violence, but less progressive in terms of political tendency and collaboration with Marxism.[94] Finally, independent and blue-collar workers in each of the organizational subgroups were less progressive than workers who were sacramental or cultural Catholics. In sum, membership in a Catholic organization appears to have diluted the progressiveness of respondents from all social classes.

Differences in terms of the residential sector in which the respondent lived and the type of parish to which he or she belonged were sharper. Catholics living in *poblaciones* were more progressive; an average of 60 percent of them gave progressive answers to the four political questions, compared to 49.1 percent of those living in "popular" (lower-middle-class) neighborhoods. Among organizational Catholics living in *poblaciones*, on the other hand, those belonging to "socially oriented" parishes were substantially more progressive (averaging 62.6 percent) than those affiliated with "spiritually oriented" parishes (48.9 percent).[95]

The progressive character of particular parishes or communities could derive from their activities and programs over the years, from the work of a particular priest or pastoral team, or from the history and characteristics of the surrounding neighborhood. The fact that the gap between more and less progressive people was widest among those involved in parish activities (organizationals and regular sacramentals) would seem to indicate that the parish, not the area, was responsible. But it could also be that Church authorities assigned their most progressive personnel to highly politicized areas, or that pastoral agents were radicalized by the communities to which they were assigned.

To further examine the interplay between ecclesial and societal influences, we can look at the experience of three parishes in our sample: one of them, Nuestra Señora de Dolores of Quinta Normal, is a dynamic, spiritually-oriented parish; the other two, Nuestra Señora de los Parrales of La Granja and Comunidad Cristo Liberador of Villa Francia, are a socially oriented parish and community, respectively.

Three Parishes

Nuestra Señora de Dolores serves some 60,000 to 70,000 Catholics in the community of Quinta Normal, on Santiago's near west side. Since 1965, it has been staffed by the Schoenstatt Fathers, an offshoot of the Pallotine order founded by German-born Father Joseph Kentenich. During the 1980s, four Schoenstatt priests attended to the spiritual needs of the main church, three satellite chapels, an adjoining parish, San Vicente de Palotti, and a nearby shrine, the Medalla Milagrosa. They were assisted by seven Schoenstatt nuns, and a large number of lay volunteers, many of whom are active in the Schoenstatt lay movement.

The parish's main church had three Sunday masses, at which the average attendance in the 1980s was around 500 people. Three satellite chapels, two of which were located in poorer sectors of the parish, had a Sunday mass as well. The parish sponsored a full range of catechism classes, retreats, and apostolic organizations for men, women, and young people. Dolores was out of step with most other parishes in the archdiocese's progressive western zone. For most of the period since

1973, it refused to use pastoral or catechetical materials produced by the zonal vicariate, judging them to be of an unduly political character. Unlike other area parishes, moreover, it was never a staging ground or arena of antiregime agitation. As it adjoins an industrial park with a number of large factories, there were occasional strikes or demonstrations in which some parish members took part, but otherwise there was little or no political activity.

Between 1973 and 1987, in fact, pastoral programs and activities (youth organizational meetings, catechetical groups, support groups, etc.) were organized and carried out almost exclusively at the parish complex. Apparently, people felt more secure there, less likely to be rousted by military or police authorities. The same fear is said to have inhibited the development of Christian base communities, although these began to spring up in 1987, following the relaxation of restrictions on "public" meetings and "political" activities.

The parish activists and nuns with whom we spoke appeared to share the pastoral outlook of the Schoenstatt priests. The consensus reflected and contributed to the effectiveness with which the parish has been managed since 1965. Mass attendance and participation in parish activities have been high, indicating a positive response to the leadership of the Schoenstatt Fathers by area residents generally.

The residential sector, Quinta Normal, may have something to do with this. It is an older, more established "popular" *comuna*, with a predominantly lower-middle-class population known for its pragmatism, social stability and upwardly mobile (*arribista*) mentality. Although its support for the No option (61.4 percent) in the 1988 plebescite, and for opposition candidates in the 1989 presidential (57.7 percent) and parliamentary (59.4 percent) elections, exceeded the averages for Santiago as a whole, it was the least pro-opposition of the areas in which our three parishes are located.[96]

Nuestra Señora de los Parrales is a working-class parish on Santiago's south side. From 1978 to 1988 it was staffed by the Sacred Heart Fathers (or "Padres Franceses"), an order known for its theological, pastoral, and political progressivism.[97] Sacred Heart priests also staffed two adjoining parishes, San Gregorio and San Pedro y San Pablo, both known for their strong community life and political activism.

Parish activities were centered in the main church, located on Santa Rosa Avenue. It held several masses on Sunday, and housed a variety of pastoral programs and activities. Three smaller chapels functioned as mini-parishes and community centers in their respective sectors. On Sunday, a priest came to each of these chapels to hear confessions, say mass, and meet with sector organizations. Ecclesial Base Communities of up to eighty people functioned in each, and people were encouraged to center their religious and organizational activity around these communities, not in the larger (and more prestigious) parish church.[98]

Parish members, particularly young people, were very active in protests and other antiregime activities during the 1980s. Protests and demonstrations would begin suddenly, be sustained for brief periods, give rise to periods of relative calm, and then erupt again, weeks or months later. The parish community was very much a part of this process, given its insertion in the life of the comuna (La Granja) and because of the main church's location at a major intersection on Santa Rosa.[99]

The Sacred Heart priests were openly sympathetic in this regard but were not instigators or activists themselves. They reminded people that Christians had civic responsibilities and should be doing something to improve current socioeconomic and political conditions. They stressed, however, that such activities were their responsibility as laypersons, and not the business of the Church or its pastoral programs.

In mid-1988, they relinquished their control of the parish to a conservative Spanish priest, although they continue to work with two of the satellite chapels (or communities). Since their departure, the main parish has continued to function sacramentally, but the level of organizational activity has diminished markedly. Apparently some activists have moved to other parishes (also run by Sacred Heart priests) whose orientation and general environment they found more congenial.

Cristo Liberador is a large Christian community located in the west Santiago neighborhood of Villa Francia. Although within the boundaries of Jesus de Nazarét parish, it has functioned as an autonomous ecclesial community since its foundation, in 1970, by Father Mariano Puga, a well-known worker-priest. Since the mid-1970s, it has been "accompanied" by Father Roberto Bolton, with the help of Sisters of the Divine Master, most of whom come from Argentina and Uruguay.[100]

Cristo Liberador functions as an ecclesial base community (CEB) with an active membership that has ranged between sixty and a hundred people since its establishment. Its liturgies are held in a chapel built on the edge of a vacant two-block lot in the middle of Villa Francia. For a number of years, it has made use of the Msgr. Enrique Alvear Center, a large complex several blocks away, where workshops, classes, and social activities are held. In addition to the CEB, the community encompasses six Christian base communities (CCBs), with memberships of from ten to twenty adults (mostly couples) each. They meet weekly, and mostly on their own, for Bible study and general reflection, with Father Bolton or one of the sisters joining them occasionally. Each of the communities sends a representative to the monthly pastoral council of the CEB, at which administrative matters and broader pastoral, social, and political questions are treated.[101]

The community has had its ups and downs. A religious community in solidarity with its neighborhood and with popular-sector groups, it has brought together as many as 250 to 300 hundred people, and as few as 30. Six or seven of its original members (couples) are still active. Others have moved on to other communities and similar commitments.[102] Some have been discouraged by either government repression or internal political conflicts, and have returned to more conventional faith lives; others have taken on full-time political commitments and no longer have time for religious commitments as such.

Villa Francia has been a hotbed of political radicalism since the late 1960s. During the years of military government, it was frequently a center of protest and opposition activity, and almost as frequently the object of sweeps and occupations by security forces. In some instances, its political activism has had an invigorating effect on the life and activities of the Christian community and the neighborhood as a whole; at other times it has generated division and disillusionment. The level of support for the No vote, and for opposition and left candidates in recent elections, was slightly higher than in the areas around Dolores or Parrales.[103]

The thinking of Catholics in these parishes broadly parallels the political character and tradition of the neighborhoods surrounding them. In their responses to political questions, members of the Cristo Liberador community held the most progressive views, averaging 50 percent, while those of Nuestra Señora de los Parrales and Nuestra Señora de Dolores parishes averaged 45.3 percent and 47.3 percent, respectively. Table 4.2 illustrates these differences.

It should be noted that Catholics in more conservative Dolores parish were more inclined to describe themselves as leftists than those from the other parishes, were more interested in collaboration between Christians and Marxists than Catholics from Parrales, were quite active politically, but were the least favorable toward the Allende government and the least involved (less than half the extent of the others) in social organizations.

To better understand the local church impact, table 4.3 compares the attitudes of *organizationals*, i.e., the Catholics most exposed to the Church, with those of other Catholics in each of the three parishes. The percentages in parentheses indicate the differences between their views and those of *sacramentals* and *culturals* on the six questions. A minus value indicates that the responses of sacramentals and culturals were higher (more progressive), a positive value that they were lower (less progressive), than those of *organizationals.*

In Dolores, a spiritually oriented parish in an area affording some upward mobility, the responses of organizational Catholics were the lowest of any on three of the five general political questions and considerably lower (by an average of 18 percentage points on these same questions) than those of sacramentals and culturals. Their views were also less progressive, although more functional to a peaceful transition, on the use of violent methods.[104] Their levels of social and political involvement, on the other hand, were also somewhat lower. These data suggest a strongly moderating influence on the part of the priests and nuns of Dolores.

The average scores of organizational Catholics in Parrales and Cristo Liberador, each of which was guided by a progressive pastoral team and surrounded by poorer and politically more militant areas, were considerably higher (50.4 percent in Parrales and 63.1 percent in Cristo Liberador) than in Dolores (41.7 percent). Organizational Catholics were marginally more progressive than other Catholics in Parrales, but markedly more so in Cristo Liberador. In both parishes, the percentage of those willing to use violence under some circum-

TABLE 4.2

Political Attitudes of Catholics in Three Popular-Sector Parishes, Santiago, 1987.

	Left Political Tendency	Best Government, Allende	Best Government, Frei	No Violence	Christian-Marxist Collaboration	Participation in Social Organization	Participation in Political Organization	Range of N
Dolores	88.0%	31.5%	42.5%	59.7%	78.1%	14.7%	39.0%	46–62
Parrales	87.0%	58.3%	27.8%	57.5%	69.2%	41.5%	42.5%	27–32
Liberador	87.5%	70.3%	27.0%	43.2%	95.0%	38.6%	16.3%	23–34

Source: Fleet, 1987a.

stances was lower among organizationals than among the other two groups. Finally, in both Parrales and Liberador, organizationals were more likely be involved in social organizations than other Catholics, but less so politically. These data point to a mixed impact on the part of pastoral agents in Parrales and a strongly progressive impact in Cristo Liberador in all areas except party involvement.[105]

Another indication of differences among the three parishes and their impacts on organizational Catholics were people's responses to survey questions dealing with religious issues. Respondents were asked about: (1) the reasons (spiritual or political) for Christ's persecution and death; (2) the importance of obedience to Church authorities, as against following one's conscience, when dealing with moral problems; (3) the propriety of the bishops speaking out on economic questions; and (4) whether, in the event of conflict, one followed his or her priest or bishop. In each parish, organizational Catholics gave higher (more progressive) responses to these questions than did other Catholics.[106] Those in Dolores were much less progressive (41 percent) than those in Parrales (57.1 percent) and Cristo Liberador (67.8 percent), however, and in both Parrales and Dolores less religiously progressive elements were often (though not always) more progressive politically.

These responses suggest that both social context and pastoral orientation were affecting Catholic attitudes. In Dolores, those most involved with the Church were also more moderate in their political views than Catholics with less contact, an indication, perhaps, that the more conservative pastoral leadership of the Schoenstatt Fathers was

TABLE 4.3

Political Attitudes of Organizational and Other Catholics in Three Popular-Sector Parishes, Santiago, 1987.

	Dolores		Parrales		Liberador	
Left political tendency	76.9%	(–23.1%)	78.6%	(– 8.9%)	93.3%	(+30.8%)
Allende best	21.4%	(–16.7%)	43.8%	(– 1.7%)	76.5%	(+31.0%)
Favor Christian-Marxist collaboration	73.5%	(– 9.8%)	68.8%	(+11.7%)	100.0%	(+15.7%)
Participation in social organization	10.8%	(– 9.2%)	33.3%	(+11.9%)	36.4%	(+11.4%)
Participation in political organization	26.5%	(–31.4%)	27.8%	(– 1.8%)	9.1%	(– 7.7%)
Average	41.7%	(–18.0%)	50.4%	(+ 2.8%)	63.1%	(+20.1%)
No violence	67.6%	(–21.8%)	64.7%	(+21.0%)	50.0%	(+ 8.3%)
Range of *N*	46–62		22–32		23–34	

Source: Fleet, 1987a.

spilling over into political matters. Catholics in the other two parishes were politically more progressive. In Parrales, organizational Catholics were not very different from other Catholics, but in Cristo Liberador they were substantially more progressive. Both Parrales and Cristo Liberador were socially oriented parishes, and the religious views of their organizational Catholics were moderately progressive and highly progressive, respectively. In view of this, it seems reasonable to characterize priests and pastoral agents in Parrales as reflecting or accommodating the concerns of the people of their area, and to credit those of Cristo Liberador with a markedly progressive impact at least in terms of people's political attitudes.

In all three parishes, political party involvement was lower among organizational Catholics, but as the data presented in table 4.4 indicate, the attitudes and behavior of those belonging to parties were quite different from those of Catholics who did not belong to them. Table 4.4 gives the responses of organizational Catholics in each parish belonging to a political party, followed (in parentheses) by those of organizational Catholics who did not belong. A minus value indicates that the responses of nonmembers were higher, and in all but the case of no violence more progressive; a negative value indicates that the responses were lower, and therefore less progressive, than those of members. Again, here, the patterns in Dolores were unique.

In Parrales and Cristo Liberador, the organizational Catholics belonging to political parties had higher scores (i.e., held more progressive views) than those that did not belong; in Dolores, however, politically involved organizational Catholics trailed those who were not

TABLE 4.4

Political Attitudes of Popular-Sector Organizational Catholics Belonging, and Not Belonging, to a Political Party, Santiago, 1987.

	Dolores		Parrales		Liberador	
Left political tendency	66.7%	(–20.8%)	80.0%	(+ 2.2%)	100.0%	(+ 7.7%)
Allende best	40.0%	(+21.8%)	80.0%	(+52.7%)	100.0%	(+26.7%)
Favor Christian-Marxist collaboration	70.0%	(– 7.3%)	100.0%	(+45.5%)	100.0%	—
Participation in social organization	—	(–16.0%)	40.0%	(+ 9.2%)	100.0%	(+70.0%)
Average	48.6%	(– 4.5%)	75.0%	(+27.4%)	100.0%	(+26.1%)
No violence	50.4%	(+22.0%)	25.0%	(+51.9%)	—	(+65.0%)
Range of *N*	25–35		14–18		15–22	

Source: Fleet, 1987a.

involved by 4.5 percentage points. In addition, their average score was much lower (48.6 percent) than those of their counterparts in Parrales (75 percent) and Cristo Liberador (100 percent). In this context, party involvement appears to have moderated the attitudes of organizational Catholics even further in Dolores, and to have made them more progressive in Parrales and Cristo Liberador.[107]

These data do not permit us to generalize about Chilean Catholics or Catholic activists. But they do suggest that in three important parishes and/or communities both social context and parish character affected the political attitudes of their people during the transition period.

Conclusions

The Catholic Church contributed to the successful transition to democracy in Chile at several key junctures. Individual bishops and segments of the lay community played key roles in the process through the plebiscite of 1988. The hierarchy pressed for a peaceful and orderly transition to democratic rule throughout the period, playing first a neutral, and then a mediatory, role.

The bishops themselves were not always agreed on all matters. Determined to show the young, the poor, and the oppressed that the Church was with them, a handful of progressive bishops were continuously and explicitly critical of the government. Moderates, on the other hand, wanted to avoid breaking with the government and abetting in any way forces advocating "all means of struggle" against it. They also wanted to deemphasize political issues because of their divisive effects on the Church. Conservative bishops, whose numbers grew steadily with John Paul II's appointments, either liked Pinochet or preferred his regime to a government dominated by the left. Along with most moderates, they wanted normal relations with authorities that included ceremonial and other institutional contacts.

Despite such divisions, the Pinochet regime treated the Church as a whole as an adversary, and allowed security and paramilitary forces to attack its pastoral agents and programs. This hostility ultimately led the Chilean Episcopal Conference to close ranks. When their pastoral workers were harassed, foreign priests were expelled, and the authority of individual bishops was publicly challenged, moderate and conservative bishops joined with progressives in a more critical and spirited defense of perennial core interests. In doing so, they helped to undermine the regime's legitimacy long before parties became an effective alternative to it.

Had the regime not been as heavy-handed in its dealings with Church people, moderate and conservative bishops might have remained on the sidelines throughout the transition process. But because it was, and because it threatened the Church as a whole, they could not. As public protests against the regime mounted in 1983, the bishops articulated a general moral framework for assessing social and political developments in cautiously critical terms. Their efforts strengthened the hand of moderates in both government and opposition camps, isolating and partially delegitimating more extreme elements. Their most significant efforts on behalf of a peaceful transition, however, came as private mediator and promoter of the National Accord and later the Social Pact.

Archbishop Fresno and his advisors failed to convince Pinochet to negotiate a settlement in the mid-1980s, but they brought opposition parties together around positions with which many government people and their supporters could live. They did not dictate the terms of these agreements, but they offered an atmosphere of confidence and trust in which to talk, assistance in helping people to see the extent of their common concerns, and finally moral legitimation to the final accords. They helped opposition forces to unite, and to become the reasonable alternative to Pinochet that they had not been to that point.

It is ironic, perhaps, that the hierarchy was most influential when it was least prophetic, when it worked behind the scenes on behalf of political compromise and not, as earlier, denouncing evil and abuse in the public arena. Whatever local Church activists might have thought, the conservative Fresno was the ideal Church leader for the time. Raúl Silva's courageous defense of human rights made him a hero to the general public and to activists within the Church. But he was profoundly distrusted by both religious and political conservatives, and could never have allayed or disarmed their fears to the extent that Fresno did.

Other elements of the larger Catholic community played important roles in the transition as well. Practicing Catholic elites, prominent in business circles and active in parties of the left, center, and right, advised the bishops, facilitated the dialogues hosted by Fresno and other bishops, and generally embodied the openness to negotiation and compromise that the bishops were calling for. The unwillingness of

Catholic business and political leaders to heed the bishops' call for negotiation in 1972 and 1973 undercut the Church's potential role in that crisis. Conversely, during the mid- and late-1980s, they acted as carriers for Church values and perspectives (moderation and compromise), and had precisely the opposite effect.

Local Catholic activists, on the other hand, were an important part of the broader opposition movement, although an unrepresentative minority among the general Catholic population.[108] As members of local Christian communities, as participants in youth groups and in classes preparing young people for the sacraments, or as activists in local, Church-sponsored meal programs or human rights defense groups, Catholic activists took part in marches, demonstrations, work stoppages, protests, and other antiregime activities. Their ties to the Church made them valued allies and occasional sources of protection for other regime opponents. For them, and for many people, they "embodied" the Church in their parish or zone.

We cannot generalize about Chile's Catholic activists or its larger Catholic population. It is conceivable that they included a substantial number of religious and/or political radicals who either supported or were active in groups like the Sebastian Acevedo Movement against Torture or the Coordinadora de Comunidades Cristianas. However, among the local-level Chilean Catholics whom we interviewed, a substantial number of whom came from some of Santiago's most progressive Catholic parishes and/or most militant neighborhoods, most of the organizationals and sacramentals were quite moderate. In particular, they displayed attitudes and practices likely to facilitate a successful transition, i.e., a willingness to work closely with leftists in a reconstituted democracy, and reluctance to consider violence a legitimate or appropriate political method.

Among Catholics living in the parishes we studied in greater depth, the views of Christian community members and other organizationals were much more progressive religiously and politically than their sacramental or cultural counterparts in Cristo Liberador, only slightly more so in Parrales, and significantly less progressive in more spiritually oriented Dolores parish. The most active and progressive among them, however, came from poorer, more highly politicized, and traditionally combative areas, to which Church authorities generally assigned progressive priests and nuns, and where they generally received a favorable response (the case, for example, in both Parrales and Liberador).

Other factors might have affected the attitudes and behavior of Catholics in other areas, but most of those living in Dolores, Parrales, and Cristo Liberador were influenced by convergent religious and secular forces.

Finally, cultural Catholics played important roles in Chile's transition to democracy at both national and local levels. Most leaders and militants of the social movements and political parties involved came from their ranks. The experience of many of these men and women in Church agencies and Church-sponsored organizations left them with a new appreciation for the values of freedom, human rights and dignity, community, and solidarity. Their contact with the Church left them more favorably disposed to it (as a social and cultural institution), to its perspectives on social and political matters, and to the need for moderation and reconciliation at this juncture in Chilean history. Secular intellectual and political currents were also pushing them in this direction, but the Church was an important additional source of influence.

Their experience with Church also made it easier for these cultural Catholics to work with Catholic political elites. The latter were not just representatives of rival parties (such as the PDC, for example), they were fellow Catholics. They were the personal friends of a bishop, or political advisors or liaisons of a Christian community, to whom one could extend trust or make concessions with greater confidence. They were people with whom they had shared special joys and sorrows, and with whom a new and stronger sense of common interest had developed.

In sum, Chilean Catholics at various levels of the Church were more than the secondary allies of which Drake and Jaksic speak in the transition process. The support for democracy offered by the Church did not cease in 1982. It resurfaced at key junctures in the mid- and late-1980s, helping to bridge the gaps among opposition groups and to ease the anxieties of former regime supporters beginning to look beyond Pinochet. And it remained significant through the Social Pact and the plebiscite of 1988.

5

. . . .

Chile's Consolidation of Democracy

Six years after being restored, Chile's democracy remains partial and uncertain. The Armed Forces continue to be arbiters of the country's political life, defining the boundaries within which civilian politicians operate. Most of the antidemocratic limitations with which the Aylwin government was saddled when it assumed power were in place when it left. Assassinations and other acts of violence periodically disrupt the relative calm of postmilitary political life, and from time to time the military engages in shows of force, reminding civilians of what it will and will not tolerate.

Nonetheless, this partial "democracy" appears to be taking root in all sectors of society except the military. With a strategy of "consensus democracy," the Aylwin and Frei governments have consulted with allies (labor unions) and adversaries (right-wing parties) on virtually all policy matters. Differences have become matters of degree or detail, not principle, in many areas, and an atmosphere of mutual respect and accommodation has arisen. Economically, the government's policies have produced sustained, high-level growth, winning over once skeptical bankers and industrialists, yet reducing the percentage of Chileans living below the poverty level.[1] Politically, General Pinochet continues to have solid backing from the army, and remains a destabilizing element. But he has lost much of his support among conservative elites and with the public at large. Most observers think that only hard-line military elements would welcome a coup at this point, and that they are unlikely to move on their own.[2]

Unlike its role in Chile's transition, the Church's impact on consolidation has been modest and at times obstructive. The bishops have devoted considerable time and energy to the issues of social and criminal justice, but with little success. Their public statements on divorce, birth control, abortion, public education, and protestant "sects," on the other hand, have challenged, and probably weakened, liberal democratic values and attitudes.

Local Catholic activists have been no more successful in influencing government policy or in challenging restrictions on democratic rule. They have grown more independent of local and national Church leaders than in previous years but are less influential politically. Their demands for justice and for socioeconomic reform have been more restrained, however, and their skepticism of parties and political institutions less pronounced than many observers anticipated. As such, they have contributed to the spirit of accommodation that has prevailed to date. Their effect on long-term consolidation of fully democratic institutions is much less clear, however.

In what follows, we analyze the impact of the institutional and grass-roots levels of the Church on the consolidation of Chile's democratic institutions. We begin with some reflections on consolidation, the impact of modernization and social democratization, and the obstacles standing in the way of each.

Completing the Transition and Consolidating Democracy

The consolidation of a democracy is the process by which a country's democratic institutions and practices become fully democratic and take deeper and firmer hold. It marks the point at which the empowerment and accountability of elected officials become complete and relatively free from credible threats to their continued existence.[3]

Consolidation is not simply the implementation of agreements hammered out in a transition period, however. Indeed, the terms of most transitions are intended to avoid or obstruct full democratization, and at some point will have to be challenged by democratic forces. Consolidation thus involves matters that opposing forces could not face or resolve during transition. This is the case in Chile, with the military's immunity and autonomy vis-à-vis civilian authorities, its substantial oversight functions, the discriminatory aspects of the country's electoral laws, and the removal of several policy areas from the jurisdiction of elected officials.[4]

In Chile, therefore, consolidation involves two distinct objectives: the completion of a less than fully democratic transition, and its institutionalization. The first of these objectives requires: a substantial reduction, if not the complete elimination, of the military's political functions and other nondemocratic features of the 1980 Constitution; clarification and resolution of responsibilities for human rights abuses since 1973; removal of antidemocratic limitations on elected executive and legislative authorities; and reform of the judiciary.

These protective arrangements were not things with which the military and its civilian allies would be likely to part willingly. They were useful buffers or defenses against public opinion and a government with broad popular support, both of which they viewed as hostile. If the Concertation government could keep economic growth at a high level, it could probably win over some skeptical entrepreneurial elements. It would be easier to meet their needs and expectations, and they would be less inclined, perhaps, to view the military's concerns sympathetically. But growth and prosperity alone would not assure the survival or extension of democratic institutions. The mistrust of both the military and its civilian allies[5] toward Concertation parties and leaders would have to be overcome.

The personal attitudes and relationships of leaders on both sides would be crucial in this regard, at least in the short term. They could make or break the process of fulfilling or extending the arrangements established during the transition. In the longer run, however, the post-military system was unlikely to become or remain truly democratic without the effective participation of rank-and-file activists. For this to happen their leaders would have to negotiate with them on more equal footing than in the past, and their own demands would have to become or remain moderate or reasonable.[6]

The modernizing and socially democratizing tendencies currently afoot in Chilean society are pushing developments in this direction.[7] But economic growth as such is neither inherently or inevitably democratizing.[8] Neo-liberal growth strategies, for example, generate fewer jobs and require stricter control of labor costs than alternative models, and can retard consolidation over the longer term. The impact of growth and modernization in Chile thus will depend on the growth model pursued (how much and what kind of employment and social-class development it generates), and on both government and private-sector initiatives in social welfare, education, neighborhood promotion, employment training, the media, and other areas. As with civil-military relations, secular leaders will be crucial in this area as well, although the access and influence that the established and emerging social forces acquire along the way will also play a role.

It is with both elites and these rank-and-file activists, we believe, that the Catholic Church has a role to play in democratic consolidation. Its moral authority with people on both sides of the political divide, and its influence at both elite and grass-roots levels, equip it to promote reconciliation, strengthen commitment to democratic principles, defend the interests and promote the empowerment of subordinate classes, help (at times) to moderate their expectations, and assist elites in overcoming their fear of their own followers and those of their rivals.

An additional problem to be faced in consolidation processes anywhere is the need for social and political forces to express and extend their "democratic" commitments concretely. The task of consolidation is not, as in transition, to achieve agreement in principle. It is to put principles into practice in specific contexts and in relation to particular issues, to practice what one preaches to others (even, and perhaps above all) in areas of concern to oneself. And here, it must be said, military, civilian, and clerical elites alike have had difficulty learning to accept the judgments and verdicts of support bases (either their own or those of their adversaries) on issues of importance to them.

In the sections that follow, we analyze the progress made in consolidating democratic institutions and practices since Patricio Aylwin's assumption of the presidency in March 1990. In doing so, we look at the impact that bishops and local-level Catholics have had on specific policy issues and on civil-military relations generally, on majority/minority relations, and on both modernization and social democratization.

The Bishops

As Chile returned to civilian rule, the Catholic Church was its most admired and respected national institution. The work of the Chilean bishops in defense of the poor and on behalf of the victims of human rights abuses won them new stature and respect.[9] At the same time, the appointment of more conservative bishops during the 1980s gave the Episcopal Conference a more neutral image, strengthening its mediatory credibility, and enhancing its impact on the transition to democracy.

The bishops were less likely to be a factor in consolidation, however. With the country returning to civilian rule, secular forces would speak for themselves, and the center of political gravity would shift from matters of principle to problems of application and approximation, on which principled people (including the bishops) could easily disagree. Additionally, with the presidency and legislature in opposing hands,[10] most policy matters would have partisan implications, which the bishops would probably want to avoid. Finally, their renewed emphasis on more traditional Catholic moral values and positions (on divorce, sexual morality, etc.) would almost certainly undercut the influence that the bishops had enjoyed with secular leftists and cultural Catholics in previous years.

The Aylwin government took office determined to move ahead on full democratization and justice for the victims of human rights abuses. The general public appeared to be supportive. Most bishops sympathized with these objectives but thought that issues of democratization were better left to the country's political parties. They would focus, instead, on justice for the victims of human rights abuses, the socioeconomic needs of the poor, and moral questions affecting personal and social life. They soon discovered, however, that these matters were no less complex and that their own ecclesial and ideological differences, and the absence of ready ears and secular carriers in the larger social arena, greatly limited their influence.

Justice for victims of human rights abuses

Chile's Catholic bishops called continuously for justice for the victims of human rights abuses during the four years of the Aylwin government. Conservative and progressive bishops alike defended the moral claims of victims and their families. Families needed to know the fate of their loved ones, the bishops contended, and they needed to know who was responsible for their fate before forgiveness and reconciliation could take place.

The bishops first addressed the issue of accountability in a pastoral letter written by Msgr. Carlos González in early June 1990,[11] as the Special Commission on Truth and Reconciliation appointed by President Aylwin was beginning its investigation of human rights abuse under military rule. González's major point was that true reconciliation was possible only if people admitted their mistakes and asked one another for forgiveness.[12] The bishops spoke collectively several days later, when a mass grave was discovered in northern Chile.[13] They issued a strong declaration through their Permanent Committee, saying that it was time to stop referring to the atrocities of the period as "alleged" or "presumed." "The country needs to know what happened in broad outlines [*rasgos más esenciales*]," they said. "It should be the responsibility of the justice system to clarify the facts and determine guilt and innocence" (and) "it is necessary to clarify things in terms of fundamentals and to refuse to justify what is unjustifiable by arguing that we were living in a state of war."[14]

Military authorities were not about to submit to the judgments of civilians in these matters, however. General Pinochet put the matter bluntly, shortly before relinquishing the presidency: "The day they touch any of my men," he warned, "[that day] the rule of law is over."[15] In fact, to preclude such a fate, he proclaimed an amnesty law in 1978 that protected military officers from prosecution for virtually all possible violations of human rights up to that point.[16] And now, with the discovery of mass graves, other top-level officers only vaguely acknowledged feeling "as much pain as all Chileans regarding some events of the past," but added that they were confronting adversaries "who share no quarter."[17]

Several months later, the Supreme Court issued a judgment upholding the 1978 amnesty law. It also ruled that the law precluded military and police officials from being tried or investigated (as well as punished) for possible offenses. This was a direct repudiation of President Aylwin, who had urged that even cases covered by the amnesty be investigated. Periodic discoveries of additional burial sites kept the issue alive during 1990 and early 1991. As trusted and resourceful public figures in their respective dioceses, Catholic bishops were frequently approached by the relatives of disappeared people asking them for help in determining the fate of their loved ones.[18]

With pressure for full disclosure and prosecution mounting, the military counterattacked. In mid-December, Pinochet ordered army troops confined to their barracks in battle dress. The move was described as a "readiness alert," but the military was letting civilian leaders know it would not tolerate investigations of its conduct and would defend itself forcefully if necessary.[19] The maneuver had its desired effect. Most critics of the military backed off; a few, like Msgr. González, pressed on, but without effect.[20]

The problem with making the military accountable for human rights abuses was that the government had neither legal nor de facto power over it. The 1980 Constitution gives the military ultimate control of the political process and is virtually impossible to modify in the present political context. It charges the military with "watching over [*velar por*] national security and guaranteeing the institutional order." It reaffirms its "institutional independence" (making it responsible for its own assignments, promotions, and retirements) and giving military courts jurisdiction in cases involving military personnel. Finally, it reaffirms the independence of a Supreme Court that Pinochet packed in 1989 with congenial appointees.

With such provisions, democratically elected governments could do only what military authorities were willing to abide. Any of these provisions could be modified or eliminated by formal amendment, but the majorities required to do so range from three-fifths to three-quarters (depending on the provision) and thus require the support of at least some right-wing deputies and senators. To date, the right has rejected all efforts to reduce the military's power. The UDI and RN parties claim to be democrats but are content to continue operating in a partially democratic system. They would rather keep their appointed senators and an electoral system that gives them equal representation with only a third (plus one) of the votes of a district, postponing the day when they would have to confront center-left forces on their own. In fact, even those who are critical of Pinochet still favor an independent military with "guardian" functions. They particularly like its moderating effect on *concertación* governments, i.e., the premium it places on its maintenance of unity and the centripetal pull that it that it exerts on left-wing forces.

But the real issue is not the right's political convenience. The military would resist its subordination to civilian authority, and any scaling back of its political role, *even if* the votes could be found. Military officers simply do not trust civilian authorities in the present context, and are determined not to place themselves under their jurisdiction. The threat that such a position poses is most intensely felt with respect to accountability for human rights abuses (here, what is acceptable to the military is not to human rights groups, and vice-versa). But it lurks below the surface of all antidemocratic restrictions.[21]

Tensions between civilian and military authorities continued to build through March 1991, when the Commission on Truth and Reconciliation issued its report.[22] It struck a middle ground between full accounting and adjudication of all violations, and an erasure of the past that would permit a fresh start (*borrón y cuenta nueva*). It dealt only with cases that resulted in death (over 2,000), and did not identify the individuals believed responsible,[23] although it indicated the state agency involved (in most instances, the National Bureau of Investigations, or DINA), and how the victim died. It discussed the context in which the abuses took place, but argued that no alleged "state of war" could either justify or mitigate them. And, finally, in limiting the number of cases treated, it sought to avoid the impression that it was lumping all military and police officers together or passing judgment on the military government as such.[24]

In releasing the report, Aylwin urged that those involved in "state terrorism" under military rule acknowledge the pain that they had caused and take steps to lessen it. He also urged civilian courts to take action, noting that many of the cases described in the report involved criminal actions that were not covered by the amnesty law. The Catholic bishops endorsed his call, stressing again that those responsible had to ask for and receive forgiveness for there to be true and lasting reconciliation.[25]

Public reaction to the report was quite favorable. The government scheduled proclamations, discussions, and collective reflections throughout the country, hoping to put additional pressure on military and judicial authorities. Pinochet counterattacked, impugning the report as "biased," resting on "weak foundations," and full of "facile and tendentious" judgments. He also charged that the report was being used for "partisan and political" purposes and that it was compromising the country's national security (in whose defense, the military might have to take action).[26] Other military spokesmen argued that in certain circumstances "harsh actions" were unavoidable, and that they actually had saved the country from infinitely worse fates (i.e., civil war and communist dictatorship).

These people were hard-liners. For them, war, even if internal, was hell, and civilians had no business second-guessing military professionals. They also believed that there were certain political tasks and prerogatives for which the military was uniquely suited, and were determined to retain them. But even moderate military officers (air force and Carabinero officials, for example) insisted on immunity from prosecution. They feared that without it the military would be set upon by hostile and vengeful adversaries. Innocent officers would be swept along with the few that had committed abuses, and the military as a whole would suffer irreparable harm.

How long the military could have resisted pressures for disclosure or submission to judgment is difficult to know. The matter became moot in early April, when two gunmen murdered Jaime Guzmán, the conservative intellectual who helped to draft the 1980 Constitution and was elected to the Senate in 1989. Guzmán's death produced a closing of the ranks among rightist forces in support of the military. It also weakened the Aylwin government politically, pushing it to adopt aggressive antiterrorist policies (with which Msgr. González and other bishops took issue), and generally strengthening the hand of intransigent elements.[27]

Things remained unchanged for the next two years. The bishops continued to urge full disclosure of responsibilities,[28] but gave the impression that they were speaking for the record and did not expect that anything would change. The military's determination to resist was too strong, they sensed, and the possibility of a coup was too high.

The actions and insinuations of military authorities nourished such fears in the months that followed. In May 1992, for example, the government proposed to reestablish civilian control over the military, and army and navy officers again rattled their sabres. Twenty-three army generals signed a statement denouncing the idea and reaffirming the military's status as an independent branch of government, while other military elements claimed to be a necessary "counterweight" to civilian authorities.[29]

At the time several civilian judges were hearing criminal cases involving military officers, but most deferred them to military courts, where they were quickly dismissed. Popular resentment mounted in the ensuing months, and when the Supreme Court directed that the officers accused of the murder of *mirista* Alfonso Chanfreau be tried before a military court, government deputies brought impeachment charges against three of the Justices and a military judge advocate. The Chamber of Deputies approved them, and sent them to the Senate, where the government won a partial victory: one of the justices was impeached, although the order was delayed pending appeal; the other two justices and the military advocate were acquitted.

The military again countered. Army generals, some in combat attire, met in May 1993 in the Defense Ministry wth a detachment of elite, black-bereted troops pointedly standing guard. The generals denounced what they termed a "campaign" to harass and discredit them. They cited as evidence a revival of the case involving Pinochet's son, the fact that civilian courts had subpoenaed and were proceeding against some military officers, and the government's most recent assault on their "rightful prerogatives and functions," and they underscored their determination to defend their honor and the country's institutional order.

What they wanted, it seemed, was a statute of limitations on the filing of charges for human rights violations committed after 1978. They also wanted Aylwin to abandon the notion of trying cases that would later be amnestied. Aylwin and Pinochet met privately on four occasions over the next several weeks, and reached a "compromise" that tilted heavily in the military's direction. A total of 1,167 cases would be adjudicated, 800 by civil courts and the rest by military tribunals. All would be heard in secret session, the names of those actually convicted would remain confidential, and sentences, if applied, would be served in military facilities.

Aylwin accepted these terms, believing that they were as good as he was going to get from the military.[30] The agreement was denounced by the relatives of victims and by several government parties as well. A number of bishops and Church leaders also spoke publicly against the agreement, while others (Msgr. Ariztía and Msgr. Oviedo, for example) endorsed it. With his own supporters divided, Aylwin saw little chance for approval and withdrew the bill.

A month elapsed before the bishops spoke as a group on the matter. When they did, they repeated what they had been saying since 1990, i.e., that there could be no reconciliation without justice and truth.

> Those people and institutions responsible for the painful acts of the past must fulfill their moral obligations to clarify what happened and ask forgiveness for their errors. They must ask God to pardon them, do penance for their sins, repair as much as possible the damage done, and facilitate the exercise of justice.[31]

Unfortunately, the people to whom the bishops directed these appeals ignored them. Military hard-liners were proud of what they had done and resented the aspersions being cast on their "heroic" mission. Moderates, on the other hand, thought it would be hard to limit inquiries once they were underway, and were not ready to submit their fates to either public opinion or civilian courts. Some civilian rightists (the UDI, for example) denied that security forces had done the things of which they were being accused, while others conceded that they had but were determined to protect military morale and integrity.

Apart from human rights organizations and the left wing of the Concertación itself, on the other hand, there were no secular forces willing to embrace and defend the Church's position in this matter. Neither government officials nor Christian Democratic leaders were willing to run the risks or pay the costs that pushing for more entailed.

Nor were the bishops willing to challenge the military either. They continued to make the appeals they had made from the outset, but did nothing else. They could have attempted something like the National Accord, i.e., discussions among parties producing a broad consensus that the military might have to accept. But recent (1991 and 1993) secular efforts along these lines had failed utterly. What was acceptable to the military was not acceptable to human rights groups, and what the latter were demanding was unthinkable to even moderate, much less hard-line, military people.

The Catholic bishops endorsed the principle of civilian supremacy and would have welcomed a reduction or elimination of the military's special functions and powers. To date, however, they have not pressed the matter either publicly or from behind the scenes. They know that they have few "listeners" in the military, and that those that might listen are unlikely to challenge their leaders. They also know that civilian rightists have political reasons for resisting as well, and they are not prepared to challenge them publicly. Finally, the bishops have not found a "secular carrier" in either the Aylwin or Frei governments willing to move more aggressively. Indeed, they recognize the intensity of military sentiment and the wisdom of leaving these matters for cooler heads at a later time.

Economic policy

A second area of concern for the bishops was the Aylwin government's economic policy. For much of the military period, they questioned Pinochet's policies and their effects on the poor. Like many of their countrymen, they hoped that the restoration of democratic rule would bring more socially conscious policy-makers and policies. But the bishops also knew that the new government would have to accommodate investors and entrepreneurs if they were to avoid potentially troublesome economic decline and political polarization.

They were probably surprised, however, when Aylwin's government chose to retain Pinochet's pragmatic neoliberal strategy and to pursue greater equity through compensatory social programs financed by higher taxes and the additional revenues that would come with higher rates of growth. These policies proved remarkably successful in generating growth, reducing inflation, and promoting exports. They strengthened investor and consumer confidence and helped to solidify relations between the state and the private sector and the government's political position generally. And, with additional employment opportunities and expanded social welfare programs, the portion of the population living in poverty and extreme poverty fell from 43 to 34 percent from 1991 to 1993.

But the fruits of the country's growth went disproportionately to the same privileged sectors that had prospered under Pinochet. Their living standards and consumption levels achieved new heights, while millions of Chileans continued to live in poverty and want. Wages remained low and were not likely to go up as long as it was their low labor costs that made Chile's principal exports competitive.[32] And the social welfare system, although more adequately financed, remained as exclusionary and segmental in impact as under Pinochet.[33]

The government's economic policies had mixed effects on consolidation. In keeping the growth rate high and in maintaining cordial public-private sector relations, the Aylwin government's policies made allies out of social and economic forces that might have become destabilizing. In addition, its promotion of small business projects in slum neighborhoods, and its expansion of health services and educational opportunities, helped to sustain and promote social forces that would strengthen and sustain fully democratic institutions in the long run. But at the same time, the restrictions on wage increases and collective bargaining, and a refusal to consider regulating industrial and commercial firms on behalf of lower-income earners, stifled the growth and organizational development of these same, largely popular, sectors.[34]

It was difficult for the bishops to address economic policy issues. Development models and policy instruments were not things on which they could speak comfortably or effectively: they were neither well-informed nor in agreement among themselves. Most of them sympathized with the poor, but while some spoke out individually others did not. An additionally complicating factor was that the people responsible for economic policy in the Aylwin government (treasury minister Alejandro Foxley, budget director José Pablo Arrellano, and labor minister René Cortázar, for example) were CIEPLAN "monks"[35] with whom a number of bishops (including Conference president Carlos González) had been personal friends for years. And these friends were telling the bishops that policies designed to redirect significantly greater benefits to lower-income groups would be resisted by economic elites, would undermine economic growth, and might have even more disastrous political consequences. Their logic apparently convinced some bishops, although not others. But even those who remained skeptical found it difficult to challenge these points publicly or privately, or to enlist other bishops in doing so.

The bishops' first extensive reference to economic matters came in a statement prepared for a special Day of Prayer for Chile in September 1991. It noted the "urgent problems" of the poor, who had a "privileged" place in the Catholic tradition, and stressed respect for their dignity as a key to eradicating violence and criminal activity. But the bishops went no further, and actually urged people to lower their expectations of material improvement in the short term. They spoke, for example, of the need to create "a climate of confidence that would promote savings, investment and creativity," and they expressed the hope that "those who have had to postpone the fulfillment of their needs, and their leaders as well, have the good sense not to ask for the impossible or to demand the immediate resolution of their problems."[36]

This was a remarkably realistic and accommodating statement. It reflected the appointment of more conservative bishops in recent years, and an evolution in the thinking of more liberal members as well. Not all bishops were in full agreement with its terms, however. In October 1991, for example, Msgr. González remarked publicly that, after a year and a half of democratic rule, "the poor were just as bad off as they were [under the dictatorship], and in some cases worse."[37] Another partial dissenter was Archbishop Oviedo of Santiago. In a pastoral letter in 1992, he spoke of the need for solidarity with those who were "suffering the costs of modernization." He termed "certain visions of economic activity, e.g., those that viewed economic competition as an all-out struggle for clients," as "anti-Christian and anti-human." And he concluded by arguing that the existence of widespread poverty was not really an economic problem, the result of laws or forces over which one had no control, but an anthropological one, for which Christian men and women were responsible.[38]

In September, the Conference's Permanent Commission referred to the problems of the poor in a document that dealt with the problem of justice for the victims of human rights abuses. "We are troubled," they said, "by the growing differences between rich and poor, the unequal opportunities available in our consumer society, and in the way in which this situation affects young people. We thus have an ethical obligation to do everything we can to settle this social debt we have with the poorest among us, in whom we recognize the face of Christ."[39]

Having ventured this far, the bishops said little for the next several years. They broke their silence in early 1996, issuing a document prepared by a commission of social scientists and social activists close to the Church. Prompted by the release of new data on poverty (which had dropped in some parts of the country, but not in others) and income distribution (which appeared to have worsened in relation to 1992), they sought to join and to extend the debate that was emerging within the Concertation as to whether or not the country's growth could produce greater equity than it was. The letter was welcomed by those (e.g., Socialists) who thought that more could and should be done, but was virtually ignored by most other political forces.[40]

The bishops thus spoke out occasionally and clearly (if not always forcefully) on issues of poverty and social injustice. With the possible exception of their most recent letter, however, one had the sense that they were speaking for the record, without hope of effecting change. Their limited prospects for influence were a function of their unwillingness to challenge the economic and political reasoning of government policy-makers, some of whom they had known and respected for years. But even had they been willing or able to do this, it is unlikely that they could have persuaded them otherwise. For both economic and political reasons, policy-makers felt strongly that theirs was the proper course, and that workers and the poor were willing to shoulder the burdens involved. Although criticism has intensified of late, most of the latter still appear to believe that the potential costs of trying to get or do more at that point were simply too high.

Moral questions

Since 1990, the Chilean bishops have devoted much of their time and energy to personal and social moral issues. These included divorce, abortion, premarital sex, educational reform, and the growth of evangelical Protestantism. Occasioned by heightened secularization in Chile since the late 1970s, the attention given to these issues also reflected the appointment by Pope John Paul II of ecclesially and theologically conservative bishops. They were more critical of contemporary secular values, and more determined to challenge them, than their moderate and progressive colleagues. And they were not as inclined to mince words or to approach their subjects in a spirit of accommodation or dialogue. Their public exhortations during the early

1990s cost the Chilean bishops some of the good will they had earned in the 1970s and 1980s. They also caused many to question the Church's commitment to the democratic principles that it was urging on other groups.

There has been vigorous debate in Chile on a number of moral issues in recent years. Women's groups, many of which first emerged in opposition to the military government, were increasingly vocal in demanding reforms in the laws and practices affecting women. In 1991, women began broadcasting from their own radio station, Radio Tierra, whose programming featured interviews and discussions calling for more liberal laws and regulations concerning sexuality, divorce, birth control, and abortion.[41]

Public opinion appeared to support their calls for reform and greater freedom in these areas. According to a survey taken in 1989, and published in December 1990, between 68.8 and 72 percent of all Catholics (and 67.1 and 71.6 percent of all Chileans) thought that infidelity, lack of love, and lack of understanding were sufficient causes for divorce, while 43.1 percent (and 42.9 percent of the sample) actually favored making divorce legal.[42] By March 1991, the percentage of Chileans favoring legalization had risen to 55.6 percent, with educated people 66.5 percent), young people (64.1 percent), upper-class Chileans (70.1 percent), leftists (64.8 percent), and rightists (62.9 percent) more adamant in this regard. Somewhat surprisingly, Catholics (54.6 percent) were more supportive of such a law than Protestants (47.2 percent).[43] This reflects the strongly conservative stance on family values of fundamentalist and Pentecostal Protestants, who now dominate Chilean Protestantism.

Opposition to abortion was more substantial, but was growing weaker as well. In 1989, 86.4 percent of Catholics believed that it was always wrong (*un delito*), although most of these were more permissive in the face of specific extenuating circumstances. By 1991, only 49.2 percent of all respondents were willing to rule out abortion under all circumstances, while 44.7 percent thought that it was justified "in special cases." As in the case of divorce, wealthier (75 percent) and more educated (55.4 percent) Chileans were more inclined to accept abortions (under special circumstances, if not on demand), and Catholics (53.0 percent) were more apt to do so than Protestants (29.6 percent).[44]

Divorce was the more politically prominent of the two issues. Chile is one of the few predominantly Catholic countries in the world where divorce is still illegal. Instead of pushing for legalization, Chilean governments have made it easy to obtain annulments on technical grounds. Each year, roughly 8,000 Chilean couples end their marriages in this manner. After paying a fee of U.S. $360, they need only "prove" that the original paperwork contained errors, which can be done by getting people to testify (however falsely) that the addresses listed in the documents were not their correct addresses at the time. For those who cannot afford the fee, an amount three times the minimum monthly wage, the only available option is de facto cohabitation, which leaves the woman without the right to property or child support.[45]

Sentiment in favor of correcting this fraudulent and discriminatory situation has been building for several years. In September 1990, women belonging to the Socialist Party and the Party for Democracy (PPD) drew up a bill to test the waters. Public reaction was generally favorable, but the bishops denounced the idea and most Catholic deputies and senators (Christian Democrats and rightists alike), even those who had remarried after obtaining fraudulent annulments, fell into line behind them. The bishops argued that divorce was contrary to both God's law and the law of nature, and that its legalization would weaken families and increase the damaging effects of dissolution. They acknowledged the existence of many "sad situations" and the need to safeguard both children's welfare and patrimonial rights and obligations, but they urged that these things be addressed in ways that would not weaken the family.[46]

A divorce bill was introduced in the lower house in early 1991. The Aylwin government opposed it, and most Christian Democratic Deputies voted against the measure, effectively killing it. It was resubmitted in July, at which time the Episcopal Conference's Doctrinal Commission sent the legislature a position paper outlining its views.[47] It was more conciliatory than earlier statements, conceding, for example, that current arrangements encouraged fraud and spared couples the bother and expense of serious judicial procedures. But instead of allowing divorce, the bishops urged that religiously celebrated marriages be annulled only when and if religious authorities had already so decided, and that in purely civil marriages solutions for "special cases" be sought in "keeping with natural morality and the common good of the person and the family." They denied wanting to impose Catholic doctrine on non-Catholics, and praised religious freedom as a positive good. But they added that they had an evangelical mission to fulfill, and that it would be a mistake (*no sería acertada*) to allow individual churches or ideologies to establish their own doctrines in these matters.[48]

In mid-1994, President Frei's interior minister proposed that a referendum be held on the issue in 1995, although he subsequently backed away from the idea. Similarly, several bills circulated in the Congress during 1994 and 1995, although neither garnered enough support from Christian Democratic and right-wing legislators to warrant serious consideration. By mid-1996, however, two bills authorizing the dissolution of civil marriages under restrictive circumstances and offering help to "second" families (thought to number in the hundreds of thousands) badly in need of strengthening, were getting favorable responses from both Christian Democratic legislators and some Catholic bishops.[50]

The opposition of Catholic legislators (from the Christian Democratic Party, the Independent Democratic Union, and Renovación Nacional) helped to defeat this proposal too. Public support for reform remained strong, however, and the issue is unlikely to fade away given the abuse of existing legislation and the injustice that it imposes on those who cannot afford a legal annulment.

Abortion is another issue being discussed more openly in Latin America in recent years.[49] No formal effort to legalize abortion has yet been made in Chile, but observers believe that it is only a matter of time before one is mounted. Women's groups are increasingly outspoken on the subject, and public opinion polls indicate growing support for a relaxation of current restrictions. According to 1989 survey data, for example, 76.5 percent thought that abortion should be permitted when the mother's life was in danger, and 48.7 percent thought so when the pregnancy was the result of a rape.[51] Neither the Vatican nor the Chilean bishops have shown any indication of willingness to reexamine their views on abortion itself or legalization. In the years to come, therefore, the Chilean Church is likely to be increasingly at odds with public opinion and with constituencies (women's organizations and progressive secular groups) with which they enjoyed quite cordial relations in the 1970s and 1980s.

AIDS and its containment is another issue looming large in Latin America. The number of reported cases in the region has gone from fewer than 8,000 in 1987 to 60,000 in 1993. This sharp upturn is generally attributed to unprotected homosexual and heterosexual activity by males, although an increasing number of married women are contracting the disease from their husbands.[52] Chile has had government-sponsored family planning programs since the 1960s, and recently began distributing condoms as part of an effort to curb the spread of AIDS, prompting a vigorous response from the bishops.[53]

Their opposition was part of a broader concern with sexual permissiveness generally.[54] In 1991, Msgr. Oviedo made sexual morality the focus of his first pastoral letter as archbishop of Santiago. Oviedo was an ideological moderate, but an ecclesiologically and theologically conservative prelate. Along with other bishops, he was troubled by indications of increasingly permissive attitudes and practices in terms of premarital sex.[55] His letter was sharply criticized by the liberal media, by secular social and political activists, and by theological progressives who wanted him to address social and political issues.

The Church was also critical of a proposed educational reform that the Aylwin government had circulated in early 1992. According to the Episcopal Conferences's secretary, Msgr. Cristían Caro, the bishops had serious reservations regarding "the conception of the human person and the moral relativism that can be seen running through the entire proposal."[56] The problem with the draft document, he said, was its failure to even mention absolute values, to link human freedom to values capable of fulfilling human potential, and to relate sex education to the family and the indissolubility of the marriage bond.[57]

At issue here was the extent to which public education should reflect official Catholic perspectives in a largely Catholic but religiously pluralist country. Bishops like Caro believed that unless the proposals were rooted in explicitly Christian or Catholic values, moral relativism was inevitable.[58] They were determined to confront the government and to demand "more Catholic" texts. Others, like Permanent Committee president Carlos González, disagreed. They recognized that a democratic and pluralist state could not tie itself to a single religious vision. They were more willing to cooperate with the government and to help the state to fulfill its ethical responsibilities without compromising its secular character.

In the end, conciliatory elements prevailed, producing a reworded document acceptable to both church and state. It helped that secretary of government Enrique Correa had spent two years in the seminary and knew a number of the bishops personally.[59] But the conflict underscored the differences that divided Church leaders and alienated some of their one-time secular allies.

A final, divisive issue since 1990 has been religious toleration.

Chile's Protestant population dates to English and German immigrants in the late nineteenth and early twentieth centuries. In 1960, Protestants constituted about 11 percent of the population, with the bulk of them belonging to the so-called mainstream (Lutheran, Anglican, Methodist, Presbyterian, and Baptist) churches. More recently, there has been a significant increase in the number of Evangelical and more fundamentalist churches, such as the Jehovah's Witnesses and the Church of Jesus Christ of Latter-Day Saints (Mormons). These denominations have engaged in aggressive proselytizing efforts, particularly among nominal, low-income Catholics. By 1985, Protestants made up between 15 and 21 percent of the total population, with non-mainstream churches accounting for almost 80 percent.[60]

The rapid growth and expansion of evangelical Protestantism has adversely affected Catholic-Protestant relations. During the Pinochet years, Catholics and mainline Protestant denominations, most of which were opposed to the dictatorship, worked closely together. In contrast, the newer denominations (Pentecostals as well as Witnesses and Mormons) were more favorably disposed to the military government, which frequently hailed them as "true Christians who stayed clear of politics."[61] Since 1990, the Church's relations with Protestant groups have been strained. In 1992, the bishops issued a pastoral document in which they differentiated among mainline denominations, Pentecostals, and more aggressive "sects," such as Jehovah's Witnesses and the Mormons. They were particularly critical of the latter, whom they saw as posing a danger for democratic society because of their "doctrinal demagoguery," their "theocratic, vertical and totalitarian structures," their insistence on possessing the "absolute truth," their control of information, their "total rejection of society and its institutions," their "suppression of individual liberties" of their adherents, their insistence that new members "break all former social ties upon entrance into the cult," and their attitude that "all those outside the group" were enemies.[62]

The bishops were less harsh in assessing the Pentecostals, who have grown more rapidly than the sects in recent years.[63] However, they lumped the two together in terms of their impact on Chilean Catholics. Their growth, they contended, had led to "confusion and doctrinal disorientation," "doubts and insecurity in the faith," "divisions in the [local base] communities and in families," and a "lack of respect for religious symbols and images of the Catholic Church." And they saw

Catholics suffering from the "pressures of proselytization," from a "feeling of inferiority," and from "grave pain over family divisions caused by some sects."[64]

The document urged active competition with these groups. It called for the training of local Catholic activists with respect to the Bible and Catholic doctrine. It urged them to function as neighborhood missionaries getting nominal or fallen-away Catholics to return to active religious practice. Finally, it advocated greater use of radio, television, and the printed media to spread Catholic teachings.[65] Thus, despite its moderate language, the document left the impression that the bishops were launching a spiritual war whose objective was the rolling back of the inroads that Protestant groups had made among former Catholics.

In doing so, they had Rome's full support. In his address to the meeting of Latin American bishops (CELAM IV) in October 1992, Pope John Paul II referred to the new generation of Protestants in Latin America as "rapacious wolves" responsible for division and discord among Catholics. He claimed that large amounts of cash were being donated (he did not name the source) in an effort to destroy Catholic unity in Latin America. He strongly endorsed the program of "new evangelization" developed in preparation for the conference, and urged an all-out effort to cultivate popular piety among the great masses of Catholics in Latin America.[66]

With the Vatican continuing to appoint ecclesially and theologically conservative bishops, the sentiments expressed in this letter will probably shape Catholic pastoral strategies for years to come. They are also likely to generate increased tensions with both Pentecostal churches and the more aggressive sects, and could lead the bishops to call for restrictions on religious freedom in the future.

The impact that the more conservative views of Church authorities on these matters have had on Catholics is difficult to assess. It could be that the bishops have kept rank-and-file Catholics on a properly moral course, but they appear to have driven many others away. Surveys taken in 1990 and 1991 indicate a drop in the level of confidence in the bishops during this period, for example. While 42.5 percent of all respondents expressed "a lot of confidence" in them in December 1990, by March 1991 the figure had fallen to 34.8 percent, although it rebounded to 37.9 percent following Jaime Guzmán's assassination. When these figures are broken down in terms of socioeconomic strata and political position, the bishops have enhanced their image with upper-class Chileans (46.3

to 48.1 percent) and have lost ground with those identifying with the left (from 57.6 to 47.9 percent).[67] The first of these shifts is difficult to relate to the issues of divorce and abortion, since rightists are among the most liberal Chileans on these issues. But their positions are probably at least partially responsible for their declining support among leftists generally, and upper- and middle-class leftists in particular.

The uncompromising manner with which the bishops approached these issues, and the fact that they did so although many Chileans thought otherwise, did little to solidify their democratic credentials or to strengthen the force of their call to others to respect the will of the people. The country was clearly in the throes of an ethical crisis, and the Church had a job to do. But as a *Mensaje* editorial put it in mid-1993, Church authorities needed to convey principles in a more Christian, i.e., a more inviting, less negative, manner. Christianity could not be imposed from above through laws; it needed to be preached in a way that was "humanizing and respectful" of the people to whom it reached out. In short, the bishops needed to accompany these people, not simply direct or regulate them.[68]

The bishops could argue, of course, that there was no compromising with the truth, that these matters could not be decided by referenda or survey research. But in arguing that the issues in these cases took precedence over the right to one's own point of view, over the need to respect the views of others, and over the need to see differences within a larger context of community and accommodation, they made it easier for the army and the political right to do likewise in areas of importance to them.

The Local Church

Most Catholic priests and nuns in Chile were not politically involved during the country's transition to democracy; even fewer have remained active since 1990. With the resumption of normal political activity, most previously active Church people have returned to conventional pastoral activities and projects. Parishes and communities appear to be more fully staffed than in previous years. This is a result of a fall-off in religious practice and an increase in the number of seminarians and ordinations during the years of military rule.[69] Most observers agree, however, that new priests are ecclesially and theologically more conservative than those of earlier generations.[70]

Reduced interest in politics and social issues has been particularly apparent in Chile's Christian communities, whose number has declined significantly since 1987. With the relaxation of political restrictions and controls, there was less need for refuges, functional distractions, political sounding-boards and organizational cells. Those continuing to operate were less likely to discuss political matters or to support political initiatives than during transition. With the recovery of public space, Catholic activists interested in politics were free to affiliate directly with organizations of their choosing, and were usually, although not always, encouraged to do so.

Most priests and sisters concerned with political affairs at the local level steered clear of direct involvement but sought to "accompany" their lay leaders pastorally, i.e., to provide support in difficult moments and to help in the reconciliation of religious and political commitments. Most with whom we spoke sought to strengthen people's sensitivity to the needs and interests of others and to help them learn to defend and promote their own in the parties and organizations in which they are active. Some of them were resolute *basistas*, determined to endorse whatever their congregations decided to do ("lo que vea la base," as some put it), while others thought of themselves as exercising a critical, practical, or moderating influence.

In general, pastoral agents and workers at the local level have had a moderating influence on popular expectations for the postmilitary period. One or two of the more than twenty priests, sisters, and lay pastoral workers we interviewed between 1987 and 1992 thought that the first postmilitary government should begin immediately to undo the damage done under the military. As a lay worker at the western zone vicariate put it:

> I think that the oligarchy has to pay for everything that it has done. I would include confiscations, expropriations, drastic measures, profound structural changes. . . . And right away. Unless we are able to do this we will be right back where we started. We'll begin to give the oligarchy a little bit of space and as soon as it has such space it will engineer another coup.[71]

But most pastoral workers were willing to settle for much less. In their view, the best way to avoid a return to military rule was *not* to challenge the military and to show business elites that they had nothing to fear from a civilian government that respected human rights and democratic liberties. A southern zone social worker, for example, agreed that just getting rid of Pinochet and reestablishing a formal or bourgeois democracy would not solve the country's problems. But he cautioned that in view of Chile's economic situation (particularly its external indebtedness), "there would be very little that even an ideal postmilitary government [in his view, a centrist coalition] could be able to do for the neediest people, for the popular sector."[72]

A nun working in Pudahuel, one of the western zone's poorer and more militant neighborhoods, had mixed feelings on these matters. She expected that government and party elites would try to manage things their way and keep the people from interfering in the delicate business of accommodation:

> The politicians are already saying that the highly sensitive, emotional mass public wants everything all at once, that it does not understand the political maneuvering one has to engage in at this point, the concessions one has to make, given what there is right now. These politicians are constantly talking about Argentina. How many of them have pointed to the case of the Mothers of the Plaza de Mayo, saying "these blessed mothers, they know nothing about politics, they are ruining us, because they operate with such emotional intensity and with an unwillingness to maneuver, to compromise and to negotiate, that they force us to assume unrealistic and maximalist positions." And so there is no hope. They [the politicians] simply confirm our suspicion that they are standing in our way.

And yet she knew that people needed to lower their expectations if disasters were to be avoided:

> I think that there is going to have to be an intervening period, a few years of transition for arranging things, so that people get used to feeling a little freer. Now I believe that after that there will have to be daring reforms, which if undertaken immediately would generate another military coup. But in five years or so someone will have to think of the major economic reforms that we might need to undertake. In the long run it [socialism] is both desirable and feasible. But in the short term, I think, unfortunately, that it would be a total failure.[73]

In a similar vein, a nurse who directed a medical clinic in the Santiago archdiocese's northern zone expected that progress in the postmilitary period would be political rather than economic. "We understand," she said, "that there are not going to be immediate improvements in economic conditions the minute this government changes." The country's current conditions and limited resources precluded meeting people's basic needs or demands that they would begin to make. She did think, however, that "small, local development alternatives . . . with minimal government support" might be an option for some. "There might not be a lot of these businesses," she said, "but at least real needs would be made clear, real interests expressed." "And right now," she added, "the most important thing is that people will be able to make demands known, and to express them in concrete ways."[74]

Finally, a nun who was part of the pastoral team for a parish in Peñalolén, another of Santiago's more militant areas, was equally pessimistic, although less inclined to blame the constraints of the post-military period. In her view, popular-sector Chileans were often obstacles to their own advancement. "I have come painfully to the conclusion," she said,

> that the poor are responsible for some of these obstacles. Right here in Nazareth, for example, we have set up a little university which is a house where we try to get people to attend meetings, take courses, and participate in formation workshops. [But] it is even hard to get workers to read the Bulletin of the Vicaría de Solidaridad. And those packages that are up there [she pointed to a shelf], they are for a course in citizenship training, and they cost a lot [to produce]. Well, all these things people ask for, but when the time comes few of them actually take part.[75]

Two additional developments worth noting in connection with the local Church were carried out shortly after Msgr. Oviedo became archbishop of Santiago. The first was the formal dissolution, in March 1990, of the Coordinating Council of Christian Communities that brought together politically oriented Catholic activists during the years of transition. The second was the dismantling of the Vicaría de Solidaridad, in November 1991. In both cases, conservative bishops succeeded in getting rid of organizations that had troubled them in the past. Without having to call their activities or their political impact into question, they could point to the availability of secular agencies or outlets capable of carrying them on in the postmilitary period. In doing so, they denied the Catholic activists in these organizations at least some of the respectability that their Church ties lent them.

Local-level Catholic Activists

The attitudes and activities of lay Catholic activists will certainly affect, though will hardly determine, the success or failure of consolidation. Lay Catholic activists can help to strengthen democratic institutions by participating constructively at the local level and by refraining from making excessive demands that would burden and destabilize still fragile political institutions. In fact, observers who had highlighted the political impact of Christian communities under military rule thought that Catholic activists would be important sources of support for democratic institutions and practices once power passed to civilian hands. Some saw natural "linkages" between human rights advocacy and democratic methods and practices. Others felt that activists imbued with liberation theology would be less likely to defer to authorities or leaders and would act as a democratizing force within whatever parties or movements they formed or joined.[76]

But other observers were not so sure. They were troubled by "fundamentalist" Christians who were suspicious of "politicians" generally and had trouble making concessions to people they did not know or with whom they disagreed. In their view, such people were more likely to play disruptive roles by refusing to accept party discipline, insisting on full discussion and debate of all issues, and making economic and political demands that would destabilize the new institutions before they had a chance to take root.[77]

Of the Christian community activists we interviewed in 1987, relatively few were politically involved during transition, most held moderately progressive political views, and most tended to defer to Church authorities.[78] Unfortunately, we do not have access to public opinion surveys that enable us to compare the political views of Catholic activists with those of other Catholics during the subsequent (post-1990) consolidation process. The best we can do is use our 1987 survey data as a basis for projections for the post-1990 period, and then look briefly at data gathered since. In both cases, we are concerned with the political attitudes, expectations, and involvement of Catholics over whom the Church exercised special influence in comparison with the attitudes of other Catholics.

1987 data

The Christian community activists we interviewed in 1987 had generally more positive attitudes toward politics than did most other Chileans. Catholics belonging to Christian communities, for example, were both more interested in politics and less cynical about them than other respondents. This can be seen in table 5.1, in which we compare the views of various subgroups of Catholics.

All Catholics surveyed had a substantial interest in politics, but Christian Community activists and those belonging to Church-sponsored social organizations were the most positive in their views. A smaller percentage viewed "politicians" favorably, although again Christian community and social organization activists were much more positively disposed.

The same pattern emerges with people's attitudes towards political parties. Here, Christian community activists are the most positive in their views towards parties and are only slightly less satisfied with relations between the parties and their social bases. Table 5.2 summarizes responses to questions touching on these matters.

The reservations of some of these respondents may reflect the difficulties that parties were having in mid-1987 (e.g., uniting against Pinochet, keeping in touch with constituents, and delivering tangible benefits to them). As with other political matters, however, the most positive subgroups were those belonging to Christian communities and social organizations, while the least positive were those affiliated with religious organizations.

TABLE 5.1

Political Attitudes of Popular-Sector Catholics, Santiago, 1987.

	Interested in Politics	Politicians Work for Good of All	Range of *N*
CEBs/reflection groups	92.6%	45.5%	68
Social organizations	95.1%	42.9%	41–42
Religious organizations	82.6%	34.8%	23
Sacramentals	86.5%	19.6%	50–51
Culturals	90.5%	34.8%	21

Source: Fleet, 1987a.

TABLE 5.2

Attitudes of Popular-Sector Catholics towards Parties and Party Practices, Santiago, 1987.

	Adequate Party/Social Organization Relations	Parties Do Not Reflect Interests of Ordinary Chileans	Range of *N*
CEBs/reflection groups	61.3%	35.9%	62–64
Social organizations	73.5%	47.5%	34–40
Religious organizations	61.9%	76.2%	21
Sacramentals	66.7%	54.5%	39–44
Culturals	57.9%	64.7%	16–19

Source: Fleet, 1987a.

Also noteworthy are the relative modest demands of people with regard to leader-base relations. Table 5.3 details responses to the question: "How, in your judgment, should the leadership and base of a political party be related?" The question was an open one, with responses clustering around three options: a leadership better informed about constituent needs and interests, better communication between leaders and followers, and equal footing between the two. This last response can be taken to represent the sentiments of so-called *basista* or militant rank-and-file elements.

Most respondents gave middle-of-the-road answers (better communications between leadership and base). A minority of each group was more assertive, wanting leaders and followers to relate on more nearly equal footing. An even smaller number in each category thought that leaders should continue to lead but should be better informed about the needs of their followers. Catholics belonging to Christian communities and reflection groups emerge in the middle of the pack; they are neither the most demanding (*basista*), nor the most deferential, of the subgroups.

Finally, Christian community activists and Catholics belonging to relection groups were highly supportive of liberal democratic institutions, as willing as most of the others to accommodate or reconcile opposing class interests, and the most supportive of the military fulfilling its "traditional" role.[79] Table 5.4 details responses to these questions.

As can be seen, very few of the Catholics we interviewed favored a popular democracy (from which the rich were excluded), and more than three-quarters wanted the standard liberal democratic regime (from which no one was excluded). On the other hand, nine of ten respondents favored accommodation of class interests (over

TABLE 5.3

Attitudes of Popular-Sector Catholics regarding Leader-Base Relations, Santiago, 1987.

	Informed Leadership	Better Communication	Leaders/ Followers on More Equal Footing	*N*
CEBs/reflection groups	6.9%	81.0%	12.1%	58
Social organizations	2.8%	86.1%	11.1%	36
Religious organizations	15.4%	76.9%	7.7%	13
Sacramentals	7.3%	68.3%	24.3%	41
Culturals	—	77.3%	22.7 %	20

Source: Fleet, 1987a.

defense of one's own). Organizational Catholics (of both types) were only slightly less supportive of these "moderate," proconsolidation positions than other Catholics, although a handful favored exclusion of some elements.

Most Catholics also favored the military's fulfillment of its "traditional" role (as defender of national sovereignty and territorial integrity), with those belonging to Christian communities and social organizations the most pronounced (74 percent) in this regard. As of mid-1987, they were thus holding to a middle-of-the-road position with respect to the military. They opposed the permanently expanded role established by the 1980 Constitution, but showed no desire to punish the armed forces either. We do not know how they proposed to strip the military of its additional prerogatives, or what they would do if it resisted, but their generally moderate views suggest that they would support a compromise or negotiated settlement at some point in the future.

TABLE 5.4

Political Attitudes of Popular-Sector Catholics, Santiago, 1987.

	Popular Democracy	Non-exclusionary Democracy	Extremists Excluded	Accommodate Class Interests	Traditional Military Role	Range of *N*
CEBs/reflection groups	5.9%	83.8%	16.5%	88.2%	75.4%	65–68
Social organizations	2.4%	70.7%	26.8%	87.5%	71.8%	39–41
Religious organizations	4.5%	77.3%	18.2%	90.0%	50.0%	20
Sacramentals	4.8%	88.2%	5.9%	94.2%	59.2%	49–52
Culturals	4.5%	86.4%	9.1%	90.5%	61.9%	21

Source: Fleet, 1987a.

1989 data

Over the next two years, Catholic activists continued to be a moderating force within the political arena. In October 1989, the Centro de Estudios de la Realidad Contemporanea (CERC) surveyed the political and moral attitudes of Chileans throughout the country,[80] making it possible to compare the views of practicing and nonpracticing Catholics with those of evangelical and other Protestants. The CERC study was based on a different sample and asked different questions; it cannot be compared directly with our 1987 survey.[81] But it can be compared with previous CERC surveys (of both Santiago and the country as a whole), and helps to reinforce the implications of our 1987 data.

CERC's data confirm, for example, that Catholics generally, and more active Catholics in particular, were a moderating political force as Pinochet was about to relinquish power. They indicate that Catholics (45.4 percent) were more "interested" in politics than evangelicals (32.7 percent) and those belonging to other religions (39.4 percent),[82] that practicing[83] Catholics (50.3 percent) were slightly more "interested" in politics than non-practicing Catholics (45.8 percent), and that younger Catholics (those between eighteen and twenty-four years of age) were the most interested in politics (53.1 percent), the most sympathetic (31.8 percent) to the left, but the least regular (21.1 percent and 31.9 percent for men and women, respectively) in terms of religious practice, and the least inclined to think of the Church's influence as beneficial.[84]

The CERC data also indicate that Catholics in general and active Catholics in particular identified primarily (38.5 percent) with the political center and had modest expectations regarding economic improvement in the postmilitary period. They further indicate, that Catholics (generally) were slightly more sympathetic to the right (23.3 percent) and slightly less sympathetic to the left (24.2 percent) than other Chileans,[85] that slightly fewer Catholics (58.3 percent) thought the existing economic system was either "quite" or "very" unjust (either *bastante or muy injusto*),[86] and that they were less inclined (62.8 percent) to think that fundamental changes were necessary or that everything needed to be changed.[87]

These data, gathered in 1989, confirm our 1987 findings. They showed active Catholics (not Catholic activists)[88] to be more politically oriented, but less progressive, than other Chileans, especially those having no religion, and that the most progressive and most politically oriented of all were younger Catholics who had the least contact with the Church. The CERC survey did not ask respondents in what kind of neighborhood they lived, and the data themselves offer no clue as to the sources of their politicization or radicalization. But it is possible, as with our 1987 survey, that the social and political attitudes of many Catholic activists were products of the secular environments in which they lived and worked.

Potential for Future Dissent within the Church

These data, drawn during the transition to democracy, suggest that Catholic activists would be a stabilizing, moderating influence in the ensuing process of consolidation. Before concluding, however, we would like to reflect briefly on their likely responses to the initiatives of the Chilean bishops relating to divorce, abortion, educational reform, and the growth of Protestantism. If large numbers of Catholic laypeople were to support the hierarchy in opposing divorce and abortion reform, or to enlist for combat in the struggle with Pentecostal and other Protestant churches, Catholicism's impact on the consolidation of democracy could be even more divisive and obstructive. But this would not be the case if rank-and-file and activist Catholics were to oppose or ignore the hierarchy's more conservative agenda. In this event, the Church itself could be in for stormy times, but would be less of an obstacle to the strengthening and extension of democratic institutions and practices.

Available survey and interview data suggest considerable potential for dissent within the Church if the bishops push their moral and religious agenda vigorously. Table 5.5 details 1989 CERC data on Catholic support for divorce and abortion reform. Respondents are divided into those who attend mass regularly, those who do so occasionally, and those who do not attend.

The table data indicate substantial dissent by all types of Chilean Catholics to official Church positions and to priorities emphasized by the Chilean bishops specifically in the opening years of democratic consolidation. Thus a significant percentage of regularly practicing Catholics disagree with key Catholic teachings on divorce and abortion, and a clear majority accepts these practices under certain conditions. With such a degree of dissent among even devout sacramental Catholics, the prospects for the Chilean bishops mustering lay support for their antireform positions in the years ahead appear limited.

Our 1987 survey of Catholics in the popular neighborhoods and *poblaciones* of Santiago also suggests this. It did not include questions

TABLE 5.5

Support for Divorce and Abortion Reforms among Santiago Catholics, 1989. *N=1,786*

	Regular	Occasional	Nonpracticing
Divorce			
unqualified	34.9 %	45.6%	51.4%
when no love	62.4%	71.2%	74.3%
Abortion			
mother's life	70.9%	79.3%	79.3%
after rape	41.8%	50.2%	57.5%

Source: Rayo and Porath, 1990.

on either abortion or divorce but asked respondents to indicate whether they would follow their conscience or Church authority on moral issues in which they disagreed with official Catholic teachings, and whether they would go along with their local priest or the bishop when there was a difference of opinion between the two levels of Church authority. Table 5.6 presents the results of these questions for popular-sector Catholics with varying degrees of involvement in the local Church.

These data indicate significant potential for dissent among popular-sector Catholics at the local level. Only organizational Catholics belonging to religious groups show a very high degree of loyalty in principle to Church authority, and more than 40 percent of these would support their local pastor over the bishop in matters of disagreement between the two levels of authority. Catholic activists in base Christian communities (CC) or in local Catholic social organizations were even more supportive of legitimate dissent within the Church.[89] Although this does not translate into open opposition to the hierarchy on specific issues, it indicates that the principle of legitimate dissent was well established among Chilean Catholics. They seem to have taken the teachings of Vatican II and the Medellín and Puebla Conferences to heart in terms of the sanctity of conscience and the importance of the local church. This could result in some open disagreements by local activists with the hierarchy in the years ahead.

These data are consistent with the strong support of regularly practicing Catholics for both divorce and abortion (under certain circumstances) reported earlier, although popular-sector Catholics hold much more restrictive views than their middle- and upper-class counterparts.[90] These practicing Catholics are not likely to line up with the hierarchy on these issues. And if local priests begin to drag their feet or resist these positions as well, they will almost certainly have the support of a substantial number of "good" Catholics at the local level.

TABLE 5.6

Potential Dissent Among Popular-Sector Catholics, Santiago, 1987.

	Follow Conscience over Church Authorities	Follow Priest over Bishop
CEBs/reflection groups	42.6%	45.5%
Social organizations	45.2%	56.8%
Religious organizations	26.1%	43.5%
Sacramentals	32.7%	52.6%
Culturals	57.1%	40.0%
	N=206	N=179

Source: Fleet, 1987a.

In addition, given their strong pluralist sentiments (reflected in table 5.4) organizational Catholics do not appear to be hostile to or anxious to compete with their evangelical Protestant neighbors. The latter have been most successful in winning over lower-income Chileans (particularly among lapsed or nonpracticing Catholics) but popular-sector Catholic activists may not be the ready source of missionary energy that the bishops are hoping they will be.

Finally, since women are often active leaders in the local Christian communities in Chile, many times outnumbering men, dissent may occur in the years ahead over their role in the Church. The Vatican adamantly opposes the ordination of women as either deacons or priests. Yet, in Chile, as in all of Latin America, women have been exercising important ministerial functions for years because of the shortage of priests. As early as 1976, in Chile, for example, more than 10 percent of the parishes in Chile (80 of 750) were under the exclusive pastoral guidance of women, primarily nuns. And with the blessing of the bishops, women are currently performing such traditionally priestly functions as baptizing, officiating at weddings, preaching, and burying the dead.[91]

There may come a point in Chile, especially in light of the growing women's movement in the country, when Catholic women—especially those who are exercising priestly responsibilities, de facto if not de jure, in local Catholic communities—will begin to question openly the Church's refusal to grant females equal pastoral ranking for equal pastoral work. The contradiction is stark; were it not for women, there would be no local Catholic organizations, religious or social, in many areas of the country, yet males dominate all positions of authority in the Chilean Church. And since this issue touches upon the innermost nature of the Church itself, over the long run it may prove even more explosive for the hierarchy than dissent on either divorce or abortion reform.

In sum, these various issues and problems forbode internal difficulties for the Chilean Church in the years ahead. The difficulties may or may not have an effect on Chilean democracy generally, but they are clearly part of the Church's own struggle with democratic values and impulses.

Conclusions

The Church's impact on consolidation has been modest and at times obstructive. The armed forces and civilian rightists continue to oppose elimination of military independence and other antidemocratic features inherited from the country's relatively smooth but incomplete transition, and neither the Catholic bishops nor Catholic activists have been willing to press the issue with a forcefulness likely to produce positive results. With the trauma of more than sixteen years of often brutal military rule still fresh in their minds, they appear to prefer partially democratic civilian rule to no civilian rule at all.

National Church authorities and local Church activists have reached an analogous conclusion in terms of government economic policy; they prefer the current preoccupation with growth and stability, with only very modest advances in equity, to more pointedly redistributive policies that might reduce growth, unsettle economic confidence, and/or arouse political adversaries on the right.

Individual bishops and Catholics active in small groups have been less patient in terms of civil-military relations, justice for the victims of human rights abuses, and the rate at which poverty and gross income disparity are being reduced. But the expectations and demands of the bishops as a whole, and of the vast majority of activist and rank-and-file Catholics have been more moderate in these areas than many observers thought they would be. The period since 1990 has hardly been tension-free. Military leaders refuse to give up the controls and restraints with which they saddled the political system initially, and periodically underscore their determination in this regard in forceful ways. An uneasy truce thus persists. Within certain boundaries, political forces are "free" to pursue their interests and those of their constituents, although most do so with a degree of accommodation seldom observed in the country's pre-1973 political history.

This state of affairs underscores the principal dilemma of Chile's consolidation process: the spirit of moderation and accommodation required for success in the long run, are obstacles to short-term progress. This is because there has been no movement beyond the terms of the original, and only partial, transition. It is because the center-left and democratic right have been sensitive to the fears and concerns of the armed forces and the civilian right, but these forces have not reciprocated. What is being accommodated, in other words, is the intransigence of one of the sides, which is effectively preventing the emergence and consolidation of truly democratic institutions and processes.

The divisions and ambivalence of Church authorities with respect to democratic principles are most apparent and costly in this context. While older, and more liberally oriented, bishops would like to be more active in supporting fuller democratization, the ever-larger contingent of ecclesially and theologically conservative bishops has made this impossible. These latter bishops, most of whom have been named since 1979, have forced the Episcopal Conference into adopting moral positions that make abundantly clear their own unwillingness to accommodate contrary opinions and concerns on matters they regard as important. And this attitude not only alienates once supportive elements of the general public but offers a model on which reluctant democrats in other sectors can justify similar behavior on issues of concern to them.

The Episcopal Conference's increasingly conservative orientation did not prevent the bishops from playing a constructive role in the transition process. Continued attacks on Church people during the waning years of military rule helped the bishops to close ranks and convinced even ecclesially conservative prelates that a return to civilian rule was in the best interests of the Church. The interest of ecclesial and ideological conservatives such as Cardinal Fresno in promoting dialogue and compromise among the civilian opposition helped to win support for the transition from rightists who trusted his instincts and integrity.

The context was quite different, however, during the consolidation period. There were no external attacks on the Church to precipitate unified episcopal support for a more fully democratic regime. Most bishops were satisfied that external pressures on the Church and serious human rights violations in society had subsided significantly. In the absence of a determination by civilian political leaders to roll back the remaining limitations on civilian rule, few bishops were willing to push on alone for the changes that democratic consolidation required, nor would they been successful if they had. The secular carriers for their views that had been available during the transition were no longer there.

More important, the ecclesially and theologically conservative bishops sensed new threats to Catholic teaching that were not as apparent or as pressing under military rule, i.e., proposals to remove the remaining vestiges of Catholic morality still embodied in Chilean law—the prohibitions against divorce, artificial birth control, and abortion—and demands for greater toleration for the country's burgeoning evangelical Protestant population. Although from a secular perspective these changes were precisely what full liberal democracy required (since they appeared to have majority support among the public at large), they were contrary to Catholic morality and therefore unacceptable to the Catholic bishops. Some progressive bishops and clerics, as reflected in the editorial positions taken by the Jesuit monthly *Mensaje*, might have been willing to accept some changes on these issues, but most were not. After all, they were appointees of a pope whose opposition to liberalization of any sort was quite clear, and they could not be persuaded otherwise. In effect, the bishops were unwilling to tolerate or compromise with opposing or divergent views if these ran counter to official Catholic positions. And they thereby turned their backs on practices essential to the consolidation of democratic institutions.

At the national level, Catholic lay activists took the same positions as the bishops on virtually all issues. In fact, their reluctance to push more aggressively for additional democratic reforms, and the credibility that they enjoyed with the hierarchy, were undoubtedly instrumental in persuading the bishops not to press more forcefully on issues of military impunity, accountability for human rights violations, and the restrictions on democratic institutions and procedures.

Local Catholic activists exhibited the same moderate political views on consolidation that they had with respect to transition. Their attitudes were characterized by a high degree of tolerance for pluralism, optimism regarding political institutions and leaders, modest expectations about the economy and issues of distribution, and support for a more restricted role for the military. While not as active in political movements as either cultural Catholics or secular Chileans (those with no religious affiliations), those closest to the Church at the local level certainly were utterly unsympathetic to extreme elements of the right or the left, and thus provided at least residual support for consolidation.

Thus, during the first several years of civilian government, at no level of the Catholic community was there forceful leadership on either removing the remaining restrictions on civilian rule or implementing a majoritarian secular democracy. Nor were there significant forces in Chilean society demanding such changes. The various parts of the Catholic community shared the satisfaction of most Chileans with what had been accomplished in the transition, and they were equally reluctant to jeopardize its continuance by pushing for additional concessions from hard-line military elements and their civilian allies.

6

. . . .

The Church and the Transition to Civilian Rule in Peru

Peru's transition from military to civilian rule was less protracted than Chile's. The Church's role was also less central. Chile's military government remained strong for a longer period of time, giving ground only with the emergence of a united opposition to which the Church contributed significantly. In addition, regime hostility helped to discourage divisions among the Chilean bishops and between them and priests and pastoral workers at the local level. In Peru, on the other hand, most political forces, including the military, favored a return to civilian rule. They were also less mistrustful of one another, and less in need of assistance in developing a democratic consensus.

The Peruvian Church also suffered from more extensive divisions than its Chilean counterpart. Conservatives could always be found in large numbers among its bishops and priests. In the absence of regime hostility, and as the prestige and influence of moderate progressive bishops like Landázuri and Bambarén began to wane in the mid and late 1970s, conservatives grew more assertive. Their pre–Vatican II spirituality, their concern with internal matters and personal moral issues, and their efforts to restrain progressive elements, had a diluting and distractive effect on the Episcopal Conference, making it more difficult for it to speak with a single voice on social and political issues.

Individual bishops, the bishops of the Southern Andean region, and the Bishops' Commission for Social Action (CEAS) were less inhibited, and spoke out forcefully from time to time. CEAS developed educational materials, conducted workshops, and provided social services to workers, slum-dwellers, and other "popular sector" groups. Its initiatives had a significant impact on local communities but little at the national level. It lacked the exposure and authority of the bishops' conference, and its efforts could be greatly diluted, if not entirely neutralized, by the public statements of more conservative bishops.

Local church leaders and activists energetically opposed the economic and human rights policies of the

Morales Bermúdez government. Priests, nuns, and lay activists actively encouraged and assisted the social and political organizations that emerged in the *pueblos jóvenes* and popular neighborhoods of most of the country's cities. They supported unions and community groups in strikes, demonstrations, and other forms of antiregime protest, as their counterparts would later do in Chile. The protest activities of which they became a part were a key factor in the military's decision to relinquish power, although they failed to persuade the Morales government to alter either its economic policies or its hard-line approach to law, order, and dissent.

Our treatment of this period covers the Peruvian bishops as a whole, individual bishops and groups of bishops, and activists at the parish and community level as they responded to the Morales government and its policies. We also stress the development of tensions between the Church and the Morales government and within the Church itself.

The Bishops

The last several years of Velasco's presidency were marked by intense polarization and conflict. The president's popular support diminished as his reforms stalled and the economy faltered. His decline and fall made it more difficult for the Church to mount a challenge to Morales Bermúdez. Bishops like Landázuri and Bambarén, who had supported Velasco and his reforms, found their judgments questioned by the public at large and by increasingly assertive conservative forces within the Church.

Morales Bermúdez had been Velasco's treasury minister, but led a coalition of progressive and conservative military officers that ousted him in August 1975. He pledged to deepen Velasco's reforms and to facilitate greater popular participation in politics, but he moved in the opposite direction, abandoning or diluting most of the changes introduced under Velasco. In the waning months of 1975, Morales weeded out most of Velasco's remaining appointees, preparing the way for an orthodox stabilization program drafted by Luís Barúa Castañeda, the civilian economist he chose to be minister of the treasury and economy. Barúa's program was designed to cut the deficit, hold down labor costs, stimulate domestic savings, attract foreign investment, and promote exports.

Morales wanted to liberalize some structures and procedures and to encourage investors and producers. He hoped to make the country's exports more competitive, stabilize its currency, and strengthen its ties with more developed economies. These policies, he claimed, would set the country on a more stable and prosperous course and help to consolidate the advances made under Velasco. In fact, he was so confident in this regard that he was willing to impose his policies whether people accepted them or not. During late 1975, the government allowed prices for basic consumer goods (including food) to rise while restricting wage increases in both public and private sectors. Workers and community organizations protested these moves, and some unions went on strike. Declaring the strikes illegal, the government had striking workers dismissed from their jobs and arrested or intimidated many of their leaders.

These policies would bear modest fruit towards the end of Morales' term. But for the first several years the living standards of most Peruvians deteriorated. The economy (GDP) grew by 2 percent in 1976, but shrank by −.1 percent and −.5 percent in 1977 and 1978 respectively. Between 1976 and 1979, real wages (in manufacturing) fell by almost 34 percent, unemployment rose from 4.9 to 7.0 percent, and inflation rose from 33.4 percent to almost 74 percent before dropping to about 60 percent.[1] Declining living standards prompted workers and poor people to mobilize against the government. Demonstrations, work stoppages, strikes, building seizures, and other protest activities began in earnest following the announcement of price hikes and wage and salary ceilings in early January 1976. Over the next several months, teachers, metalworkers, copper miners, and newspaper workers went on strike. In some instances (with newspaper workers and miners, for example), labor demands were met; in most, however, the strikes were declared illegal and their organizers arrested. A state of emergency was declared in July, after students and workers rioted to protest a 44 percent devaluation of the sol and sharp increases in prices for food, clothing, utilities, and public transportation.

The bishops' conference and its Permanent Council stayed silent for most of this period. In late 1975, they authorized the circulation to Christian communities of a previously drafted letter (it was actually written while Velasco was still president) urging the military to allow more popular participation in its revolutionary process, and challenging its claims that this would jeopardize national security.[2] For most of 1976, however, the conference said nothing, even though living standards of low-income Peruvians continued to deteriorate and labor and popular organizations continued to be repressed. Individual bishops like Bambarén, the Episcopal Conference's Social Action Commission (CEAS), which he headed, the bishops of the Southern Andean region, and the priests' organization ONIS encouraged and defended popular organizations that were opposing the regime and its policies. But while their initiatives helped to sustain groups in villages and *pueblos jóvenes*, they had little impact on either elite or public opinion at the national level.

The conference as a whole spoke out in October 1976, in a document entitled "Reflexiones de Fe sobre el Momento Actual," in which they denounced the "unrestrained liberal capitalist" model, argued that unless employers paid more equitable salaries (instead of forcing people to work for less) worker anger would build up and violence would intensify, and urged the government to respect the rights of workers to organize and defend their interests. Beyond this, however, they would say only that if sacrifices were going to be asked of lower-income groups, the upper- and middle-income people making these decisions should set an example by living austerely themselves.[3]

When not simply ignoring such appeals, government spokesmen lamented the bishops' failure to appreciate the potential benefits of their policies. This angered some bishops who wanted to press matters more forcefully, but conference conservatives refused. They believed they had been wrong earlier, when they had issued statements that appeared to endorse Velasco's policies, and they were now unwilling to criticize those of his successor. The Church should outline principles, they thought, but not dictate how they should be applied in particular instances.[4] According to conference conservatives, bishops who had access to policy-makers might express their concerns discretely but should not do so publicly. The Church should neither endorse governments whose policies conformed to Catholic social teaching nor oppose those whose policies did not. And because warnings of the risks or possible consequences of a government's policies would implicitly challenge its economic and political judgments, the Church ought to avoid issuing these as well. Finally, while some conservatives argued that the Church should avoid embracing or calling for social reforms because of their inherently divisive character, others stressed its lack of competence in such matters.

Landázuri could sense resistance to his leadership growing within the conference. From time to time, however, he pulled more conservative bishops along by controlling the conference's agenda and the context in which issues were discussed. Catholic left political leader Henry Pease recounts one such instance in early 1977. Having been invited by Msgr. Bambarén to address the bishops' assembly on the subject of human rights, Pease arrived only to learn that "some bishops" opposed his speaking. Not wanting to cause problems, he was about to leave when Landázuri appeared, greeted him effusively, and led him aside to ask about the health of Pease's father, who was seriously ill at the time. Chatting quietly as he did, Landázuri led Pease down a hallway. But then, without warning, he suddenly ushered him into the room where the assembly session was about to begin, introduced him to the bishops, and directed him to proceed with his talk, which Pease did.[5]

Changes in political context also affected the willingness of bishops to speak out on social and political questions. In August 1976, Morales told several bishops of his decision to relinquish power to a civilian successor "in the not too distant future."[6] The decision did not appear to be forced upon him by his fellow officers. Indeed, a number of them opposed the idea strenuously, although the social unrest of the first year in power made the prospect of remaining on much less attractive to Morales himself.

Morales claims to have been committed to the restoration of democratic rule from the outset of his period in power.[7] His meeting with the bishops came after several months of unrest in early 1976 but in advance of the intense conflicts of late 1976 and 1977. It made the bishops more reluctant to criticize his government or its policies. Conservatives argued that he deserved a respite for his "patriotic gesture," while some progressives were anxious not to give him reason to change his mind. Other progressives, who wanted to press matters further, had to proceed on their own.

The most prominent of these were the bishops of the Southern Andean region, who had continued to support the antigovernment activities of peasant and other popular organizations under Morales as they had under Velasco. Most of them had been appointed under Paul VI and had worked closely with the laity to build a vibrant and socially progressive Church in the region. The U.S.-based Maryknoll Fathers, who established the Andean Pastoral Institute and fostered priestly vocations among young *serranos*, provided them with valuable pastoral and material assistance. The Andean Church remained strong and socially involved through the 1980s, despite the hostility of local authorities and the efforts of conservatives to rein it in.[8]

In a pastoral letter issued in July 1977, the Southern Andean bishops denounced the violent repression of striking teachers and textile workers, and blamed the climate of violence pervading the country on "an economic development model and a social and political system that does not take majority interests into account and rests on a doctrine that subordinates people to the state and forces them to serve its interests."[9] The following year they issued a similar document which featured quotes from local parish and community members complaining of their circumstances and their mistreatment at the hands of government officials. To their testimonies, the bishops added biblical texts, excerpts from Church teaching, and their own commentary. The government attempted to prevent the publication and circulation of statements like these, but they found their way into the public arena anyway, invariably creating a stir.[10] The Southern Andean bishops made an impression on Morales as well. Recalling this period in a 1992 interview, he boasted of having maintained "cordial" relations with the Church as a whole, but characterized the Andean bishops as "troublesome."[11]

The impact of a regional group of bishops, however forceful and cohesive, was not as great as that of the Episcopal Conference as a whole, however. Their remarks and statements were often followed by demurrals or rejoinders from more conservative bishops, greatly diluting their effect on the general public. In most instances, they gave stimulus and support to the people involved in the conditions or issues they were addressing, and were well received by those already sympathetic to them. But they probably did not persuade or convince people not already on their side.

Luis Bambarén, the popular bishop of the *pueblos jóvenes*, was also a prominent critic of Morales and his policies. Bambarén had been close to Velasco and to generals Rodríguez, Fernández Maldonado, and Carpio Becerra, who were members of Velasco's inner circle.[12] It was a personal blow to Bambarén when Morales had Velasco placed under house arrest in September 1975, when he forced Rodríguez into retirement in October, and when he dismissed Fernández Maldonado as prime minister in July 1976. In early 1976, Bambarén publicly endorsed the land invasions carried out by homeless migrants on the outskirts of Lima, which he blamed on the government's failure to develop an adequate housing policy for the popular classes. He also lent his authority to CEAS statements that criticized the government's economic policy and its persecution of dissident popular groups.

Bambarén's political credibility had declined in recent years, however. His criticisms of Morales Bermúdez were applauded by progressive Catholics but were resented by others who were hostile to Velasco and favorably disposed to Morales. Fearing that he might be turning people against Morales and his policies, political and ecclesial conservatives launched a campaign to discredit him. Newspaper and magazine columnists linked him with Marxists and other subversive types, while militant Catholic traditionalists termed him a "dangerous radical."

Traditionalists claimed that Bambarén wore a ring on which a cross was intertwined with a hammer and sickle.[13] Bambarén denounced these and other charges as part of a CIA campaign against the Church, and offered to let reporters see for themselves that he wore no such ring. Whatever the truth of this charge,[14] Bambarén was neither ecclesially nor ideologically radical. He was a practical man who had little time for theoretical discussions, Marxist or otherwise. He thought that people deserved to live better than they were living (hence his support for Velasco), and that the Church's concern for their material well-being was an integral part of its mission. But he was no radical democrat nor was he an advocate of greater lay protagonism in a "people's Church." In these and other respects he was a traditional churchman, supportive of moderate reforms in society but moderate-to-conservative on issues of hierarchical authority, clerical control, Church prerogatives, and personal (sexual) morality.

Bambarén's greatest impact was through the work and publications of the Bishops' Commission for Social Action (CEAS), which he had headed since the late 1960s. CEAS pronouncements and educational materials carried the seal of approval of the Episcopal Conference, even though many bishops disagreed with its positions on particular issues. With parties and political organizations still banned, even conservative bishops recognized that the Church had to assume a surrogate political role, that it had to represent and defend social and political forces (primarily the poor) who could not speak for themselves. They could also see that CEAS could perform these functions in ways that would make it unnecessary for them to commit themselves individually. And if and when political space opened up, and parties and movements began to reemerge, the conservative bishops were confident that they could limit the organization's involvement and visibility at that point.

Until then, however, CEAS would be very busy. On May Day, 1976, it condemned investors who were withholding capital and thus denying workers the opportunity to support themselves and their families. It also expressed alarm over the rumored disappearance and torture of dissident trade unionists, and further argued that while "protest movements and demands for higher wages were interpreted by certain private- and public-sector groups as threats to an order that must be maintained," they should be viewed "as a questioning of conditions generated by an order that should be transformed and overcome."[15] Another CEAS document focused on the economic model that apparently was guiding government policy. It called attention to what it termed its excessively high social costs and the oppressive and virtually totalitarian concentration of power in Morales' hands.

These documents bore the mark of the commission's director, Father Ricardo Antoncich. Antoncich was an extraordinarily "diplomatic cleric"[16] who could phrase critical judgments in language that even the most timid bishops could embrace. Many of the lay economists and sociologists[17] who worked under him had been disciples of Gustavo Gutiérrez while at the Catholic University, and were part of the informal community attached to his Bartolomé de las Casas Institute.

Under Antoncich's watchful eye, CEAS generated a substantial number of statements, reflection pieces, and educational materials. With Bambarén's blessing, they were circulated to bishops, pastoral agents, and Church-sponsored organizations and programs, providing them with information and analysis on issues of the day. In doing so, CEAS people had a significant impact on the content of Church policies and programs. By introducing regularized procedures, consultative processes, and multilevel discussions of issues, they also enhanced the administrative and organizational capacities of individual dioceses, although this had the effect of eroding some of the bishops' once discretionary authority and control in social matters.

The priests' organization ONIS was also critical of the Morales government, attacking its economic policies and its repressive practices. It challenged its economic policies for placing additional burdens on those groups (workers, peasants and the slum-dwelling poor) least capable of bearing them. And it flatly rejected the notion that worker rights (to protest and mobilize) had to be suspended in order to prevent their being manipulated by "outside interests."

The Local Church

The greatest amount of Church opposition to the Morales government came not from its leaders or agencies but from local-level Catholic activists and communities. Priests, nuns, and lay volunteers and activists in *pueblos jóvenes*, popular neighborhoods, and rural communities helped to form social organizations and movements throughout the country, and assisted them both in pursuing their own immediate ends and in relating to the explicitly political groups that were beginning to reemerge.

Most of these groups were unknown and untrusted in the *pueblos jóvenes*, where only APRA and the military government (with its many programs and activities under Velasco)[18] had established an identity and presence. Some of the organizers that the leftist parties sent to the *pueblos jóvenes* were long-time militants without much feel for slum-dwellers or their problems. But those belonging to one of the "revolutionary" groups that had formed during the early 1970s were more effective because they knew the priest or one of the lay activists, or because they themselves had worked for a time as popular educators or community organizers.

As socioeconomic conditions worsened, leftist groups developed substantial followings in Lima's poorer neighborhoods and *pueblos jóvenes*. Ideologically, Catholic political activists ran the gamut from advocates of radical revolution to timid reformers. Some were of a moderate bent, and saw their anti-Morales activities as a means of restoring civilian rule and the possibility of competing with other forces within a "democratic" political context. Others were firm believers in direct political action and in the possibility of overthrowing the existing economic and political systems.

Some of these radical Catholics were Catholic in the cultural sense only; others continued to practice their faith and were active in Christian communities. As university students in the mid-1960s, they had abandoned *aprismo*, Acción Popular and Christian Democracy for revolutionary Marxism. Alienated by the concessions that Haya de la Torre was willing to make in exchange for restored legal status, and by Belaúnde's backsliding on agrarian reform,[19] they were taken with the

Cuban revolution, and with Marxism's ability to cut through "bourgeois" fears and hesitations in matters of analysis, strategy, and tactics. Some were willing to renounce their "bourgeois" faith in order to become full-fledged revolutionaries,[20] but many others remained believing, and even practicing, Catholics. In defending their option, they invoked Father Gutiérrez' "distinction of planes" argument, according to which religion and politics belonged to different levels of action, and turned on different judgments.[21]

In the view of an older Catholic radical, socially sensitive Christians were forced to choose one or another of the Marxist splinter groups emerging at the time. The country's Christian Democratic Party never amounted to more than "a handful of provincial, but urban, middle-class intellectuals in a desperately poor, highly unequal, and still largely rural country."[22] And yet the students viewed the country's Moscow-oriented communists with as much disdain as they did its middle-class parties. The communists, after all, had backed Velasco and supported Morales during the first eighteen months of his presidency. Most student radicals therefore ended up in Vanguardia Revolucionaria (VR), the Movement of the Revolutionary Left (MIR), or one of the smaller Trotskyist, Maoist, or Castroite parties that sprang up in the mid or late 1970s. These were invariably led by dynamic vanguard figures who talked fast and forcefully but had difficulty working or sharing authority with others. Their limited prospects for "taking power," it seems, merely intensified the sectarian tendencies of these groups, and over the next several years the splinterings and divisions into which they were drawn left them powerless and demoralized.

With the revival of protests and social mobilization under Morales, the political climate changed abruptly. State and governmental structures suddenly appeared fragile and inconsequential, and both activists and observers began to think in terms of "revolutionary breakthroughs." Under these circumstances, young Catholic radicals experienced a revival of interest in political parties and organizing. Most chose to remain politically active "independents," working closely with social movements and popular organizations at the local level. Of those who affiliated with political parties, most chose the Partido Comunista Revolucionario (PCR), headed by Manuel Dammert, a one-time UNEC activist who remained on good terms with many Church people although he was no longer a believing or practicing Catholic.[23] Others identified with the Partido Socialista Revolucionario (PSR), a party established by left-wing Catholics who had worked in the Velasco government.[24]

Some Catholic radicals tried to keep up their apostolic work even though they were politically active. Not all were able to, but most kept their responsibilities distinct and resisted the temptation to use their religious credibility to recruit people to their parties. Few, if any, of the priests or nuns working in popular neighborhoods identified with a particular party or group, although most were generally supportive of laypeople who did, and were more favorably disposed to groups on the left than to APRA or the rightist parties.

Priests and other pastoral agents provided meeting space, good offices, and other assistance to local workers, neighborhood activists, and even partisan political organizers. According to Klaiber,

> the Church did take on the role of representative . . . in the absence of political space. . . . That would be the example where people would be holding political meetings in Villa El Salvador in the late 1970s in parish halls. Ah, that bothered the military [government] no end. But . . . that happened everywhere. What happened is that the bishops warned priests and other pastoral agents not to do that too often, but as a matter of fact it did happen in every place.[25]

Party activists and organizers interested in penetrating *pueblos jóvenes* and popular neighborhoods knew that they would have to win the approval of local priests, nuns, and lay leaders; people trusted them more than they did the parties or other outside forces. For their part, pastoral agents knew that these people were looking for "bodies" with which to bolster their bids for influence at the national level, and tried to help local activists from being manipulated or taken over by them. But they also knew that if local groups were to be effective they would need to develop linkages with regional and national organizations; and they were willing to work with both parties and unions, mediating between them and local organizations, and helping parishioners relate to groups with which they shared objectives.

The years under Morales thus witnessed the mobilization of large numbers of Peruvians living in the shantytowns and popular neighborhoods of larger cities. Many had been introduced to or guided in their social involvement by pastoral agents (priests, nuns and lay workers) in their local communities. Some moved on to partisan political involvement. Most chose not to affiliate with any organization but were willing to work with parties in pursuit of common goals. During this period,

Christian communities, self-help or survival organizations, and local strike-support or political groups were often difficult to differentiate.

Divisions within the Church

Not all Catholics were unhappy with Morales Bermúdez's government. In fact, Morales's seizure of power was a windfall for conservative Catholics who had opposed Velasco but were unwilling to challenge Bambarén and Landázuri. Right-wing business and political elites, most of whom were nominal Catholics, had mixed feelings toward Morales. They approved of his stabilization policies but objected to their limited access to policy-makers and to the still substantial economic role that his government assigned the state. Sodalitium Christianae, a new conservative organization modeled on the older Opus Dei, was more categorically supportive. It defended Morales' austerity measures and condemned strikers and protestors as "trouble-makers." In 1977, with the sympathy if not actual backing of several conservative bishops, it mounted a public campaign against liberationist elements in the Peruvian Church, charging that they had politicized its mission and that Marxists and other "rapacious wolves" had made inroads through CEAS and ONIS.

Right-wing newspapers and magazines played up the "story" of an infiltrated Church, attacking both Cardinal Landázuri and Bishop Bambarén. One of the issues they seized upon were the educational reforms that Velasco had introduced in 1972 and that Morales decided to review in 1977. As earlier, these reforms were viewed by conservative critics as an assault on private education. But the critics were also upset with what they termed the "radical" content of the proposed religious textbooks.[26] Parents' groups and the head of the Consortium of Catholic Schools expressed these judgments publicly at every opportunity. They were backed up by conservative bishops,[27] and by similarly minded newspaper editors and columnists.

In an effort to placate critics, Cardinal Landázuri asked a number of Catholic theologians and educators to assess the materials. They concluded that the materials were theologically unobjectionable, although somewhat abrasive in style. A second committee appointed by the Episcopal Conference's Education Commission concurred. But these reassurances did little to appease critics or to shake their belief that the Church had been infiltrated by subversives. Clearly, the bishops and the broader Catholic community were badly divided on important matters. The controversy over education made this abundantly clear and undercut the Church's role as teacher. Already reluctant to speak on social issues, the bishops grew more so for fear of provoking one or another Catholic group, and further polarizing matters.

Church-State Tensions

Morales's government continued to face determined opposition on a variety of fronts. In January 1977, the president publicly expressed his willingness to relinquish power to a civilian successor in "three or four years." In February, he issued his Tupac Amaru plan, which called, among other things, for "democratic elections" within three years. Morales hoped that the prospect of civilian rule would weaken the resistance to his economic policies and undercut social and political mobilization generally.[28] It did not. Dissident elements continued to stage strikes and protests during the first half of 1977. These frequently ended in pitched battles with security forces and the arrest of most of the activists involved. In many instances, CEAS statements or documents endorsed the demands of the unions, while local Church groups, particularly Christian communities in poorer areas, provided emotional and material support for members and their families.

Morales' Tupac Amaru plan retained Velasco's agrarian reform but targeted the social property sector for elimination and liberalized policies with respect to foreign investment and internal markets. The communists denounced the plan, broke with the government, and directed their labor activists to join with other unions in resisting wage freezes and price hikes, leaving APRA and the PPC as the only parties still supporting Morales. Strikes, building seizures, and protest demonstrations dominated national life for the next several months. The unions and popular organizations showed considerable determination and resilience; as one strike ended, forcibly broken in most instances, another began.

Morales grew impatient with such defiance. Labor leaders were thrown in jail and remained there. Former military colleagues were arrested and expelled for impugning the character of the "revolutionary" government. Statements by Church leaders even mildly critical of his government were prevented from reaching the general public. Churchmen wishing to express concern over specific incidents visited Morales or his ministers individually, but were regarded as naive meddlers who were being manipulated by leftist union and political leaders.[29]

As it confronted these protests, the government was negotiating with the International Monetary Fund. An agreement with the fund was crucial if Peru was to avoid defaulting on its external debt, but the institution's terms were very stringent. Intragovernment conflict led Barúa to resign in May, and his successor to follow in early July, at which point another wave of strikes, led by students and bank clerks, was launched. Their repression prompted a sharply worded pastoral letter from the Southern Andean bishops (there had been student strikes in Cuzco and Arequipa), and a call for a general strike in July by communist, Christian Democratic, and normally progovernment unions.

Rapidly declining socioeconomic conditions assured the strike widespread support. Lima and several other cities were effectively paralyzed. Six people (students and workers) were killed by patrolling soldiers, as were another twenty in clashes between troops and slum-dwellers in the *pueblos jóvenes.* Archbishop Landázuri and other bishops decried the loss of life and urged that the country's "crisis" be solved cooperatively (presumably with both sides making concessions) and without hatred and violence. The government responded that the strike was "political" and "subversive," and authorized the firing of all public and private sector workers who had participated. Landázuri issued a statement critical of the decree, but the newspapers (under either government or conservative control) refused to publish it. The bishops as a whole said nothing, and only a few pressed matters in private meetings with regime officials.

This context prompted the Church to become a surrogate provider of social services through the Bishops' Social Action Commission (CEAS). Bishop Bambarén was approached by relatives of workers who had been arrested or dismissed from their jobs. He sent CEAS staffers to the jails and commissaries in search of those who had been arrested. The police were not always happy to see them. One CEAS staffer recalls being accused of "helping the communists," but she and her associates were almost invariably allowed to carry out their inquiries "because they were Church people" and because "Bambarén's name still meant something."[30] As requests for such assistance increased, Bambarén assigned additional people to investigate the complaints of dismissed and arrested workers and their families.

Later that year, CEAS set up small workshops where fired workers could earn money or acquire the skills needed to get another job. The government took visible offense. In its view, the Church was lending moral and material support to its enemies. The extent of the Church's social service efforts remained modest, however. There were never more than five or six people providing or overseeing services in Lima, and little was being done in other parts of the country. Many people, including CEAS staffers, believed that the Church could have done more, but its impact remained largely symbolic. Its assistance programs cast the Church in a quasi- or mildly adversarial light but were woefully inadequate to the level of actual need.[31]

More might have been done more if CEAS had not lost its executive secretary and if Msgr. Bambarén had remained in Lima. Father Antoncich's departure in mid-1977 for a position with the Latin American Council of Religious Men and Women (CLAR) in Medellín, Colombia, reduced the commission's influence with moderate and conservative bishops. Bambarén's appointment as bishop of the northern coastal city of Chimbote in June 1978 made it more difficult for him to oversee the organization's activities and reduced the Church's influence in both military and political circles.

Also limiting the expansion of CEAS programs and services was the reluctance of most bishops to challenge the government. According to people who worked for CEAS at the time, the Church's efforts in the areas of human rights and social service were "primitive," but not for lack of resources; no proposal for assistance was ever rejected for lack of resources. The problem, rather, was the desire of most Peruvian bishops to get along with those in power. Something on a scale of the Chilean Vicaría could have been mounted. The necessary resources could have been obtained, as they were in Chile. But the bishops would have had to challenge the government on a national scale, and most were unwilling to take this step.[32]

Early 1978 was dominated by a teachers strike that closed public schools from May to July, and by preparations for the Constituent Assembly election in June. Most teachers, including priests and nuns who staffed state-financed Church schools, supported SUTEP, the United Peruvian Education Workers Union, which the government refused to recognize and the pro-Chinese Communist Party of Peru, better known as Patria Roja (Red Fatherland), with which SUTEP was closely linked. SUTEP officials, aided by priests, nuns, and laypeople from low-income areas served by state schools, persuaded Landázuri and other bishops to endorse the teachers' wage demands. In July, on Landázuri's urging, the Church's education office and the Consortium of Catholic schools helped to broker an end to the strike in a series of bargaining sessions between SUTEP leaders and military authorities.

The election of a Constituent Assembly that would draft a new constitution was the first step toward the restoration of civilian rule. It was the product of negotiations between military "institutionalists" (those supporting Morales Bermúdez and willing to relinquish power) and center (APRA) and right-wing (PPC) political groups. Fernando Belaúnde, the country's last democratically elected president, refused to take part in the elections, arguing that they ought to be choosing a democratic government and not just an assembly to draft another constitution.

The left, whose support among the country's poor and working classes had increased dramatically during Morales's presidency, was also divided. Some groups were willing to participate in order to demonstrate their appeal and to help to determine the content of the new constitution. They were willing to suspend direct-action initiatives for the moment, and to spend their time shaking hands, making promises, and trying to differentiate themselves from other political forces. Others refused, arguing that elections were a dangerous distraction of political energies that should be directed at bringing down a bourgeois order on the verge of collapse. In their view, the last two years of strikes, protests, building seizures, Church occupations, and street clashes had produced a revolutionary moment that should be seized. Instead of turning to "politics," therefore, they maintained their direct-action campaign right up to the day of the balloting.

The election was held in mid-June 1978. With Belaúnde out of the picture, the 100 seats were apportioned in roughly equal shares among left, center, and right forces. APRA led the way, winning 37 seats. Leftist parties won a total of 30 seats, with the largest number going to an independent Trotskyist party. The pro-free enterprise (PPC) won 25 seats, picking up the votes of many who would have supported Belaúnde. The election strengthened the government's hand in dealing with its critics and opponents. It went a long way toward deflating the country's political crisis, heading off the development of a "revolutionary situation" in the country and helping to isolate the more radical unions and parties. It also enabled Morales to concentrate on the economy, where signs of renewed health were beginning to show up in statistical data and academic assessments.[33]

Although the left surprised most observers by capturing a third of the vote, it could be outvoted by the combination of *aprista* and PPC delegates. With his party holding the largest number of seats, APRA's Haya de la Torre was elected Assembly president. He resisted the left's urging that the Assembly's mandate be broadened, and played the role of balancer between the government and its critics within the Assembly until his death from cancer in early 1979.

Assembly delegates were organized into commissions and subcommissions dealing with different aspects of the Constitution. Church authorities had little impact on discussions of economic or political matters but sought to influence individual delegates and subcommittees on matters affecting religious education, the family, and the rights of citizens, organized labor, and the unborn. They were fairly successful in doing so, although the reaction of the country's Catholic population was mixed. Conservatives cheered their call for a constitutional prohibition of abortion, for example, but were less attentive when the rights invoked were those of workers or political dissenters.

Also taking up the bishops' attention during this period was the negotiation of a new concordat defining the Church's relationship to the Peruvian state. Early in 1979, they approached the papal nuncio, Msgr. Mario Tagliaferri, asking for permission to discuss the possibility of a "friendly separation" with the Morales government.[34] Apparently, most Peruvian bishops, including several leading conservatives, considered the existing church-state relationship, in which the latter exerted influence in the naming of bishops and other matters,[35] a "bother" or "inconvenience."

The bishops may have taken this initiative in order to preempt the possibility of separation under less favorable terms at a later date (e.g., under civilian rule). The Church was on reasonably good terms with the Morales government, and church-state relations were not likely to be better under a successor civilian regime. Moreover, Protestantism was still a relatively insignificant force in Peru, but both evangelical Protestant denominations and radical Judeo-Christian sects (i.e., the Mormons and Jehovah's Witnesses) had expanded dramatically in recent years.[36] A number of bishops believed that pressures for full disestablishment would continue to mount, and some felt that it might be better to broach the subject of modifications sooner rather than later.

The Vatican initially refused to consider the matter but later changed its mind and authorized exploratory talks. These concluded rapidly with terms that were favorable to the Church. It became autonomous with respect to its internal affairs. Other religious denominations were granted rights and protection under the constitution. But Catholicism remained "an important element in the historical, cultural, and moral formation of Peru," and continued to enjoy special legal status and a host of both symbolic and substantive privileges, including state subsidies for its primary, secondary, and post-secondary educational institutions.[37]

The election of a Constituent Assembly helped to create an atmosphere of openness and trust in which church-state relations could be redefined in mutually satisfactory terms. It also led to a reduction in the political involvement of the Church's progressive elements. In permitting the resumption of partisan political activity, and in raising the prospect of presidential elections in 1980, the Assembly relieved both individual bishops and organizations like CEAS and ONIS of their surrogate political responsibilities. With citizens able to speak politically for themselves, and political groups competing for their support, the bishops could withdraw from the political arena.[38]

The military government faced two additional challenges before leaving power, however. The first was the draft constitution proposed by the Constituent Assembly after almost a full year of deliberations. The second was another strike called by the national teachers' union. The draft constitution contained provisions (involving the death penalty, human rights, and the jurisdiction of military courts) that the Morales Bermúdez government refused to accept, and it ordered the Assembly to reconsider. The Assembly rejected any limitation on its authority, however. Instead, it reaffirmed its original text, and voted to disband, hoping to force the government's hand. But Morales refused to go along, and simply promulgated a version from which the offending provisions had been removed.

In June 1979, SUTEP called another strike, accusing the government of failing to live up to the previous year's agreement. As president of the Permanent Council of the Episcopal Conference, Cardinal Landázuri directed the conference's Educational Commission to explore the possibilities of mediating the dispute. The bishops were not as united in their views as they were the year before. Bishops Bambarén, Vallejos, and Quinn, and the priests' organization ONIS, issued statements in support of the teachers. But other bishops argued that the union was more interested in confrontation than in teacher welfare and noted, among other things, its ties to the pro-Chinese communist faction, Patria Roja, and the latter's denunciation of the Constituent Assembly. In its report, the Bishops' Commission also questioned SUTEP's real objectives and accused it, not the government, of breaking the terms of the 1978 agreement.

The ensuing months were unpleasant for all concerned. Most of the religious orders and parishes that staffed state-supported schools expressed solidarity with their striking teachers, but few were willing to endorse SUTEP or the strike as such. Those who did were fired by state administrators or by their religious directors (in schools run by more conservative Catholic orders). As relations between the government and SUTEP deteriorated, so did those between the Church's National Education Office and the more conservative Consortium of Catholic Schools. The major issue in dispute was the determination of some religious superiors to dismiss employees who had denounced their refusal to support the strike or otherwise challenged their authority.

As the conflict wore on, the ranks of the strikers and their supporters dwindled. In the end, the government broke the strike, dealing a blow to SUTEP in the process. It was aided by the recent upturn in the economy and by the adoption, with modifications, of a democratic constitution in July 1979.

General elections were held in April 1980. Former president Belaúnde ran as the candidate of his Acción Popular Party. He was opposed by Luis Bedoya Reyes of the PPC, Armando Villanueva of APRA, and five different leftist candidates running on separate tickets.

Following more than three years of relative silence, the Episcopal Conference issued a surprisingly strong pastoral letter on the eve of the election. In it, the bishops expressed the hope that the victorious candidate's government would adopt a development model that would focus on the problems (hunger, health care, housing, employment, and education) afflicting the poor.[39] With Belaúnde, Villanueva, and several leftist candidates all expressing concern for the plight of the poor in their campaigns, their plea carried no particular electoral implications. Belaúnde won a surprisingly decisive victory, gaining 45.4 percent of the votes in a multiple-candidate field. Bedoya and Villanueva trailed badly, at 15 percent and 20 percent of the vote, respectively, while the leftist candidates managed only 15 percent among them. The left might have made more of an impact had its various parties supported a single candidate, but their inability to do so blunted their appeal.[40]

Belaúnde's election amounted to a repudiation of both military rule and political sectarianism.[41] And yet, in the end, Morales Bermúdez obtained the kind of transition he sought from the beginning. He was not particularly popular, but his economic program was intact, bearing at least modest fruit, and a relatively peaceful and controlled transition to civilian rule had been achieved despite formidable obstacles and skepticism. Moreover, the military regime achieved this pretty much on its own, without significant help from the Church, which spread itself thinly in a number of directions and was unable to speak or act as a single entity.

Conclusions

The Church did not have much of an impact on Peru's transition to democracy. The process of transition was negotiated among opposing tendencies within the military,[42] and with center and right-wing parties. Priests, nuns, and lay activists played significant roles in the popular opposition to Morales' government. They provided meeting space, good offices, and other forms of assistance to workers, neighborhood activists, and partisan political organizers. The protest activities which they helped to sustain put additional pressure on the military to withdraw but had little effect on either its economic policy or its political tactics once that decision was made.

The bishops, on the other hand, were unable to speak clearly or effectively during the transition period. Their earlier ambivalence toward Peruvian democracy (e.g., at the time of Belaúnde's overthrow in 1968) did little to strengthen their credentials as allies or advocates of transition. Their internal divisions were more pronounced than those of the Chilean bishops, and grew more apparent and obstructive as the influence of people like Landázuri and Bambarén declined. Divided on both ecclesial and ideological matters, the Peruvian bishops lost their collective voice on social and political issues. The terms of Catholic social thought prevented the conservatives from endorsing or defending the existing order, the socioeconomic or political agenda of right-wing groups, and the government's human rights abuses (except at the outset and at the very end of its term), but they refused to attack the government or challenge it publicly. In this connection, it is worth noting that the Peruvian bishops were treated with greater deference and respect by military authorities than their Chilean counterparts. As a result, they had less cause to confront the regime and more reason to rely on direct, behind-the-scenes contacts with authorities.

In the absence of statements or pronouncements from the Episcopal Conference, those of individual bishops were often featured in the national media and had a significant impact in some instances. But these usually prompted clarifications or rejoinders from other bishops, which had the effect of underscoring conference divisions and diluting the impact of both individual members and regional groupings.[43]

Finally, in contrast to the Chilean case, issues such as educational reform, family values, and separation of church and state intruded on the transition process in Peru, reducing the time and energy available for socioeconomic and political concerns. This was partly due to timing and partly to the character of Peru's transition, in which most military officers were in favor of relinquishing power. But it also reflected the more substantial foothold that conservative bishops had within the Episcopal Conference, and their greater concern for core Catholic interests.

The Peruvian bishops were thus neither the articulators of a broad consensus supporting democratic rule nor the mediators among political actors that their Chilean counterparts would be. The Peruvian Church's influence on the transition to democracy was greater at the local level, where its pastoral agents worked closely and supportively with popular movements and political organizations. Priests, sisters, and laypeople "representing" the Church in popular neighborhoods supported the efforts of neighborhood organizations and trade unions, rallying opinion against the military government and its policies. But even here, their impact was felt less in terms of concrete achievements than in the bases laid down for future organizational activity.

Moreover, social and political involvement by local church leaders and some bishops may actually have intensified conflicts within the Church, further reducing its willingness and ability to speak even minimally on social and political matters. Beginning with Velasco, and continuing under Morales Bermúdez, priests, nuns and lay activists were caught up in the commitments and aspirations of their peasant, worker, and slum-dweller congregations. This brought "the Church" into conflict with both political authorities and established social forces (landowners, factory owners, etc.), and was troubling to conference conservatives, who had reluctantly accepted the changes introduced by Landázuri. But as his political stature diminished, conflicts of perspective and interest long-contained came to the surface.

A substantial number of conservative Catholic laypeople supported the Morales government's policies of austerity and repression of dissent. The statements and actions of local clergy or progressive bishops drew their wrath and prompted efforts to discredit Catholic progressives. In Chile, where right-wing influence within the Church was less substantial, Cardinal Fresno, a conservative himself, helped to allay conservative fears, thereby easing an otherwise difficult transition. Conservatives were more numerous in Peru, and the views of moderate progressives like Landázuri merely heightened their anxieties, thereby intensifying the conference's divisions and reducing the episcopacy's capacity for effective influence.

Finally, Catholic political activists in Peru were also more fragmented than their Chilean counterparts, and their impact on political events was accordingly diluted. Rather than acting as a moderating leaven for compromise and tolerance, Catholic laypeople could be found at virtually all points along the political spectrum, including the polar extremes.

7

. . . .

The Church and the Consolidation of Democracy in Peru

On April 5, 1992, President Alberto Fujimori declared a state of emergency in Peru, dissolved the national Congress and Judiciary, and began to rule by decree. These moves were necessary, he said, to deal with the country's economy and guerrilla insurgency. Since this *autogolpe* (self-inflicted coup), Fujimori has overseen the election of a compliant Constituent Assembly, convinced it to draw up a constitution greatly enhancing his powers and permitting his reelection, and won election to a second term, in which he enjoys substantial majorities in both legislative houses. This sequence of events dealt the process of democratic consolidation that began in 1980 a major, if not fatal, setback.

In some ways, consolidation should have been easier in Peru than in Chile. Peru's generals left office without control over their own promotions and retirements, and without immunity from prosecution for crimes committed while in power. Nor were rightist or promilitary elements in Peru given control of either the Congress or the Supreme Court. In other respects, however, consolidation faced serious obstacles in Peru. Never particularly stable or democratic, Peruvian politics were in greater disarray than at any time in the country's recent history. The broad consensus that Chile's civilian elites had pieced together to unseat the military had not been needed, and did not exist, in Peru. Indeed, the country's parties and party factions were in shambles, incapable of offering either effective or credible representation for the country's social forces.

Economic conditions and capabilities also boded ill for the Peruvians. Chile's strong economy could continue to grow even as some resources were redirected to low-income sectors. In Peru, however, structural and other weaknesses left the economy incapable of sustaining growth or improving the lot of lower-income elements as it did. In short, prospects for redistributive growth that might have enhanced regime legitimacy were limited.

In terms of civil-military relations, the Peruvians also faced difficult challenges. Most of the human rights abuses in Chile had occurred in the early years of military rule and were things of the past. In Peru, on the other hand, the consolidation process followed immediately upon a period of recurrent and extensive abuses. Antimilitary sentiment was thus of more recent vintage, and not likely to diminish with the proclamation of a people's war by Shining Path guerrillas in April 1980. Indeed, that war would become the vortex of national political life for the next twelve years, underscoring the fragility of national institutions and making it difficult for civilian governments to curb military power and abuses of power.

These obstacles hampered the consolidation process from the outset. They also limited the potential contribution of the Peruvian Catholic Church. There was little that the Church could do with a willful and unaccountable military, a chronically stagnant economy, and unrepresentative and untrusted parties. However, there were other things that would limit the Church's impact for which it was much more responsible. One was the declining influence of Cardinal Landázuri, without whose moderately progressive leadership Church unity and leverage diminished markedly. Divisions among bishops and between the hierarchy and local elements had been apparent during transition but intensified with Landázuri's decline. Conservative, moderate, and progressive bishops differed sharply over ecclesial and political matters, making it difficult for the Episcopal Conference to take public stands and to speak forcefully when it did. Additionally, the failure of most bishops to encourage or support the social and political efforts of grassroots groups confined the impact of these organizations to the local level. Finally, the Peruvian bishops' preoccupation with certain moral and ecclesial issues greatly diminished their credibility and authority. The aggressive and uncompromising fashion in which they addressed the issues of abortion, divorce, birth control, liberation theology, and the growth of evangelical Protestantism, cast them in an antidemocratic light. As it had their Chilean colleagues, it cost Peruvian Church authorities some of the good will that they had won earlier in liberal and progressive circles, and blunted the thrust of their encouragement of compromise and consensus-building.

In the sections that follow we trace the evolution of the Belaúnde, García and Fujimori governments. We analyze the Church's response to each, and its impact on the consolidation of democratic institutions and practices. As in previous chapters, we distinguish the efforts of national authorities from those of bishops, pastoral agents, and activists at the local level.

The Belaúnde Government (1980–85)

Prospects for successful consolidation were poor as Belaúnde took office. For one, the country's political forces had drawn contradictory lessons from the years of military rule. The right believed that the military's reforms had "riled up" the masses and introduced disorder; they wanted the next government to concentrate on growth and leave redistribution for later. The left, on the other hand, did not think that the reforms had been substantial enough and insisted on bridging the gap between rich and poor immediately. Within the left, moreover, were social democrats, Maoists committed to a "peoples war," and others for whom democratic institutions were instruments of capitalist domination.

Belaúnde failed to manage these conflicting pressures. Pushed and pulled by neoliberal advisors and party cronies and allies, he led the country into recession and decline. His efforts to reduce budget deficits, contain inflation, and expose the economy to competitive pressures were undercut by costly public works projects, extravagant military spending, and tax breaks and subsidies that benefited certain producers and consumers. Following an initial period of modest growth, recession set in and most Peruvians sank deeper into poverty and despair. Belaúnde's mismanagement of the economy hurt him politically and placed the country's democracy in unfavorable light. A formidable guerrilla insurgency and the corrupt practices of intermediate- and upper-level government officials further complicated matters. By the mid-point of its period, Belaúnde's government had lost its way. As political forces turned their attention to the 1985 elections, APRA and Izquierda Unida moved quickly to the forefront, while the right looked to the future with growing alarm and anxiety.

Neither the bishops nor local-level Church organizations affected political developments during the early 1980s. The bishops left politics to the politicians, even as they pulled in contradictory and antagonistic directions. Locally, Catholics were active in a variety of neighborhood organizations housed in parish facilities and encouraged by area priests and nuns. But they were more concerned with the provision or distribution of material services (meals, medical supplies, access to water, land titles, etc.) than with political matters as such. As conditions worsened, many of these activists came to identify with the loosely structured Izquierda Unida coalition.

The bishops

Peru's return to civilian rule strengthened the resolve of Peruvian bishops to redirect their attention to more traditional moral and spiritual concerns. They had been moving in this direction for several years, prompted by the emerging divisions in their own ranks and the prodding of Vatican officials and the Latin American Bishops' Council.

On the eve of the 1980 presidential election, the bishops issued a pastoral letter urging adoption of a development model that would focus on problems (hunger, health, housing, employment, and education) afflicting the poor. In his campaign, Belaúnde had stressed such things as well, but, once launched, his government introduced modified, neoliberal policies designed to make the economy more competitive. Tariffs that protected domestic industries and jobs were cut in half. Income taxes were reduced, and credits to the private sector, particularly to exporters, were extended. The idea was to promote growth, from which, in the long run, rich and poor alike would benefit. Some industries (e.g., textiles) would retain their protection to preserve existing jobs; some foods and a few basic necessities would remain subsidized to keep the number of people falling below the extreme poverty line from rising sharply. But except for these measures, the poor would bear the short-term costs of liberalization.[1]

Belaúnde's own party (Acción Popular) reluctantly embraced this program. So too did the Popular Christian Party and APRA, at least at the outset. Only the Izquierda Unida opposed the plan, arguing that it would lead to greater dependency on foreign powers and to hunger and oppression for the working class.[2] Belaúnde's policies seemed to work for a while. During the last half of 1980 and for most of 1981, the economy grew at about 4 percent per capita per year (the rate achieved at the end of Morales's term). However, prices continued to climb, open unemployment and underemployment remained high, and the real wages and purchasing power of most Peruvians fell. In this context, the government's popularity began to decline. In municipal elections in late 1980, Acción Popular candidates won only 36 percent of the national vote, down from their candidate's winning 43 percent the previous May.

Although the government had ignored their appeal on behalf of the poor, the bishops were unwilling to challenge it directly. In a statement issued in early 1981, they decried the "growing breech between rich and poor, the precarious circumstances in which poorer Peruvians were living, and the incidence of violence and unemployment." But then they blamed these conditions on underlying "moral deficiencies" (drug trafficking and use, official corruption, pornography, sexually explicit advertising, abortion, euthanasia, artificial birth control, and other "immoral practices") which they went on to discuss at length. As far as the social problems themselves were concerned, they urged people to "be good," to abide by the principles of Christian fraternity and solidarity, and to rely on faith in Christ, frequent prayer, and an unshakable desire for justice and peace.[3]

These exhortations failed to move policy-makers, and economic conditions continued to deteriorate. The drop in investment in 1981 set the stage for recession the following year, when per-capita GNP fell by 2.3 percent, real minimum wages dropped to 78 percent of 1980 levels, average real wages by .3 percent, and unemployment remained at roughly 6 percent.[4] Belaúnde's political support continued to erode. APRA went into opposition at the end of 1981, coordinating some of its actions with the left. The Popular Christian Party was also critical of government policy but stayed with Belaúnde. Nationalist military officers made threatening noises in response to the letting of oil exploration contracts to foreign companies and other issues.

As economic conditions and living standards deteriorated, the bishops continued to avoid the issues and policy questions involved. Their January 1982 statement focused again almost entirely on moral issues: "permissiveness," the "breakdown of family values," and "such immoral practices as drug trafficking, consumerism, criminal violence, bribery and corruption" were preventing Peruvians from achieving their full human dignity. They noted the "grave economic situation" (the shortage of basic necessities, rising unemployment and malnutrition, and general social deterioration) that prevailed, but pointed to "public agencies," intermediate organizations, and the private sector as having the responsibility to "eradicate these evils and assure the common good of the society." Their role, they said, was "to work for the transformation of men and women from within, by means of an integral education in faith that made family and social responsibilities clear."[5]

This retreat from social and political issues was partly a function of the declining influence of Cardinal Landázuri. Landázuri was neither an ideological nor an ecclesial progressive, but he often embraced the judgments and advice of people who were. He used his stature as cardinal and primate, his close ties to Rome under Popes John XXIII and Paul VI, and his congenial and accommodating style to control the Episcopal Conference from its inception in the late 1950s. Under his leadership, it became one of the most progressive in Latin America, although many of its bishops were theologically and ecclesially quite conservative.

As the 1970s came to a close, Landázuri began to retreat from his progressive positions of earlier years. Many Church people viewed the Velasco period as disastrous, and he was anxious to mend fences with them.[6] In addition, the death of Pope Paul VI cut deeply into his Vatican contacts and connections. Finally, John Paul II's election as pope in 1978 led to the appointment of additional conservative bishops.[7] Since twenty of Peru's fifty-five bishops were missionaries from remote prelatures and not really interested in conference business, the conservatives gained the upper hand in most proceedings fairly quickly.[8] Without challenging Landázuri directly, they were able to neutralize his leadership. Landázuri remained conference president until 1988, and continued to name commission presidents and to help set conference agendas. But by the early 1980s, he lost his ability to line its members up in support of statements and documents drafted by his advisors and consultants.

Deteriorating economic conditions prompted Belaúnde to shuffle his cabinet in late 1982. Pedro Schwabe replaced Manuel Ulloa as prime minister, and Alfredo Rodríguez Pastor became minister of the treasury. They freed up prices, lowered the exchange rate, and cut credit and price supports further. The recessionary effects of these moves were greater than anticipated, and conditions went from bad to much worse. Per capita GNP fell 14.1 percent in 1983 and rose by only 2.1 percent in 1984. Investment fell by 33.2 percent in 1983 and 9.9 percent in 1984. Real minimum wages recovered slightly (to 80 percent of what they were in 1980) in 1983, but dropped to 62.3 percent in 1984. Average real wages fell by 20.7 percent in 1983, and by an additional 9.1 percent in 1984. Unemployment, finally, jumped to 9.0 percent in 1983, and held at 8.9 percent the following year.[9]

The downward trend led to strikes, protests, and a further erosion of the government's political position. The principal beneficiary of voter discontent was APRA, which captured 33 percent of the vote in the November 1983 municipal elections. Leftist parties, united in a single front, the Izquierda Unida (IU), were a close second at 30 percent. It brought together Moscow-oriented communists, Maoists, Trotskyists, "vanguard" Castroites, pro-Velasco socialists, and Catholics who had been active in union and popular-sector struggles in the late 1970s. Some of these Catholic leftists played important roles in the IU's formation and early success, and in 1983 several were elected to seats on municipal councils and as mayors of districts of Greater Lima with high concentrations of *pueblos jóvenes.*[10] The IU's leading figure, Alfonso Barrantes, a self-proclaimed "non-believing Catholic" (*católico no-creyente*), was elected mayor of Lima in 1983 and immediately became a force to be reckoned with for the 1985 presidential election.

In the face of deteriorating economic circumstances, the bishops supported the "survival" efforts of local parishes and communities (soup kitchens, food co-ops, child-care centers, etc.), but remained aloof from the country's social and political debate. From time to time they expressed their solidarity with those experiencing economic difficulty, with the victims of natural disasters (drought and flooding), and with those affected by both insurgent and counterinsurgent violence. But they could only urge people to pray and be patient. They said nothing of the need for redistributive policies, employment-generating programs, or popular participation in decision-making at the local or national level. Nor did they offer to mediate among antagonistic parties and forces in search of consensus.

They were also unwilling to condemn Sendero Luminoso directly. In August, the bishops addressed the increased violence in the central highlands around Ayacucho. They condemned terrorism and the use of torture in pursuing or interrogating "suspected terrorists." But with few exceptions they were loathe to mention the Shining Path by name.[11] Some did not want to appear to side with the military, which until 1987 was responsible for a far larger number of deaths. Others believed that if they did not directly confront Sendero, it would be more willing to negotiate with Church people or less likely to bother them.[12]

In any event, the bishops had little impact on anyone. The military and Belaúnde government officials viewed them as naive, while politicians who were troubled by the government's social policies and counterterrorism were looking for practical ideas and initiatives, not general principles. The bishops might have helped party leaders, political and military elites, and social organizations and movements to discuss and resolve their differences, but they did not. They were more concerned, it seems, with the training of Catholic priests and the dangers of liberation theology. These were issues on which the Vatican had begun to push them, and in which conference conservatives had a particularly keen interest.

Concerned with the pastoral and intellectual formation of future priests, Vatican authorities set out to rid seminaries of inappropriate influences. They sent instructions regarding curriculum and faculty appointments at the archdiocesan seminary of Santo Toribio, where secular (diocesan) priests studied, and at the Catholic Faculty of Theology at San Marcos, where most religious orders sent their seminarians. These moves resulted in the removal of progressive faculty members from both institutions and prompted the religious orders (e.g., Sacred Heart, Dominican, and Franciscan priests) to set up their own seminary, the Instituto Superior de Estudios Teológicos (ISET), with a curriculum and intellectual atmosphere more to their liking.

The other issue that consumed the bishops' attention was the writings of liberation theologian Gustavo Gutiérrez. Conservative Catholics had been challenging Father Gutiérrez's views for some time but had failed to persuade either Msgr. Landázuri or the Vatican of Paul VI that they were heretical.[13] With John Paul II's election in 1978, and with his appointment of Cardinal Joseph Ratzinger to head the Sacred Congregation of the Faith (the Holy Office), the anti-Gutiérrez campaign was given new impetus.

Ratzinger undertook a review of Gutiérrez's writings. He began in 1981, and later requested that the Peruvian bishops review and assess them as well. Msgr. Durand and others leaped at the chance but failed to get a majority of their colleagues to support either rejection or condemnation.[14] In 1984, the Peruvian bishops submitted divergent position papers on Gutiérrez's work to Ratzinger. The German cardinal was not pleased. His Office had just issued a formal "instruction,"accusing liberationists of "immanentism" (believing in the possibility of a perfect world here and how) and of being insufficiently critical of Marxist analysis.[15] But the document did not name names, and thus allowed the Peruvian bishops to embrace it without repudiating Gutiérrez. They were called to Rome in October to explain their "inadequate" response. Unexpectedly, Vatican officials sympathetic to Gutiérrez persuaded the pope to review his work personally before approving a revised statement that Ratzinger had prepared for the bishops to sign. After spending most of a night reading Gutiérrez's work, the Pope is said to have suggested that Ratzinger's text be softened and its references to Gutiérrez eliminated.[16]

The document that resulted, and that the Peruvian bishops issued in their name, was nonetheless a conservative one. It insisted on the intellectual freedom of theologians being exercised within the bounds of the Church's magisterial teaching. Additionally, the historical examples of liberation to which it pointed were clearly of a spiritual, not socioeconomic or political nature. Finally, it defined true liberation in unambiguously spiritual terms:

> It is important to underscore the essential aspects [of faith] that liberation theologies tend to ignore or leave out: the transcendence and gratuity of liberation in Jesus Christ, true God and true Man, the sovereignty of his grace, the true nature of the means of salvation, and in particular the sacraments. One should also keep in mind the true sense of ethics, for which the distinction between good and evil should not be relativized, the true sense of sin, the need for conversion, and the universality of brotherly love. One should guard against the politicization of existence that in ignoring both the specificity of the Kingdom of God and the transcendence of the person ends up sacralizing politics and using popular religiosity to benefit revolutionary enterprises.[17]

At the time, this was as far as the pope would go. When he visited Peru in May 1985, he brushed aside the grumbling of those who had wanted Gutiérrez condemned. He also endorsed the aspirations and struggles of the popular sectors with which Gutiérrez was closely identified. He would address the issue of liberation theology again during a return visit in 1988. But for the moment the matter was set aside.

Ironically, the pope's 1985 visit prompted the Peruvian bishops to issue a a rather progressive statement on the eve of that year's presidential and congressional elections. The conference had said nothing of social or political import since its August 1983 message on violence. They had said nothing about the social costs when the Belaúnde's government adopted more neoliberal policies in late 1982. Nor did they react in May 1984 when it began pumping money into politically motivated "development" projects in a blatant display of patronage politics. Apparently, they either had nothing to say or could not agree among themselves if they did.

This statement broke new ground, however, condemning poverty, injustice, corruption, social insensitivity, and violence in language they had not used since the late 1970s. They even cited passages from the pope's recent speeches. One passage condemned unemployment and underemployment as "evils and at times a true social calamity" and insisted that "the first concern of all responsible officials ought to be to provide work for all." The other referred to the country's external debt:

> In making available the necessary financial assistance, [international agencies and institutions] should work with the developing countries to define conditions [of repayment] that will enable these countries to leave their poverty and underdevelopment behind. . . . [They] should renounce conditions that over the long run instead of helping these countries to improve their circumstances cause them to sink even deeper, and perhaps to such despair as to produce conflict of incalculable magnitude.[18]

These remarks linked the Church to other government critics on the eve of a national election. Both APRA and the United Left had been critical of Belaúnde's neoliberal policies and vowed, if elected, to challenge international financial institutions on questions of government spending and debt repayment. With economic and political conditions as they were, the election of one or the other of these candidates seemed likely. And on these issues, at least, the Church appeared to align itself with them.

Local churches

With the Episcopal Conference retreating during this period, some bishops spoke individually in their dioceses or collectively in their regions. Bishops Dammert and Bambarén, Auxiliary Bishop Schmitz of Lima, and the bishops of the Southern Andean region, were forceful defenders of popular causes during the early 1980s. Bishop Schmitz actively encouraged and supported priests and pastoral agents who worked with survival organizations in *pueblos jóvenes.* The Southern Andean bishops supported the demands of area peasants for land, and were critical of police and military forces.

But these progressive bishops had little impact at the national level. The media readily distinguished conference positions from those issued by individual prelates. And in case they did not, conference conservatives were quick to note that individuals and regional assemblies spoke for themselves, not for the conference (or the Church) as a whole.[19] The remarks and actions of individual bishops were important to the people working or living within their jurisdictions; but they went unnoticed in larger arenas and were ignored by those not already in sympathy with them.

As socioeconomic conditions worsened, Catholic groups became more active at parish and neighborhood levels. They were less involved politically than in previous years, however, because of changes in the nature and focus of popular organizations. During the Belaúnde years, confrontational activity (strikes, demonstrations, street blockages, building seizures, etc.) declined in relation to the late 1970s. Survival organizations and relief activities (milk programs, soup kitchens, women's support organizations, etc.) increased proportionally, although activists had to fend for themselves in most instances. The industrial and employee unions with which they had been linked in the 1970s had been weakened during 1978 and 1979, as many union leaders lost their jobs and went to jail. With unemployment and underemployment running high,[20] they were scarcely able to remain afloat, much less help to organize slum-dwellers in *pueblos jóvenes.*[21]

Priests, sisters, and lay pastoral workers supported block and district organizations if their bishops encouraged or allowed them to do so, and in some instances even though they did not. Parish facilities and resources were made available, and priests and sisters frequently helped people solve problems or get around obstacles. The organizations became visible and respected fixtures in their neighborhoods, providing valuable social experience and both personal and spiritual development. Some priests who worked in the *pueblos jóvenes* of Lima's southern cone, encouraged the political development of their lay activists, helping them to evaluate political groups and opportunities and to integrate their political values and commitments with their Christian faith.

Many of these local Catholic activists joined either APRA or one of the leftist parties in the Izquierda Unida, particularly following the November 1983 municipal elections. According to Klaiber, most pastoral agents supported popular "issues, messages, and symbols, and never parties or people," but leftist candidates were always well received in any *pueblo jóven* parish in Lima.[22] In some areas, even nonpartisan political efforts led to confrontations between Christian activists and government or military authorities.

Wherever the local bishop would invite it in, the Episcopal Conference's Social Action Commission (CEAS) helped bring priests and parish or Christian community activists together to discuss issues affecting local residents. With more people displaced by guerrilla violence, the natural disasters (droughts and floods) of 1983, and an influx of financial help from abroad, CEAS expanded its operations. Its projects usually included material assistance and were more acceptable to conservative bishops for that reason. But they also offered opportunities for the organization's staffers to give workshops and training sessions dealing with Church social teaching, contemporary social problems, and other subjects of political import.

In Lima, Catholic activists in popular neighborhoods and *pueblos jóvenes* worked directly with programs (like the "glass of milk" program) initiated by the Barrantes (IU) administration. Most of the "outside" or "secular" people with whom local Catholic activists worked belonged to one or another of the Izquierda Unida parties. In the process, many came to identify politically with the left. Its ties to the local Church (to its priests, sisters, and lay personnel) helped the IU roll up large electoral majorities in most of Lima's *pueblos jóvenes* in the 1983 municipal elections.[23] Barrantes became one of Lima's most popular and effective mayors ever, thanks to his honesty and his initiatives in low-income housing, nutrition programs, and expanded public services.

In the 1985 presidential election, however, he could not overcome either his rival's charismatic appeal or the fears that Marxist parties still aroused among middle- and popular-class Peruvians. García won an impressive 46 percent of the vote to Barrantes's 23 percent, capturing all but a few outlying departments, and outpolling Barrantes in most of Lima's middle-class and popular districts and in *pueblos jóvenes* as well.[24]

The García Government (1985–90)

Bishops and local Church leaders and activists continued to work at cross-purposes politically during García's presidency. For its part, the Episcopal Conference remained preoccupied with spiritual and ecclesial matters, leaving economic and political problems to state officials and political leaders, whose inability to function responsibly or effectively grew increasingly clear over time.[25] Some bishops, priests, nuns, and lay activists, on the other hand, defended the rights and interests of their people at the local level, helping them to develop their political capabilities and to oppose policies and programs that adversely affected them. Their efforts had a significant impact on the people immediately involved. But the Church's impact on national politics and on the strengthening of democratic institutions remained limited. The bishops themselves were averse to political issues and involvement, the country's political forces were hopelessly fragmented and divided, and there was no viable reform party that could represent the Church's concerns politically. Grassroots Catholics were a moderately progressive and potentially stabilizing force during the period, but they too were hampered by the reticence of Church officials and by their own skepticism with respect to most political parties.

The bishops

García adopted a "heterodox" development policy in place of Belaúnde's episodic neoliberalism. He repudiated the IMF and debt-service commitments of previous governments, thereby freeing funds for development projects and other state expenditures. Private firms were obliged to purchase government bonds, inflation was contained through price controls, and domestic industries were protected and subsidized in hopes that they would produce and invest more. The idle industrial capacity inherited from Belaúnde enabled García to jump-start the economy while cutting inflation during his first year in office.[26]

This generated jobs, and an increased sense of prosperity and national pride.[27] García's popularity remained high for the rest of 1985 and most of 1986. In June 1986, however, civil guards and prison officials massacred more than 300 Shining Path prisoners in penal facilities in the Lima area. García promised a thorough inquiry, assuring the general public that he had clear evidence and that those responsible would not be able to "hide in the barracks of any institution."[28] But the matter was then handed over to military courts where it languished for several years before a lieutenant and a colonel were finally given relatively light sentences. García's handling of the affair gave the impression of a president without strong commitments and less than fully in charge of his government .

The tragedy prompted the bishops to speak. It was the first time they spoke publicly on a political matter during García's presidency, although they had earlier questioned the methods of security forces.[29] Their statement was perfunctory and evasive. Ignoring the issue of civilian control of the military altogether, it merely lamented the loss of lives, and argued against fighting violence with violence.[30] It failed to defend the public's or the victims' families' right to know the circumstances under which the prisoners had been killed, something that the Chilean bishops did on several occasions during the post–1990 period.

García's early economic success ran its course by late 1986. The previous year's growth fueled demands for imports, provoking a balance of payments deficit that cut deeply into foreign exchange reserves. Savings and investment levels remained low, nonindustrial production stagnated, consumer prices began to push up, and workers started grumbling about declining purchasing power. The November 1986 municipal elections gave García an impressive political endorsement. APRA captured 40 percent of the votes, winning control of most of the country's municipalities, including Lima. But even as he hailed his victory, his economic program continued to unravel, and following the elections he authorized price increases for food, consumer goods, and transportation, hoping to discourage hoarding and to ease inflationary pressures. He also forced private firms to purchase treasury bonds, hoping to reduce the budgetary deficit, shore up the country's currency, and undercut the run on dollars that was beginning to develop.

These measures antagonized financiers, industrialists, workers, and consumers generally. Additional price increases in April prompted the calling of a one-day general strike in May. Against the advice of his cabinet ministers, García gave in to worker demands for higher wages, thus driving up labor costs and inflationary pressures. Potential investors continued spending their money on consumer goods or converting to dollars for transfer abroad. With the previously idle industrial capacity again in use and savings and investment at all-time lows, recession and the compounding of inflationary pressure were inevitable.

Surprisingly, García responded by naming even "bolder" advocates of heterodox economics to his cabinet and expropriating privately owned domestic banks. The bank expropriation was justified as a means for putting local capital to more productive use. But it was poorly designed and implemented, further straining relations between the government and the private sector and raising questions about García's judgment and stability.

The bank takeover also brought the divisions within the Church to light again. Moderate and progressive bishops issued individual statements in support of the move. Landázuri termed it "an important and positive" step,[31] while more than 300 priests signed a statement that criticized the corruption of economic and political institutions, unrestrained pursuit of self-interest, unequally distributed wealth, the social responsibilities of property ownership, and raised other issues. Their declaration marked the first time since the late 1970s that a group of priests had spoken out on a controversial public issue.

Conservative bishops were more critical of the takeover. The statement that they pushed through the conference took no position as such but outlined the criteria that ought to govern such cases. These included consistency with the Constitution, avoidance of confrontation and polarization, and respect for private property and initiative.[32] A month later, the conference secretary, Msgr. Vargas Alzamora, described this as the Church's "official" position, and said that all others were simply the opinions of their authors.

Other matters divided the bishops as well. Conservative bishops grew more assertive in their dioceses and in conference proceedings. When the former military ordinary Alcides Mendoza was named archbishop of Cuzco, for example, he reclaimed lands that the Church had given to local peasant communities and sought to restrict the activities of the Andean Pastoral Institute, which the Southern Andean bishops had supported for many years.[33]

In early 1986, Archbishop Richter of Ayacucho vetoed the establishment of a Human Rights Commission in his city, despite a sharp increase in the number of civilian deaths attributable to the army's counterinsurgency operations.[34] In other dioceses, CEAS programs assisting people displaced by guerrilla violence and in need of social services were blocked by bishops who looked askance at anyone or anything critical of the military.

As the number of conservative bishops increased,[35] episcopal statements also placed greater emphasis on personal moral issues. A case in point was the conference's April 1987 message. Issued amidst growing economic difficulties and signs of serious problems ahead, it was devoted almost entirely to the government's birth control program. Only one of its twenty-two paragraphs dealt with the "harsh reality" of extreme poverty and misery. A second praised the government's refusal to meet prescribed debt-service payments but then chided it for failing to resist outside pressures on the issue of population policy.[36]

Finally, conservatives renewed their attacks on liberation theology. Msgr. Durand published a lengthy critique of Gutiérrez's works (Durand, 1985), insisting that they were among the "dangerous ones" to which Ratzinger's 1984 instruction had alluded. In early 1986 the Vatican issued a second instruction that praised the positive features of "properly conceived" liberation theologies but warned against "extreme" and "insufficiently careful" versions.[37] Although not mentioning Gutiérrez by name, it provided his critics with additional arguments and legitimacy.

An additional blow to moderate and progressive elements was the forced restructuring of the Instituto Superior de Estudios Teológicos (ISET). During a six-month "visit" in 1987, a Vatican-appointed representative imposed a new and conservative rector and director of studies. He also placed all nonclerical (female) and non-Catholic students into a separate program (except for students preparing for ordination), and required that faculty members sign formal pledges of support for the Vatican instructions on liberation theology. The result, according to Klaiber, was a less open institute, although the students themselves remained "pretty progressive." "What happens," he said, "is that the students and most of the teachers are still rather progressive, but certain people, mainly Gustavo Gutiérrez and people like that, are just simply *persona non grata.* They could never come there, even though they are personally welcome by most of the teachers."[38]

Focused, as they were on internal matters, the bishops were less than attentive to other developments. On the economic front, conditions and prospects worsened dramatically during 1987. Trade deficits mounted, foreign exchange reserves all but disappeared, the rate of tax pressure hit an all-time low, and economists began predicting deep recession and hyperinflation. In December 1987, García abandoned his "heterodox" policies. He devalued the currency by 40 percent, hoping to promote exports, discourage imports, and restore confidence in the economy. These moves, coupled with price hikes (literally *paquetazos,* or "package hits"), failed to contain inflationary pressure but did provoke worker unrest. A general strike was called in late January and another in July, amid rumors of a military coup. GNP fell by more than 9 percent during 1988, and inflation reached more than 1,700 percent by December.

The bishops did little beyond lamenting these developments. In November 1987, they urged a social pact between labor and management and the adoption of a "serious and realistic" economic plan, implying that the government's was neither.[39] But they did nothing to lead or assist people in moving in these directions. Conference conservatives believed that economic and political problems were the consequence of moral deficiencies, and were willing to speak about them.

But moral regeneration was a long-term project, on which people would have to work for many years. In the meanwhile, there were steps and initiatives to be taken (e.g., helping political forces work together for common objectives) that could mitigate or avert further disaster. The conservatives preferred to leave such matters to "the experts," however. They thought that their colleagues had been excessively concerned with economic and political issues in the 1970s, thereby confusing "good" Catholics with different sympathies or convictions.

Pope John Paul II's May 1988 visit bolstered the conservatives' position in the conference. The new papal nuncio, Msgr. Luigi Dossena, orchestrated the visit in concert with papal staff people. He placed the Peruvian section of Opus Dei in charge of liturgical events, and the equally conservative Sodalitium Christianae in charge of the press office. Dossena also had a hand in drafting the pontiff's speeches, several of which embraced the concerns of conference conservatives. Lack of respect for human dignity, the denial of transcendence, and a willingness to resort to violence, he told the bishops, were at the root of the country's social and political ills. But so too were

> the growing and extreme poverty in which many families live, social ills attributable to drug trafficking, the spread of religious [Protestant] sects, and *the obstinate persistence in doctrinal and methodological positions that sow confusion among the faithful and undermine Church unity.* (our emphasis)

And unfortunately, according to the pope, there were priests

> whose persistent error brings them to such a clouding of reason that they are deaf to the calls and warnings, as though they were being directed to others.[40]

The pope was referring to Father Gutiérrez and thus aligning himself with Gutiérrez's critics. When asked about the Vatican's instructions previously, Gutiérrez had insisted that his work was free of the errors to which they referred and that Rome must have someone else in mind. But with these remarks, it would be more difficult for him to continue to evade the issue.[41]

Following the visit, conservative bishops assumed formal control of the Episcopal Conference. Landázuri decided not to stand for another term as conference president, citing his impending retirement, but also conceding that new leadership was needed. This came in the person of antiliberationist Ricardo Durand, a one-time Landázuri ally who had grown more conservative during the 1970s and early 1980s, and who defeated his progressive-sector opponent by a narrow margin.[42]

Conservatives were named to head most of the conference's posts, including the Social Action Commission (CEAS), where Bambarén was replaced by Miguel Irízar, bishop of Yurimaguas and a reputed conservative.[43] Progressive lay activists reacted with alarm. Some of them approached the European foundations that had funded CEAS for years, asking for money to continue operating independently. One of the foundations, Misereor, responded favorably, providing money for a parallel organization, CEAPAZ, which was launched in late 1988. This proved an unnecessary precaution, as Bishop Irízar turned out to be fully committed to CEAS programs and projects.

Under explicitly conservative leadership, the Peruvian bishops addressed the nation on six occasions between July and December, 1988. Their messages and exhortations followed the noncommittal pattern of previous years. An August statement by the conference's Permanent Council urged the creation of more jobs and the development of soup kitchens. An October pastoral letter was critical of certain "imprudent" economic decisions, and of politicians who cover up problems with demagoguery.[44] But the bishops saw their role as encouraging others to live their faith, to have confidence in Jesus Christ, to reinforce people's "confidence and serenity in the face of difficulty," and to help all "to rectify their conduct and commit themselves to the promotion of the common good." The solutions to economic and political problems, they said, could come only from the country's political elites, its military officials, and its labor and business leaders, even though their deficiencies and irresponsibility were increasingly clear.[45]

Another factor in the bishops' unwillingness to take greater initiatives was their own division over how to express or apply Christian values in socioeconomic and political matters. In fairness to them, it might be noted that by late 1988 no one in Peru could agree on what could or should be done. A centrist government had failed, and its right- and left-wing critics offered opposing explanations and prescriptions. Right-wing parties wanted to reimpose law and order, take drastic measures to curtail government spending, repair relations with the international financial community, and unleash market forces. The left was split between radicals who saw possibilities for generalized insurrection and moderates who held out hope for democratic socialism. In this context, it would have been difficult for the bishops to rally others around a moderate, centrist position, had they wanted to. And as they were divided between moderate progressives and conservatives, left- and right-wing alternatives were even less appealing or promising. In effect, the bishops were caught in the same divisions and paralysis besetting the country as a whole.

Local churches

Things were not much better at the local level. Lay Catholics were as divided as the bishops. They could be found at all points along the party and political spectra, but none of the parties to which they belonged was capable of representing or defending "Catholic values" in the political arena. As we shall see, most Catholic residents of popular neighborhoods and *pueblos jóvenes* in Lima identified with the left generally or with one of the Izquierda Unida parties. The same was true of peasant communities in the Southern Andean communities of Cuzco, Puno, Ayaviri, Sicuani, and Juli. But the relatively few who participated actively were generally outmaneuvered or outvoted by other groups or factions. Moreover, the Izquierda Unida itself was anathema to Catholics who supported the Popular Christian Party, those still loyal to Belaúnde's Acción Popular, and to Christian Democrats who had stayed with the PDC or drifted in and out of various political groupings since the mid-1970s.

Some bishops and Church agencies worked actively with local Catholic communities and popular organizations during the García years. They helped them to articulate and defend their interests and to develop organizationally. They helped them to withstand the attacks of political and military authorities and the Shining Path. In these ways, local Church people contributed to the country's long-term democratic development. In the short run, however, their encouragement of popular organizations may have set consolidation back by alienating local- and national-level elites and making consensus more difficult. This was the case, for example, with the efforts of Andean bishops to help peasant communities to regain ancestral lands, with their support of popular civil defense groups known as *rondas*, and with many of CEAS's projects and programs.

Under Velasco, large, privately owned estates had been expropriated and restructured as government-managed cooperatives. In the highlands, peasants who had been stripped of their land in the nineteenth century were excluded from these cooperatives. Encouraged by sympathetic bishops, priests, and Church-sponsored organizers, they filed claims and pressured the government for action. Puno (in the Southern Andean region) became a center of such efforts under Belaúnde, who wanted to reprivatize the cooperatives, and viewed the peasants and their Church supporters as obstacles to this end.

The peasants continued to press their claims with the García government. Like his predecessors, García valued the cooperatives because of their export potential but continued to subsidize food imports for sale to the urban population. The amount of land available for distribution to peasant communities was thus limited, and food prices, on which most peasant incomes depended, were relatively low. When their protests failed to get results, the peasants usually occupied the properties involved, only to be ousted by police and military forces. Conflicts of this sort were common in the Southern Andean region, whose bishops called for peaceful resolution of all disputes, but usually sided with the peasants against local authorities and elites. They continued to take prophetic stances in defense of the poor despite the recent changes in their ranks.

The *rondas campesinas* were security patrols which peasants in the Cajamarca region began using to defend their communities from cattle rustling, assaults, and other antisocial activity in the late 1970s. They kept records, detained "suspects," tried them in special courts, and administered punishments (usually whipping). They also designed and carried out irrigation projects and mediated disputes involving land use and water rights. Over the next several years, the *rondas* spread to other departments and regions, and to the shantytowns of larger cities. By the mid to late 1980s, there were *rondas* in thirteen of the country's twenty-five departments, organized into provincial and departmental federations.

Most *rondas* identified politically with the Izquierda Unida. Their leaders were routinely harassed and occasionally jailed on charges of thievery, assault, and "agitation." Some *aprista* mayors and military commanders accused them of "training terrorists" and providing support for the Shining Path. This was ironic in that the *rondas* were often the only local force willing to resist Sendero, and were viewed by *senderistas* as formidable rivals. In many locales, they targeted *ronda* leaders for assassination; in others, they tried to make it appear that the *rondas* were collaborating with them, the idea being to preclude neutrality and force the *rondas* to choose between the government and Sendero.[46]

Archbishop Dammert and most bishops and priests in the Southern Andean region identified closely with the *rondas* in their areas.[47] Most of these were genuinely representative of their local communities, and dealt effectively with local problems (the need for land, corruption in local judicial institutions, excesses or abuses on the part of security forces). In the long run, they would be crucial building blocs in the country's democratic development. In the meantime, however, they inspired fear and hostility in both local and national elites on whom further democratization depended. *Ronderos* were not always disciplined or tactful, and occasionally they provoked conflicts that might have been avoided. Their protest marches, land seizures, and direct-action methods made them seem closer to Sendero than they really were.[48] But even where their leaders and activists were more circumspect, local political elites viewed them with alarm. They saw them as rivals bent on defying or displacing them and dismantling the existing social order. Local socioeconomic elites opposed their demands, resisted reforms that might have strengthened their hand, and in many cases endorsed their repression by local authorities and security forces.

Moreover, the bishops did little for the *rondas* at the national level, where viable and lasting agreements would have to be worked out. In fact, neither the *rondas* nor the bishops were comfortable operating beyond their local communities. The *ronderos* were not comfortable working with groups from other areas, whose conditions, interests, and dynamics might be different; they were also averse to making concessions to adversaries and found it difficult to trust fellow peasants, much less political or military elites, that they did not know.

Nor were the bishops eager to broker agreements at the national level. These were too "complicated," and people might accuse them of exceeding their proper sphere of action or expertise. Local-level involvement (e.g., their criticism of local judicial authorities or security forces) was easier for most bishops to defend as part of their pastoral ministry. They knew the people involved and the issues at stake seemed simpler and their moral dimensions clearer.

A third area of Church support for popular-sector groups were the programs and projects of the Episcopal Commission for Social Action (CEAS). CEAS had grown steadily from 1983 on, as the need for its social services increased,[49] and as the European foundations became more aware of conditions in Peru. Between 1983 and 1989, the number of its employees jumped from 60 to 120 as it attempted to assist refugees from war-torn areas, communities developing survival and other organizations, and people who were out of work or needed legal representation. As conditions worsened and state services were cut back, conservative bishops came to view its welfare services more favorably. But they continued to distrust its human rights monitoring, its refugee support, and its social formation programs, denying them access to many dioceses.[50] Political and military elites in these areas were similarly hostile to CEAS efforts and to Church authorities associated with them.

CEAS never approached the Chilean Vicaría in either its scale of operations or extent of political impact. One reason was the Peruvian Church's lack of resources. Apparently, it was easier to get international help for the victims of Pinochet's dictatorship than for people caught between Sendero and the Peruvian military but living under ostensibly democratic rule. But considerations of money aside,[51] Peruvian Church leaders also preferred to keep CEAS operating on a modest scale and confined to relief activities, in the interest of maintaining better relations with the government.[52]

Grassroots Catholics in Lima

Local-level Catholic activists had an important impact on the country's fragile democratic institutions and processes. They were progressive in their religious and political views, more tolerant and accommodating than other Catholics, and they expected less of their political leaders. In these respects, they were ideal citizens for a consolidation process. But they were neither active in nor supportive of party politics. Had they been so, they might have strengthened the country's democratic forces, and with them the consolidation process. But their aversion to partisan politics prevented this, and in the end other, less democratic forces prevailed.

In July and August, 1987, we interviewed 485 people in the seven districts of Greater Lima with a preponderance of *pueblos jóvenes* and popular neighborhoods. Our analysis looks at the attitudes and activities of these grassroots Catholics, and then assesses their impact on the consolidation. Because of the purposive sample on which our survey was based,[53] we cannot generalize about all activists or Catholics. But our findings will have illustrative value.

The popular-sector Catholics we interviewed were quite progressive politically: 85.2 percent considered themselves leftists; 59.3 percent favored "popular" as opposed to liberal democracy; 27.0 percent thought leftist (Marxist) ideas were either "good" or "important"; an additional 63.7 percent "saw value in certain of their aspects"; 57.2 percent saw the country's social problems primarily in structural or class (rather then in contingent political or personal) terms; and more than 90 percent favored close ties between Christians and "leftist" parties.[54] Their religious attitudes were similarly progressive. Many (31.3 percent) thought that Christ was killed because he was a political revolutionary. Almost half (47.4 percent) thought that the Church's principal pastoral concerns should be the country's socioeconomic problems, not people's personal religious or spiritual needs. And 40.2 percent thought that they should follow their own conscience rather than Church authorities if they disagreed with official teachings on moral issues.

It is tempting to credit the local Church with fostering the progressivism of these Catholics, since Catholic parishes and communities have been fixtures in popular neighborhoods and districts for many years. As in Chile, however, respondents who were more involved in the Church (the organizationals) were less progressive in both their religious and political views.[55] Table 7.1 compares the political attitudes of Peruvian Catholics with varying relationships to the Church.

Organizational Catholics held generally progressive views on the first two attitudinal questions. On leftist political tendency they lagged behind sacramentals and culturals but still were in the 80 percent range. On structural/class problems they gave more progressive answers than the first of these groups, while on the third question (leftist ideas) they were slightly more positive than sacramentals, but less so than culturals (another 65.6 percent, 64.1 percent, and 56.6 percent, respectively, saw some value in them). The most significant difference between organizational and other Catholics, however, was reflected in their social and

TABLE 7.1

Political Attitudes of Popular-Sector Catholics, Lima, 1987.

	Leftist Political Tendency	Structural/Class Problems	Leftist Ideas Good/Important	Neighborhood Organization	Political Organization	Range of *N*
Organizationals	81.1%	69.7%	24.8%	14.0%	38.5%	227–300
Sacramentals	90.9%	60.9%	23.1%	27.5%	72.5%	33–40
Culturals	97.5%	69.8%	35.2%	41.5%	70.2%	79–96

Source: Fleet, 1987b.

political involvement. As the last two columns indicate, organizationals were substantially less involved in both neighborhood organizations and political parties than sacramental or cultural Catholics.

Catholics living in districts with a tradition of social and political activism were more progressive and more involved than those living elsewhere. Within each district, sacramental and cultural Catholics were generally more progressive and more involved than organizational Catholics. Several things might account for this. Church organizations could have attracted already moderate people, because of their moderate image or because moderate and conservative Catholics gravitated toward official Church groups and activities, whatever their image. On the other hand, the pastoral orientations and materials that guided these organizations could have exerted a moderating impact on their members. And finally, the commitments of time and energy that local Church organizations required could have precluded exposure to other, more radicalizing influences.

Without knowing what these people were like before becoming involved with the Church, it is hard to say which of these alternatives is more likely. It is possible to better understand the interaction of pastoral and social environments, however, by looking more closely at three of the seven districts (Villa El Salvador, San Juan de Miraflores, and Comas) included in our survey.

Three Parishes

El Salvador is the "official" parish for Villa El Salvador, a district of more than 300,000 inhabitants located on Lima's desert-like southern coastal plain. "Villa" was established by the Velasco government in 1971, and has grown from a single-sector *pueblo jóven* to an agglomerate of seven sectors. It has been a highly successful self-development community thanks to its dedicated social activists, its honest and efficient municipal administration, and its extensive network of religious, cultural, and political organizations. The pastor of El Salvador until the mid-1980s was an American priest of Polish descent who was a good organizer, popular with his people, and adept at dealing with local officials and political types. As new areas of Villa opened up, sector churches and their networks of satellite chapels (some seven or eight apiece) were established, under the auspices of priests of the Maryknoll, St. James the Apostle, and Irish Dominican orders. These priests, together with foreign (particularly Spanish) and Peruvian nuns, ranged from moderately conservative to radical in terms of their ideological and pastoral orientations.

They worked collaboratively with the lay Catholic activists, government functionaries, and party militants that were drawn to the area. With social and community leadership coming from these groups, priests and nuns could concentrate on liturgies, the sacraments, religious education, parish administration, the preparation of catechists, and advising Christian communities and reflection groups.

Christian communities have been one of the vehicles through which the Church has promoted lay involvement in social and political affairs in Villa El Salvador. Left to themselves, however, many residents appear to prefer sacramental Catholicism or the more traditional spirituality of their religious brotherhoods. As of 1988, there were virtually no Christian communities in the first sector, where the main parish is located. In the second, third, and fourth sectors, however, communities have played an important role in social and political affairs in the district.[56]

With the resumption of electoral politics in 1980, Villa El Salvador became a bastion of support for the Izquierda Unida. It provided IU candidates with greater backing than any other district, and the IU led its municipal government between 1983 and 1992. This loyalty may have been an expression of appreciation for a job well done. Working with neighborhood organizations, Villa officials were very effective at the community level and in obtaining services and resources from the national government.[57]

El Niño Jesus is the largest of three parishes in San Juan de Miraflores, another district in Lima's southern cone. It was established in 1960 by Maryknoll priests, and ministers to a population in excess of 100,000 people. It began as a conventional, U.S.-style parish, offering a full complement of sacramental and apostolic activities from a large, multipurpose parish center. Over the years, the construction of satellite churches and smaller chapels in the district brought priests into closer contact with community residents. They continue to celebrate Sunday masses (with an average attendance of roughly 1,000 people) and other liturgies in the main church. But since 1986, they have lived in small houses or rooms in other parts of the parish, spending most of their time in the satellite churches and chapels, and serving as advisors or spiritual consultants to groups of young people, brotherhoods and other traditional organizations, family catechetical groups, and reflection groups and Christian communities.

Most of the priests in El Niño Jesus have been progressive, socially committed men determined to help their people know God, obtain justice, and live peaceful and dignified lives. Their efforts to promote the Christian communities have attracted social and political activists, many of whom have wanted to deepen their faith, inform their activities and commitments, and share with others the religious significance of those commitments.[58] The priests have been less successful with other residents. Members of local *hermandades* (religious brotherhoods) are often more interested in their own spiritual lives, the once-a-year fiestas they organize, or the prestige that comes with their role in the *hermandad*, than in working on behalf of others. The priests and nuns have learned to be patient, and to work within the context of people's traditional sentiments and practices, even as they attempt to expose them to new dimensions of the Gospel.

Socially and politically, the area around the parish offers a mix of conditions and orientations. On balance, San Juan de Miraflores is the least impoverished of Lima's seven *pueblo jóven* districts. The areas close to the main thoroughfare and regional market are relatively safe and reasonably well-serviced lower-middle-class neighborhoods, although conditions in Pamplona Alta, where most recent expansion has taken place, are as bad as any in the city. Politically, the district has favored the candidates of the United Left by narrow margins in municipal elections, but it supported Belaúnde in the 1980 presidential election, and Alan García in the 1985 presidential election.

El Señor de los Milagros parish is the largest of two parishes in Comas, a district on Lima's north side that was invaded and settled by ex-hacienda workers in the late 1950s. Occupying land in the foothills above the Pan American Highway, residents named their new settlement after the El Señor de los Milagros, a popular Peruvian devotional figure.[59] The parish was established in 1962 by Canadian priests of the Oblates of Mary Immaculate order, who were assisted by Dominican sisters, also from Canada. The pastor since 1992 has been a Peruvian Oblate who grew up in the parish and was ordained in 1987.

Since 1962, two or three priests and seven to ten sisters have ministered to an area with roughly 40,000 to 50,000 inhabitants. The Oblate Fathers have provided a full sacramental ministry, and have encouraged lay participation in a variety of apostolic groups (Legion of Mary, the cursillos de Cristiandad, etc.). They have also advised and supported neighborhood organizations (often led by parishioners) in their struggle to obtain water, sewers, and other basic services from government authorities.

Lay people with leadership potential were encouraged to go through cursillo training, and to help form large Christian communities (similar to Chilean CEBs) in different sectors of the parish. During the late 1980s, more than 200 lay pastoral agents were at work in thirteen such communities, each of which sent representatives to the parish pastoral council. Lay men and women routinely conducted liturgies, gave catechism classes, visited the sick, provided leadership for the Christian communities, and even performed baptisms.

The area around the parish has been known for its social and political activism for many years. In terms of living conditions, it ranks among the more "prosperous" of Lima's *pueblo jóven* districts, although the areas into which it has expanded in recent years are among the city's most deprived.[60] The northern cone, and Comas in particular, have been the best organized and most effective in Lima in pressuring the government and providing support for popular causes. During Morales Bermúdez's government (1975–80), it was an area of active mobilization and frequent clashes between demonstrators and government troops. With the resumption of electoral politics in 1978, and for most of the 1980s, it supported candidates of the United Left.

All of our respondents in Villa El Salvador were members of El Salvador parish. Most of our respondents in San Juan de Miraflores and Comas lived within the confines of Niño Jesus and Señor de los Milagros parishes, respectively, and those few who did not may be treated as if they did without distorting the sample unduly.[61] Table 7.2 compares the political views of Catholics in the three parishes.

TABLE 7.2

Political Attitudes of Catholics in Three Popular-Sector Parishes, Lima, 1987.

	Leftist Tendency	Structural/Class Problems	Leftist Ideas Good/Important	Neighborhood Organization	Political Organization	Range of *N*
El Salvador	83.7%	65.1%	29.1%	16.3%	50.3%	75–104
Niño Jesus	68.8%	66.7%	28.2%	28.6%	33.3%	23–33
Milagros	92.3%	67.9%	40.3%	20.2%	52.4%	24–31

Source: Fleet, 1987b.

Not surprisingly, the Catholics in Señor de los Milagros parish held the most progressive views and were the most involved politically, while those of Niño Jesus were the least. To appreciate the impact of the local Church, however, we need to compare the views of different subsets of Catholics. In one of the parishes, Niño Jesus, there were not enough sacramental or cultural Catholics, so we can only look at the other two. Table 7.3 places the political views of organizational Catholics in relation to those of other Catholics in Villa El Salvador and Milagros. Negative values in parentheses indicate that the percentage preceding it is lower than that for other Catholics; a positive value, that it is higher.

On the three attitudinal questions, the responses of organizational Catholics in El Salvador averaged 58.2 percent, some 3.3 percentage points lower than the parish's sacramental and cultural Catholics.[62] The political views of organizational Catholics in Milagros were higher, i.e., more progressive (they averaged 61.3 percent), than those of their Villa counterparts, but slightly lower (3.7 percentage points) than those of other Catholics. In both parishes, organizationals were less involved than sacramentals and culturals.

Organizational Catholics were thus more moderate and less active than other Catholics in both El Salvador and (though to a lesser extent) Milagros. Without knowing what people thought previously, it is not clear whether the organizations were appealing to already moderate people or actually moderating the views of progressives. Traditional religious groups like the Legion of Mary, the Christian Family Movement, and the various religious brotherhoods were probably attracting already moderate or conservative Catholics. But it is more difficult to imagine Christian communities or more socially oriented organiza-

TABLE 7.3

Political Attitudes of Selected Organizational Catholics in Relation to Other Popular-Sector Catholics, Lima, 1987.

	El Salvador		Milagros	
Leftist political tendency	80.0%	(–20.0%)	87.5%	(– 7.2%)
Structural/Class problems	67.9%	(+21.8%)	67.7%	(– 4.1%)
Leftist ideas good/Important	26.8%	(–11.7%)	28.6%	(– .3%)
Political organization	12.6%	(–25.8%)	16.1%	(– 1.8%)
Neighborhood organization	47.4%	(–14.4%)	32.3%	(–31.8%)
Average	58.2%	(– 3.3%)	61.3%	(– 3.7%)
Range of *N*	86–103	11–13	24–31	34–46

Source: Fleet, 1987b.

tions doing this, given their more progressive religious and political image. Neither the religious nor the political views of Catholics belonging to Christian communities differed greatly from those of people belonging to these traditional religious organizations.

As far as group activities were concerned, these could affect members directly, through discussions of political issues and the taking of positions within the group, or indirectly, by promoting religious values that had political implications. The case for indirect influence would be stronger if the religious views of organizational Catholics differed markedly from those of other Catholics, and were congruent with their political views. These conditions were met in two of the three parishes (there were not enough sacramental or cultural Catholics in the third for us to test the case there). The religious views of organizational Catholics were much less progressive (22.2 percentage points) than those of other Catholics in El Salvador, but only slightly less progressive (2.7 percentage points) in Señor de los Milagros. In all three parishes, the religious and political views of organizationals were generally congruent; those with the least progressive political views also held the least progressive religious views, making it more plausible to argue that religious and political views were causally connected. In El Salvador, the moderating effect was substantial, given the markedly progressive religious views of other Catholics. In Milagros, the views of organizational were more in line with prevailing sentiment in the area.

In parishes like El Niño Jesus it seems likely that the people being drawn to these organizations were already moderate, and remained so despite the efforts of pastoral agents to get them to think in new ways. But in highly politicized parishes like Villa El Salvador and Comas, Church organizations probably exerted a moderating influence because of what they were teaching, i.e., love of one's neighbors, reconciliation, nonviolence, respect for dissenting opinions, etc.

The fact that the most progressive members of any subcategory of Catholics were those who were union or party members, but that organizational Catholics were the least likely to belong to either, provides support for thinking that the Church's moderating effect may have stemmed from its capacity to shield people from other (politically radicalizing) influences, not from its pastoral ministry as such. Table 7.4 indicates the differences in the political views of organizational Catholics who were party members and those who were not. Again, a negative value in parentheses indicates that the preceding value is lower than that for those not belonging; a positive value, higher.

Taken together, the political views of party members in El Salvador averaged 69.2 percent, 23.6 percentage points higher than those of non-party members. In Milagros, party members averaged 65.0 percent, only 4.5 percentage points higher than nonparty members. El Salvador party members were more likely to be active in neighborhood organizations, but those in Milagros less so. The discrepancy between party-member and non-party-member views, and the fact that moderate and relatively apolitical people greatly outnumbered radicals and progressives, might mean that people were not being politicized by their experience in Church organizations but were being drawn to them having been politicized earlier or independently. Had these organizations not attracted such people, they would have been even more moderate.

The moderate Catholic activists we interviewed contributed to democratic consolidation in several ways. In a country like Peru, where growth and equity have rarely coincided, successful consolidation of democratic institutions requires limited expectations, a willingness to accommodate other groups and interests, and an extraordinary amount of patience. The attitudes with which most of the country's social and political forces approached the 1980s augured poorly for successful consolidation in these respects. Owners and workers drew sharply divergent lesson from the twelve-year period of military rule, as did right-wing and left-wing forces. But in the late 1980s, one of the few hopeful signs in Peru was the modest expectations and prodemocratic attitudes of Catholic activists in Lima's popular neighborhoods and *pueblos jóvenes.* These emerge in their responses to open questions concerning Peruvian society ("Do you think that the Peruvian society in which we live is a just one?") and the overcoming of socioeconomic

TABLE 7.4

Political Attitudes of Selected Organizational Catholics Belonging, and Not Belonging, to a Political Party, by Parish, Lima, 1987.

	El Salvador		Milagros	
Leftist political tendency	76.9%	(– 3.7%)	75%	(–15.0%)
Structural/Class problems	92.3%	(+27.9%)	80%	(+14.6%)
Leftist ideas good/Important	46.2%	(+22.4%)	40%	(+13.9%)
Neighborhood organization	84.6%	(+43.0%)	20%	(–14.6%)
Average	69.2%	(+23.6%)	65%	(+ 4.5%)
Range of *N*	13	62–90	4–5	20–26

Source: Fleet, 1987b.

differences ("Do you think that the differences between rich and poor will ever disappear?").

Responses to the first question ranged from progressive (people referring to "structures of domination" and "class differences") to right of center (those complaining that "the laws aren't being obeyed" or "the wrong policies are being followed"). In our view, the latter responses (that avoided structural or class issues) were less likely to frighten or alienate established economic and political forces, and were thus more favorable to democratic consolidation. Responses to the second question ranged from left (those saying yes because of "historical necessity") to right (those saying no because "God wills them"). In this case, we consider responses that avoided both extremes to be the most functional with respect to consolidation.

Table 7.5 indicates the percentage of respondents (from all seven districts) expressing proconsolidation views. Unlike previous tables, it differentiates Christian community members and those belonging to social organizations from Catholics active in religious organizations.

The differences in Catholic thinking on these matters were minimal. Nor does the Church appear to have had much impact, since the views of sacramentals were much like those of Catholics belonging to religious organizations. It is possible that the Christian communities and social organizations were attracting already more progressive members to their ranks. But whatever the reason, those closely involved with the Church held more moderate views and were thus not a polarizing force.

On procedural matters, Catholic activists were much more supportive of the consolidation process than other Catholics. This emerges from their responses to the questions treated in table 7.6. The first of these (nonexclusionary democracy) asked the respondent to choose

TABLE 7.5

Percentage of Popular-Sector Catholics giving "Moderate" Responses to Questions Dealing with Political Analysis and Expectations, Lima, 1987.

Groups	Ills Due to Non-Structural Problems	Differences between Rich and Poor Will Disappear	Range of *N*
CEBs/Reflection groups	28.1%	58.4%	113–28
Social organization	34.6%	40.3%	107–33
Religious organization	23.1%	43.8%	32–39
Sacramentals	38.4%	56.2%	32–39
Culturals	32.2%	60.1%	80–93

Source: Fleet, 1987b.

from among the following regime types: a military regime, a civilian government that excluded violent or nondemocratic groups, a pluralist democracy from which no one would be excluded, a popular or socialist democracy, or none of these. The second question (accommodate class interests) asked whether one ought to defend his or her own class interests or accommodate the interests of all social classes. The third question (followers behind leaders) asked what was more important in a social or political movement: the leader, the rank and file organized behind a leader, or an autonomous and organized base and leaders capable of serving it. The fourth question (residents, residents/authorities responsible) asked (without indicating options) who ought to solve the problems of the neighborhood (responses ranged from the government and the mayor to the party, residents and authorities together, and the residents themselves).

Table 7.6 indicates the percentage of respondents (again from all seven districts) choosing the alternatives most supportive of democratic consolidation, namely, nonexclusionary democracy, accommodation of class interests, an organized rank and file behind its leader, and residents or residents and authorities together.

In three of the four questions (nonexclusionary democracy, accommodate class interests, and followers behind leaders), a greater percentage of Christian community members and social organization activists held proconsolidation views. In the fourth (residents, residents/authorities responsible), they stood only slightly behind religious organization activists but substantially ahead of both sacramental and cultural Catholics. People who held mayors or governments responsible might appear to be less demanding and more accepting of their institutions, but we believe that those determined to proceed on their own or

TABLE 7.6

Percentage of Popular-Sector Catholics giving "Moderate" Responses to Questions Involving Political Practices and Relationships, Lima, 1987.

Groups	Nonexclusionary Democracy	Accommodate Class Interests	Followers behind Leaders	Residents Residents/ Authorities Responsible	Range of *N*
CEBs/Reflection groups	37.4%	64.6%	18.9%	64.5%	113–28
Social organization	35.5%	71.1%	24.5%	71.5%	115–31
Religious organization	27.8%	52.6%	21.1%	72.9%	36–38
Sacramentals	33.3%	55.0%	20.5%	45.0%	36–38
Culturals	14.8%	54.3%	17.3%	40.0%	88–92

Source: Fleet, 1987b.

to work with authorities are doing more to strengthen institutions in both the short and long runs.

Among the people we surveyed, Church activists were thus stronger in terms of procedural values and commitments than other Catholics. Unfortunately, their political impact was much less than it might have been. During the late 1980s in Peru, the political center (APRA) was rapidly unravelling along with the García government, while radicals and moderates vied for control of the unwieldy Izquierda Unida. Had Catholic activists been willing to enter partisan politics, they might have reinforced the positions of the moderates. But they were not, and did not.

Their aversion can be seen in table 7.7. While "interested in politics," Catholic activists were the least likely to be party members, were skeptical of partisan involvement, and had little interest in working with leftists (Marxists) in a single party (interestingly enough, cultural Catholics were even less open to this themselves).

The Christian community members and social organization activists we surveyed claimed greater interest in politics and in partisan political involvement than religious organization activists, but were less involved politically. This could reflect the constraints of time and energy. Christian communities met frequently, and their members, and those of Church-sponsored social organizations, were busy with parish ommunity projects. It may also be that they were open to partisan political activity generally but not willing to affiliate with the Izquierda Unida under current circumstances. Had they been, they might have strengthened the moderate left, perhaps enabling it to fill the void left by Belaúnde's failed liberalization and by García's disastrous populism. Had they ventured beyond their neighborhoods, reaching out to other groups, forming and extending political ties, they might have become a

TABLE 7.7

Percentage of Popular-Sector Catholics Active or Interested in Partisan Politics, Lima, 1987.

Groups	Interested in Politics	Party Member	Partisan Involvement	Single Party	Range of *N*
CEBs/Reflection groups	74.2%	14.8%	13.2%	10.2%	126–28
Social organization	63.2%	11.3%	16.2%	6.3%	128–33
Religious organization	59.0%	20.5%	7.7%	2.6%	38–39
Sacramentals	57.5%	27.5%	25.0%	5.0%	40
Culturals	83.0%	41.5%	25.8%	8.6%	93–94

Source: Fleet, 1987b.

political force. They might have helped to develop a progressive democratic alternative around which popular-sector Peruvians could have rallied. "Christian" elements in the Izquierda Unida (e.g., Rolando Ames, Manuel Piqueras, Henry Pease, and Javier Iguíniz) favored such a position. But the Catholic activists who were their natural constituents and supporters were of little help to them.[63]

In some cases, they resisted the urgings of their priests or pastoral agents to become politically involved. In others, they joined and then left political organizations frightened or disillusioned by their experiences. During the early stages of the García government, large numbers of progressive Catholic leaders abandoned the MIR, the PCR, the UDP[64] and other leftist parties to become "independent" (i.e., non-affiliated) members of the IU; others left the alliance entirely, unhappy, it seems, with its political opportunism and with the growth in some circles of support for armed struggle. As one of them put it in a 1987 interview,

> There was a moment, it must have been at the end of 1985, in which many Christians [at intermediate and top leadership levels] abandoned the left. Not all, but many resigned from their parties, and regrouped as independents. . . . I think it was a sort of tiredness. I left a little earlier, because I could not take it any more, no? I could not take the absolute immorality of political calculations and the pursuit of votes. [I could not take] the presence of *caudillos* in Peruvian political life, which is a problem of the left as well as the right. [I could not take] the development of interest in dialogue with Sendero and in taking up the armed struggle.[65]

Another Catholic argued that Church authorities had "disarmed" Christians as far as social and political action was concerned. He was referring to the hierarchy's condemnation of collaboration with Marxists, which he believed kept many middle-class Catholics from embracing the left, and thereby weakened it politically:

> the response that we have to give is political and not just social, and it [the Church's prohibitions] therefore influences those who are most susceptible to intellectual blackmail, which are the middle sectors. And so it has stopped right there, and the Church is unfortunately responsible for many people who could be politically and social useful to the country not being so, because of that message.[66]

Local-level Catholic activists that we interviewed thus had a mixed impact on the consolidation of democracy in Peru during the García presidency. Their moderate attitudes and aspirations were helpful in developing the kind of grassroots participation on which democratic development depended. But their limited involvement in partisan political activity made it easier for less conciliatory forces to prevail within the Izquierda Unida. Unable to pull together, the IU split into rival (radical and moderate) factions, leaving a vacuum that was filled by forces from the right.

The Fujimori Government (1990–95)

The ease with which President Alberto Fujimori assumed dictatorial powers in April 1992 sharply underscored the failure of the consolidation process begun in 1980. Elites and nonelites alike apparently agreed with Fujimori that neither the Congress nor the judiciary were "working" properly, and that he needed virtually unchallengeable authority to resolve the country's economic and security problems. A sharp reduction in the rate of inflation and the capture of top guerrilla leaders has bolstered Fujimori's political standing since then. A strong presidential constitution permitting the president's reelection was approved in November 1993, although by a narrower margin than most observers predicted.[67] In April 1995 Fujimori was elected to a second term with more than 60 percent of the vote, while most of the country's traditional parties failed to get the minimal 5 percent to retain their legal status.

The Church has played a marginal role in political developments under Fujimori. The bishops remained focused on nonpolitical matters through most of 1991, and when they began to address socioeconomic and political matters, their potential for impact was greatly reduced by changes in the political context. The efforts of individual bishops, and of local-level priests and pastoral workers, had some impact within their respective jurisdictions but were of limited national impact. In fact, as if to underscore the Church's diminished influence, Fujimori has recently launched an aggressive campaign on behalf of sterilization and other forms of birth control opposed by Catholic teaching.

The bishops

Church-state relations under Fujimori got off to a poor start over the alleged influence of Protestants in the political newcomer's campaign organization. Fujimori, who emerged from political obscurity to defeat novelist Mario Vargas Llosa in a runoff in June 1990, was himself a

Catholic. But the fact that evangelical Protestants held important positions in his Cambio 90 movement, and were actively promoting his candidacy, was of considerable concern to bishops already alarmed at the recent growth enjoyed by Protestant denominations in Peru.

Although Protestants constituted only 3 percent of the Peruvian population in the mid-1980s, their growth rate of 470 percent from 1960 to 1985 was one of the highest in Latin America. If that rate were sustained through the year 2010, Protestants would then comprise 14 percent of the national population.[68] Pentecostal Protestants and other fundamentalist denominations (the Mormons, the Jehovah's Witnesses, the Seventh Day Adventists) accounted for the great bulk of this increase. All were engaged in aggressive face-to-face proselytizing efforts in Lima's *pueblos jóvenes* and were most successful among uprooted migrants from the countryside, to whom they offered strong community ties, discipline for their personal lives, and some economic and social assistance. The millions of Peruvians who have migrated to Lima and other coastal cities since the 1960s, together with inadequate numbers of priests, sisters, and lay catechists, have left many nominal Catholics "unchurched," and open to proselytization by evangelical Protestant denominations.[69]

One of the bishops alarmed by the rise of Protestant "sects" in recent years was Msgr. Augusto Vargas Alzamora, the conservative Jesuit who succeeded Cardinal Landázuri as archbishop of Lima in January, 1990.[70] Vargas allowed himself to be drawn into the presidential campaign by publicly accusing evangelical groups of distributing anti-Catholic propaganda between the first and second rounds. In doing so, and in then organizing a public reparation in the form of an off-season procession in honor of the Lord of Miracles, he created the impression that the Church opposed Fujimori and supported Vargas Llosa.[71]

A week before the runoff, the new archbishop publicly stated that the Church could coexist with an agnostic (as Vargas Llosa described himself during the campaign), as distinct from an atheist (the term he had used to characterize the writer's beliefs before the campaign). Vargas Alzamora's handling of the affair provoked comments and criticisms from both Catholic and non-Catholic circles. In the end, far fewer of Lima's Catholics turned out for the procession (to reject "Protestant aggression") than he had hoped, and he actually may have helped Fujimori to defeat Vargas Llosa in the runoff by provoking an anticlerical backlash.

Of Japanese descent,[72] Fujimori was seen as a "popular" alternative to the patrician Vargas Llosa and his orthodox ("shock") approach to the country's economic ills. However, once in office, and following talks with Peru's creditor banks in New York, Fujimori was converted to neoliberal economics. On "Black Thursday," August 9, 1990, he authorized price increases and an austerity program as far-reaching as anything Vargas Llosa had proposed.[73] The bishops complained that these measures would impose additional sacrifices on those that could least afford them, but did not press matters.[74]

In fact, hoping to improve relations with the new government, the bishops agreed to participate in the Programa de Emergencia Social (PES), a program in which religious and private development groups would help to distribute government relief funds to poor neighborhoods and villages that were suffering the effects of Fujimori's shock policies. Its involvement made it appear that the Church accepted the government's plan and its attendant social costs. This, and the fact that only a fraction of the originally committed funds were actually released ($100 million of $515 million), caused the bishops to withdraw from the program in February 1991 and to channel assistance (that they received from other sources) through their own organizations (Caritas, CEAS, etc.).

Other, noneconomic issues were distracting the bishops' attention as well. One of these was the announcement, in October 1990, of a government program providing free family-planning advice and products to all citizens wishing them.[75] Another, introduced the following January, was the proposed decriminalization of abortion. The bishops strongly condemned both initiatives. They opposed "artificial" means of birth control, arguing that couples had to remain "open to the possibility of new life," and that only "natural methods" of contraception (e.g., abstinence during fertile periods) could be used. In rejecting decriminalization of abortion, on the other hand, they invoked both the 1979 Constitution, which gave unborn children the rights of living persons, and the Civil Code, according to which human life began with conception.

At the time, growing numbers of Peruvian women were asking for access to birth control and the legalization of abortion under some circumstances. But for the Peruvian bishops, conservatives and progressives alike, immutable moral principles were at stake, and could not be bent to accommodate popular sentiment. Progressive bishops believed that population control programs were assaults on the poor and a means of avoiding needed structural changes. They also believed that poverty caused overpopulation, not vice-versa.[76] But in these cases, the bishops were determined to defend the poor (from population control) whether the poor wished them to or not. They were saying, in effect, that the issues involved were not matters for public opinion or democratic process. And, of course, their own unwillingness to submit to democratic constraints would make it difficult for them to criticize others (e.g., Fujimori) for refusing to do so either.

Fujimori's austerity policies extended the recession that had begun under García, imposing serious hardships on lower-income-earners. The GDP shrunk by 14.4 percent in 1990, rose by 2.0 percent in 1991, and then fell again by 2.1 percent in 1992. Open unemployment and underemployment rose, as did the percentage of people living below the poverty line (17 percent to 41 percent), while only 5 percent of all wage earners, as against 53 percent at the start of the period, earned enough to purchase the standard food package for a month. Between 1990 and 1991, the infant mortality rate went from 9 to 12 percent, and the school dropout rate went up as well.

People remained amazingly supportive of the government despite these conditions. Opposition parties were in disarray, and attempts to organize antigovernment strikes and demonstrations failed to attract significant numbers of workers or other popular-sector forces. The forceful image that Fujimori projected was partly responsible for his appeal. So too was his success in getting inflation under control.

Church-state relations were dominated by issues arising in connection with the National Peace Commission, which was attempting to develop a plan for ending the guerrilla war peacefully. García had reactivated it during the last month of his presidency, and the Church agreed to participate when it began meeting in mid-1991. The preliminary sessions were chaired by the Church's representative, Msgr. José Dammert, and focused on which groups should be invited to participate. Subsequently, Dammert was offered the council's presidency. He was willing to accept but ended up declining when the Episcopal Conference's Permanent Council and Assembly refused to endorse the project.

The commission itself never got off the ground because of conflict over its makeup. Some members insisted on vetoing others. Had the Church persisted, these problems might have been overcome and a consensus reached on how to fight the guerrilla war more humanely and effectively. Ending it entirely would require the agreement of both the Shining Path and the Peruvian military, neither of which was willing to lay down its arms. And it was also hard to imagine Msgr. Dammert or anyone else brokering a peace formula that conservative right-wing forces (Vargas Llosa's Movimiento Libertad) and revolutionary Marxist-Leninists (like the PUM, the PCP-BR, or the PCP-PR)[77] would have embraced. In any event, Dammert was denied the chance and the war continued on its destructive course.

Individual Church leaders were publicly critical of Fujimori's economic policies during this period. The conference itself said nothing, but began to address social issues in the second half of 1991, once progressive Archbishop José Dammert of Cajamarca replaced Msgr. Durand as conference president. Dissatisfaction with Durand created a vacuum into which Dammert quickly moved. Apparently progressives and conservatives alike came to the conclusion that the country's political elites were not going to provide the necessary leadership, and that the Church would have to play a more active role.[78]

Under Dammert's leadership, Church authorities paid more attention to social and political issues. For one, they became more explicitly critical of Sendero Luminoso. The organization began targeting civilians in 1987, quickly replacing the army as the principal cause of noncombatant deaths. It also went after priests and nuns (previously considered untouchable), trying to eliminate those with a following among the rural and urban poor.[79] Finally, with political institutions and parties in evident disarray, Church authorities saw the need to show Catholics and Peruvians generally that they were not alone and that Sendero could be challenged.

This more aggressive attitude was also apparent in their pastoral letter, *Paz en la Tierra*, issued in December 1991. It was drafted with the help of lay and clerical advisors who had worked for the conference in the 1960s and 1970s. It criticized the government's disregard for citizen rights, called for policies that took the burden of economic recovery from the poor, spoke against plans to put health care and education on a pay-for-service basis, and defended the integrity and autonomy of the *rondas* and other popular organizations that the government was trying to discredit or control.[80]

In late 1991, conflict broke out between Fujimori and the Congress. At issue were presidential proposals eliminating subsidies and tax exemptions for farmers and granting the military carte blanche (in treating "suspected subversives") in emergency zones. In December, Fujimori issued multiple-decree laws dealing with these and other matters, but the Congress refused to ratify them within the thirty-day period. Fujimori counterattacked, calling for the restructuring of the "do nothing" Congress and the blatantly corrupt judicial system. In early April 1992 he declared the two branches of government "dissolved," and, backed by the military, he assumed "full emergency powers."

In effect, Fujimori overthrew the other branches of government, and proceeded to rule by decree, unchecked by either legislative approval or independent media. The state of emergency that he imposed also restricted the CEAS human rights programs. It could no longer distribute pamphlets outlining the rights of detainees, and its lawyers were prevented from representing more than a single person arrested as a "suspected terrorist."

The outside world unanimously condemned the *autogolpe*, but most Peruvians supported it, and virtually no one was willing to defend the "democracy" that it overthrew.[81] The bishops were uncertain how to respond, and consulted with Cardinal Landázuri. He urged them to condemn the move forthrightly, arguing that "if they did not insist on the restoration of democracy, they would lose the high regard [*prestigio*] in which they were held by most Peruvians." Even he had mixed feelings, however. He granted that the country had never had a "true democracy," that the Congress talked a lot but did very little, and that the judicial branch of government was "completely immoral." He also acknowledged that the Peruvian people "were not fully prepared for democratic citizenship," that their "cultural level was low," and that "many did not possess the proper standards [*conceptos rectos*]" with which to discern political matters.[82]

In May, the conference's Permanent Council issued another lengthy document, *Tarea de Todos*. It lamented the breakdown of democratic institutions and urged the government to begin meeting with opposition groups to discuss their rapid restoration.[83] It also addressed the country's social problems, stressing their institutional and socioeconomic, as well as moral, roots, and underscoring the connection between social justice and political stability. The bishops again urged that the burden of economic recovery be lifted from the shoulders of the popular-sector Peruvians, and that the government rethink its approach to the problem of violence as well, as "the lack of coordination and coherence [on the part of security forces] had contributed greatly to a sense of disillusionment and helplessness on the part of the population as a whole."[84]

Finally, in a spirit of self-criticism that was rare for the Catholic hierarchy, the bishops acknowledged their own quota of responsibility for the current state of affairs. Admitting that they should have said and done more in the past, they wrote: "The destructive conflicts and confrontations and other negative situations that we lament today would not exist, or would be notably more manageable, if we had acted in time and dialogued [with other groups] with the valor required by the circumstances."[85]

Their honesty and their appeals had little impact on either the government or Peruvian public opinion. Fujimori's disdain for the judicial and legislative branches of government had struck a chord with the general public. His demands for unrestricted powers to deal with Sendero made sense to people exhausted by more than a decade of violence and terror. The efficiency with which he seemed to be proceeding was a welcome departure from the ineptitude of the Belaúnde and García governments. In the words of one observer, "The theme of democracy and respect for human rights is not that important. The No. 1 desire is for peace, that Peruvians can go out in the street and not fear that they will be killed by a bomb."[86]

The capture of MRTA leader Victor Polay in June 1992, and of Shining Path headman Abimael Guzmán the following September, vindicated Fujimori's logic in extraordinarily timely fashion. Senderistas had placed bombs in public buildings in Lima and other cities in May, June, and July, killing large numbers of people and demoralizing the general population. But when Polay and then Guzmán were taken into custody, thanks to efficient, if somewhat fortuitous, police work, the government's image (as determined, forceful, and effective) was greatly enhanced and Fujimori's critics found it difficult to challenge his policies or methods. In short, his government's performance seeded to prove his point: authoritarian rule was precisely what the country needed in its present circumstances.

Fujimori's successes and his favorable standing in public opinion polls blunted the impact of the Church's new, more critical orientation. The bishops persisted, but they had little impact on national political sentiment or dynamics. The absence of a congenial and broadly based social or political movement that embodied the values and concerns which they espoused left their Catholic followers without a political alternative.

Another sign of the bishops' declining public influence was their marginal influence on the draft of the new constitution to be submitted to a referendum in late 1993. The bishops vigorously opposed the document's provision of the death penalty for terrorists, and Bishop Bambarén led a public campaign on behalf of amendments prohibiting abortion and sterilization. He also organized the collection of 50,000 signatures in defense of "natural" forms of birth control and against the use of condoms. The constitution approved in November 1993 authorized the death penalty for terrorists. It was silent on the issues of abortion and sterilization, but left the door open for future initiatives in these areas.[87]

During 1992 and 1993, the bishops' political impact was confined to the Church itself, and to small-scale activities directed or advised by local pastoral agents. The resurgent influence of progressive and moderately progressive bishops had little effect on developments in individual dioceses. Authorities in Lima and fifteen other dioceses, for example, continued to ban CEAS workshops (on topics like national reality, human rights, and the social teaching of the Church), to reject its educational materials as "too radical" and "too political," and to invite it only if and when appreciable material benefits were involved.[88] In Lima, "progressive" materials occasionally found their way into pastoral activities thanks to the initiatives of individual priests, but official programs (like the *semanas sociales*) used materials developed by Msgr. Vargas's own advisors and consultants.

Archbishop Vargas's conservatism was also apparent in his handling of Father Gutiérrez. Shortly after his appointment in 1990, he ordered Gutiérrez to publicly clarify his thinking and to submit future work for review prior to publication. Gutiérrez dutifully complied. In 1991, in a preface to the twentieth-anniversary edition of *A Theology of Liberation*, he acknowledged "unfortunate ambiguities" in his earlier work and characterized his current thinking as "more mature" and clearer on several matters. Vargas acknowledged these "rectifications," expressing regret, however, that the original text was nonetheless intact.[89]

Gutiérrez's submission to his bishop was typical of his desire to avoid confrontation with Church authorities. A founding father of ONIS, he quickly stepped aside, leaving others to direct its efforts and speak for it publicly. An astute political observer, he has long defended the need for lay Catholic activists to act and speak for themselves in political matters. A thinker who has always stressed the interplay of religious and political life, his own writing has taken an increasingly spiritual direction since 1980. Without yielding intellectual ground or making more than marginal concessions, Gutiérrez has tried to minimize conflict, preferring to burrow or fight from within rather than publicly.

His cautiousness has not won over his critics, who see it as cleverness. But it has frustrated some of his admirers. A Peruvian writer who thinks that the priest has been too careful complained to us of Gutiérrez's unwillingness to endorse a political declaration that other priests (and other Christians and progressives) had signed:

> That really got me. I have said to him more than once: "Look, Gustavo, stop screwing around. It's not any big deal. You can sign and no one is going to accuse you of being the Black Pope." [But] he is very careful. You have got to take what he says with a grain of salt, because he's just not . . . he is sincere, but he does not say everything he is thinking. He's sincere, what he says is solid, but he is wily. But that's his problem, right? He has had some successes. He has managed to get the Pope to say that liberation theology contains many positive elements. That's a triumph. He has gotten them to treat him much better than they treated Boff. But. . . ."[90]

Gutiérrez continues to direct the Bartolomé de las Casas Institute, which provides advice and educational materials to parishes, communities, and popular organizations. In addition, since the late 1980s, he has been active in the Fraternidad Sacerdotal de Lima, an organization that seeks to promote a better sense of identity and collegiality among archdiocesan (secular) priests from Lima, particularly those ordained in recent years.[91]

Church-state relations reached their lowest level in years in mid-1995, as Fujimori completed his first term and began his second. Fujimori struck the first blow in June, pushing an amnesty bill through the Congress (literally in the evening of its last day in session) that exonerated all military personnel implicated in war-related human rights cases. Among these were the famous Barrios Altos murders in 1991 and the National Education University (La Cantuta) students and their teacher, whose bodies were discovered in 1992. Convictions (of lower-ranking personnel) had been won largely on the strength of information leaked by dissident junior army officers, and had been applauded by the bishops as necessary and appropriate. The amnesty, on the other hand, was thought to be the work of senior officers who pressured or forced Fujimori into obtaining blanket amnesty, thereby protecting themselves from similar charges in the future.[92]

The second blow came as the president began his second term by proposing a population control program that included easier access to birth control products and sterilization. In an obvious reference to Peru's Catholic bishops, he vowed to overcome the opposition of certain "sacred cows."[93] When the Congress approved the bill in September, Archbishop Vargas answered in kind, warning of "the power of darkness" and decrying the "cowardly positions" of Catholic legislators. Fujimori upped the ante, pleading his cause to a world audience at the International Women's Conference in Beijing, and in a *New York Times* interview in which he urged other Latin American governments to unite "to break the Vatican's influence in terms of birth control and family planning."[94] Observers attributed the unusual acrimony of the exchange to the issue's importance to each side,[95] and to the mutual antipathy of the two leaders, neither of whom liked having his thinking challenged by anyone.

The local churches

The bishops' new, more aggressive stance had little effect on either Fujimori or Peruvian public opinion. Nor were there parties or persons willing or able to defend the Church's concerns in the political arena. The encouragement that some bishops gave to priests, nuns, and others working with popular organizations had an indirect and limited impact at the local level. But the lack of broader support, and the absence of a political vehicle into which these forces could feed, prevented them from mounting a challenge to Fujimori's government.

As in previous years, the support that some bishops gave to the *rondas* and other popular social organizations had mixed political effects. Fujimori's government was able to persuade some *rondas* to accept training and arms and to cooperate with military authorities. Others refused. Having lived under the control of economic and political elites their entire lives, their members enjoyed the sense of empowerment that these organizations provided. As earlier, however, they were viewed hostilely by local elites and authorities, on whose good will the strengthening of democratic institutions depended in the short run.

The *rondas* also antagonized Sendero Luminoso. The Southern Andean region, and places like Ciudad de Dios and Villa El Salvador on Lima's southern edge, became arenas of struggle between Sendero and local organizations led by Catholic activists. As conflicts arose, and local resistance efforts grew stronger, Sendero targeted local (and largely Catholic) political activists for assassination. In early 1992, the vice-mayor of Villa El Salvador, Maria Elena Moyano, was killed, and an attempt on the life of the mayor, Michel Ascueta, narrowly miscarried.[96]

Like most other segments of the Peruvian electorate, Catholic activists grew weary of the factionalism of the United Left, and were drawn to alternative political camps in the late 1980s and early 1990s. We cannot trace their electoral behavior directly, but in the 1989 municipal elections many abandoned the left in favor of Lima television personality Ricardo Belmont, a political independent.[97] In the first round of the 1990 presidential elections, voters in these districts favored Fujimori (44.7 percent) and Vargas Llosa (22.6 percent) over both leftist candidates, who together won but 14.7 percent of the votes.[98] In December 1993 they voted solidly (over 60 percent) in favor of the Constitution drafted by the pro-Fujimori Constituent Assembly.[99]

Potential for Future Dissent among Catholic Activists

The Church activists we interviewed were moderate progressives in the political context of that time (1987). The views of organizational Catholics were less radical than those of other Catholics but in no sense conservative or reactionary. Indeed, they were generally favorable to the consolidation of democracy (i.e., more tolerant, more willing to accommodate class differences, etc.).

Our survey also probed religious attitudes. It asked about the mission of Jesus (whether he died for being a political dissident, or for moral and spiritual reasons) and that of the Church (whether it should focus primarily on people's personal spiritual needs or should help the poor materially or organizationally). They also asked if respondents would obey Church teachings even if they disagreed, and whether or not the poor had an obligation to love the rich. Formal Church teaching holds that Jesus died to free men and women from sin, not because he was a political dissident. The Church also defines its mission as helping persons with their moral and spiritual, not their social, problems. And it has always stressed that faithful Catholics must follow

Church teachings, not their own consciences, whenever the two conflict. Finally, the Church emphasizes the universal obligation of Christian love of one's enemies, citing the precedent set by Christ himself. As table 7.8 indicates, a significant number of Catholic activists we interviewed dissented from official Church teachings, although they were less likely to do so than other Catholics.

Among the Catholics we surveyed, those who were less involved with the Church (sacramental and cultural Catholics) were the furthest removed from official teaching on these matters. More importantly, however, a relatively large number of organizational Catholics (religious organizationals included) dissented from formal Church teachings. Over two-fifths of all organizationals showed a preference for a socially engaged Church, and close to a quarter wanted a Church that preached solidarity with the poor rather than spiritual reconciliation between rich and poor.[100] Between one-third to two-fifths supported following one's conscience when it conflicted with Church teaching. And even a sizable number viewed Jesus' death in political rather than religious terms.[101] In general, sacramental Catholics exhibited a higher degree of dissidence on these questions. It is clear, therefore, that those with the most contact with the Church are not docile laypeople at the beck and call of the hierarchy for whatever agenda the bishops choose to pursue within or beyond the Church.

The importance that popular-sector Catholic activists and sacramentals place on social issues and personal choice may be the result of general secularization, the progressive religious and social positions of their pastors, or both. For whatever reason, many of those whom we interviewed have internalized the attitudes of Vatican II and Medellín regarding a Church in solidarity with the poor and in which laypersons make

TABLE 7.8

Dissident Religious Views of Selected Lima Catholics, 1987.

	Christ a Political Revolutionary	Church's Economic Role	Conscience over Authorities	Poor Need Not Love the Rich	Range of *N*
CEBs/Reflection groups	24.0%	43.2%	43.0%	23.6%	125–28
Social organization	34.6%	40.3%	36.4%	22.5%	127–33
Religious organization	18.9%	41.4%	35.9%	38.4%	37–39
Sacramentals	40.0%	55.0%	38.4%	48.7%	35–40
Culturals	41.3%	69.2%	37–39	57.6%	91–94

Source: Fleet, 1987b.

responsible decisions on their own. Their independent-mindedness could lead to tensions within the Church. The hierarchy's conservative approach to socioeconomic problems since the early 1980s, its opposition to CEAS programs, and its lack of support for Church-sponsored human rights commissions in some instances, all ran counter to the views of local Catholic activists. Were they to persist in such positions during the 1990s, particularly if socioeconomic conditions do not improve, the bishops may well encounter dissent and resistance.

In fact, differences between rank-and-file Catholics and their bishops may be one of the reasons why the latter have not been able to deter Fujimori's population control initiatives. Public opinion polls indicate that close to 80 percent of all Peruvians favor artificial means of birth control.[102] Had Catholic activists and sacramentals been more in line with their bishops, or more willing to follow their lead (and not their own consciences), the voice of the bishops might have carried more weight. However hostile Fujimori may be to current Church leaders, he would have to be more deferential in matters on which most Catholics had common views.

The Catholic activists in our Peruvian sample are thus quite different from the loyal cadres trained in Catholic Action circles in the 1930s, 1940s, and 1950s. The former were frontline troops in the defense of Catholic values and interests in society. They went along with the bishops in fighting divorce reform in the 1930s and in acquiescing to authoritarian regimes accepted by the bishops in the 1940s and 1950s. Today's hierarchy cannot count on Catholic activists to defend Church policies in the public arena. Contemporary lay Catholics are more independent-minded, committed to the reforms of Vatican II and increasingly supportive of liberal and democratic values, including tolerance for moral and religious differences. Efforts by the bishops to mobilize them to oppose the access of women to birth control or abortion, to challenge the rights and freedoms of evangelical Protestants, or to abandon, as Church members, the pursuit of social justice, are likely to generate disappointing results.

In fact, they might drive some Catholic activists and sacramental Catholics away. In either case, the Church's ability to compete with evangelical Protestants, and its influence on society and politics through the actions and choices of practicing lay Catholics, would further diminish. Indeed, because they need Catholic activists to give the

Church a presence in districts and areas that are without priests and nuns, even conservative bishops are likely to be more tolerant in the years to come.

Conclusions

Buffeted by continuous crisis during the 1980s and early 1990s, Peru's democratic institutions never took root. The country's social and political forces continued to value their own interests and objectives above democratic principles or practices. When Fujimori suspended the Congress and assumed emergency powers in April 1992, most Peruvians supported him. Except for the opposition political elites who were the direct objects of his assault, only the Peruvian bishops spoke out in defense of the country's democratic institutions.

The bishops lamented the dismantling of those institutions. In line with recent Church pronouncements on the subject, they stressed the congruence between democratic and gospel values. They took their stance after the fact, however. From 1980 through mid-1991, they paid scant attention to the country's developing political crisis. They bemoaned the socioeconomic problems that people were facing, and helped those mired in poverty or threatened by violence. But they never offered guidelines for addressing the structural factors underlying socioeconomic problems, and they never proposed criteria for choosing among alternative policies. Leaving such matters to politicians and public authorities, they focused instead on ecclesial affairs and personal moral issues.

Their retreat was a function of Cardinal Landázuri's declining influence and the growing number of conservative bishops appointed after 1978. Social Christians had been a majority in the Chilean Bishops' Conference from the late 1950s on, but in Peru, only Landázuri and two or three other bishops embraced this perspective. More conservative bishops deferred to him into the late 1970s, but with the election of a new pope and the appointment of additional conservatives, his influence waned. By the time (1988) he stepped down as conference president, the Peruvian bishops were focusing almost exclusively on personal moral issues (divorce, abortion, birth control), questions of religious orthodoxy, and the growth of Pentecostal and other fundamentalist denominations.

This more conservative conference was not always tactful or understanding in dealing with these matters. Its positions alienated some in the Church while failing to convince or effectively pressure policymakers. In fact, the uncompromising spirit with which the bishops addressed most issues (leaving little room for differences of opinion) added to the divisiveness in Peruvian society at a time when efforts at consensus-building were needed. In effect, the Peruvian bishops reverted to positions that they held in the 1940s and 1950s. At that time, most bishops were political and ecclesiological conservatives. They focused largely on personal moral and spiritual issues, leaving social and political matters to lay Catholics and political elites, whether democratic or not. In the next several decades, however, as socioeconomic and political conditions worsened for most Peruvians, the Church broke with tradition and placed its moral authority and mediating assistance at the service of the poor and less fortunate.

A similar revival took place in the late 1980s and early 1990s, following another period of social and political retreat. The bishops came to oppose Sendero Luminoso directly and explicitly, denouncing its terrorist methods and appealing for tolerance, accommodation of differences, and other democratic values. When the economy collapsed under García, they called for an accord among major groups. And when Fujimori applied the same neoliberal policies against which he had campaigned, they criticized the imposition of additional sacrifices on the poor. At this juncture, however, they could not agree on criteria for choosing among alternative solutions to these problems, nor did they promote or offer to mediate dialogues or possible compromises among contending political factions.[103]

Moreover, by the time the bishops began addressing socioeconomic and political issues again, it was too late. The moment of opportunity had passed. The vacuum that had developed was filled by Fujimori, and there was no political space in which to rally people, no secular carrier (a Christian reform party or moderate, centrist coalition, for example) to which these people could turn.

During the 1980s and early 1990s, Church people were socially active at the local level, although more in some areas than in others. CEAS staffers, priests, nuns, and lay pastoral agents came to the aid of workers, slum dwellers, peasant groups, and religious communities beseiged by guerrilla violence, counterinsurgency measures, and grinding poverty. They helped people survive difficult circumstances, and encouraged them, not always successfully, to defend themselves politically and to play more active roles in community affairs.

Their efforts were most visible and successful when they were endorsed and materially supported by diocesan and archdiocesan authorities, as in the Southern Andean region, the departments of Cajamarca and Chimbote, and the *pueblos jóvenes* and popular neighborhoods of metropolitan Lima. Church support helped the *rondas* and other grassroots organizations to survive the efforts of local political and military elites to neutralize or eliminate them. Over the longer term, these popular organizations might have made an important contribution to the development and strengthening of Peruvian democracy. At the time, however, their impact on consolidation was negative. They generated political support for the United Left, thereby heightening alarm and hostility among traditional political elites (e.g., APRA, rightist parties, and military officers).

In fact, the obstacles to successful consolidation of democracy were far greater in Peru than in Chile, and made it harder for Church leaders (at either the national or local level) to positively affect the outcome. The Peruvian economy failed to generate or sustain growth and to attend even minimally to the employment, housing, and health-care needs of popular-sector elements. The latter became more and more alienated from the economy and the political order, as did financial, commercial, and industrial elites, on the other side. Secondly, a sizeable part of the political left in Peru remained committed to a radical socialist agenda, embracing democracy only to the extent that it was useful in advancing that project. In this respect, the Peruvian left was closer to the Chilean left of the Allende period than to the chastened and moderate Chilean left of later years. Thirdly, the governments of this period proved unable or unwilling to control the Peruvian military, whose human rights violations escalated in the middle of the consolidation. In contrast, the worst violations of human rights by the Chilean military were over before consolidation began. Seeking a measure of justice for victims of past abuses was a formidable task, but less difficult than curbing ongoing violations under virtual wartime conditions. Finally, the Peruvian political right, unlike its Chilean counterpart, wanted nothing to do with even the moderate left. As a result, there was little interest in Peru in the kind of moderately progressive alternative around which the Church might have helped to rally political forces, as it had in Chile. In fact, with centrist projects (Belaúnde's moderate neoliberal experiment and García's radical populism) having been tried and found wanting during the 1980s, political forces were moving toward the political extremes.

These obstacles might have been insurmountable whatever the Peruvian bishops said or did. There were things that the hierarchy could have done that would have favored consolidation, however. The bishops could have pushed more forcefully for dialogue and compromise among the leaders of contending parties. They could have done more to bring grassroots activists, party leaders, and public officials together to discuss common concerns. They could have insisted that "collective social actors" (unions, neighborhood associations, etc.)[104] be recognized as the legitimate representatives of popular groups, and that political elites deal with rather than denounce them. In short, they could have been more active and resourceful as moral tutors for democracy and as mediators helping to strengthen its institutions and processes.

Unfortunately, only a handful of Peruvian bishops were willing to consider such functions. Most embraced democracy abstractly, but found politics distasteful. Most supported limited government, due process, human rights, and civil liberties, but questioned the ability of ordinary Peruvians to judge political matters thoughtfully. Few realized that unless people strengthened themselves politically, they would continue to be manipulated and demoralized. The Chilean bishops (Cardinal Silva and the moderate and progressive majority that supported him) learned this early on, and used the Vicaría de Solidaridad and other Church resources to defend people's interests and to prepare them to act on their own once democratic institutions were restored. But the Peruvian bishops, most of whom viewed politics as a distraction from, if not a corruption of, religious activity, did not.

Their unwillingness to mediate among contending forces or to take on socioeconomic and political issues had a stifling effect on Catholic activists at the local level. The positions and practices of party leaders had negative effects as well. But without the hierarchy's moral leadership and encouragement, Catholics in local, church-sponsored organizations were less inclined to political involvement than they would have been otherwise. And, thus, yet another of the Church's resources, the leavening impact of an aroused and committed laity, was further diluted.

Given their relatively modest numbers, it is doubtful they could have broken the stalemate that led to Fujimori's election and subsequent *autogolpe*. However, the limited participation of local-level Catholics was more damaging to the consolidation process in Peru than in Chile. In Chile, moderating currents were dominant in most political parties, and progress towards consolidation was possible without high levels of Catholic involvement. In Peru, on the other hand, moderating influences were in shorter supply, and Catholic disinterest strengthened polarizing tendencies by default.

During the transition in Peru, grassroots Church groups played a significant role in opposing the Morales Bermúdez government. They did so without the support of higher Church authorities largely because coordination with others (at regional and national levels) was less crucial with a government that was willing to relinquish power. In the case of consolidation, however, the key factor was the capacity of democratic groups to work together at the national and local levels, and here the absence of Church encouragement and support was more costly.

It would be wrong, however, to blame the Church for the failure of Peruvian democracy. Its authorities could have been more supportive of political involvement and struggle, and broader Catholic participation might have strengthened the hand of moderate and democratic political forces. But the fact remains that significant portions (both upper- and lower-class elements) of the population had never trusted national institutions, and that, with the failure of recent neoliberal and populist reform experiments, the country's social and political forces were growing more, not less, polarized. In this context, the Church's own lack of resolve was less an obstacle than the sociopolitical environment's hostility to the moderately progressive perspectives and policies around which the Church might have rallied its people.

8

Conclusions

During the 1970s and 1980s, the Latin American Catholic Church's efforts to minister more directly and integrally to popular-sector Catholics won it acclaim as a resourceful and caring institution. They also helped it to win back many lapsed or fallen-away Catholics who had strayed from its ranks in previous decades. During this period, however, U.S.-based evangelical and Pentecostal Protestant churches, most of whose converts came from these same popular sectors, grew dramatically throughout the continent. The Catholic Church's pastoral reforms and social and political initiatives gave it renewed credibility and influence with many Latin Americans, but the new Protestant churches were more successful, it seems, in appealing to and accommodating their religious sensibilities.[1]

Moreover, the Church's reform efforts produced tensions and divisions within its own ranks. While generating greater vitality and commitment at all levels, they came at the expense of institutional authority and coherence, as increasing numbers of Catholics have begun making moral decisions based on the dictates of their own consciences. Catholics committed to liberation theology and/or politically active through their local Christian communities (usually in opposition to military governments) have run afoul of conservative authorities determined to distance the Church from "politics" and to reaffirm their institutional prerogatives. Pope John Paul II, members of the Roman Curia, and most of the bishops appointed to head vacant dioceses in Latin America since 1978, are leading this "restoration" movement.

The pope has replaced retiring bishops (many of them moderate progressives named by Paul VI) with theologically conservative successors. He has publicly criticized the "sectarian" tendencies of some Christian communities and has insisted on stricter compliance with official Church teachings, especially in the area of sexual morality. Finally, he has approved written warnings against liberation theology, the interrogation of several of its leading exponents, and the temporary silencing of Brazilian theologian (and now former priest) Leonardo Boff.

These moves have encouraged conservative elements in most Latin American Churches. Some bishops have moved liberation-oriented priests out of their popular-sector parishes, and have sent progressive foreign missionaries home. Some liberal seminaries have been closed, and the faculties and curricula of others, along with lay-leader training programs, have been restructured to emphasize prayer, biblical scholarship, Church history, canon law, and personal counseling.

The long-term effects of these efforts at retrenchment are not yet clear; nor are their implications for the support that the Church has given to social justice and democratic politics in Latin America over the last thirty years. Students of the Latin American Church have differed in their assessments of the phenomenon and its likely effects. Some expected the pullback to continue and the Church to renew its ties with socioeconomic and political elites.[2] A second potential scenario calls for an increasingly polarized Church moving toward de facto, if not formal, schism.[3] A third possibility is a period of adjustment in which the Church places renewed emphasis on its primary religious mission while continuing to address social issues through its teaching authority.[4]

We think that an adequate assessment of the meaning and impact of Vatican efforts at restoration requires a more comprehensive approach to religious and political change than is offered in any of these studies. In our view, authors identifying with the first and second scenarios give too much weight to the Church's domination by class or social forces, while those embracing the third credit it with excessive independence from them. We think that both societal and religious factors must be fully considered. More specifically, we argue that changes in the Church's conception of itself and its mission (its ecclesiology) affect its understanding of its interests and the means selected to defend them in particular social contexts. In effect, the strategies forged by Church leaders to protect religious interests in the face of contemporary political challenges have made it into a supporter of democratic processes. We also believe, however, that the Church remains wedded to perennial religious perspectives (e.g., belief in a common good that takes precedence over individual prerogatives and interests, and commitment to universal moral laws that are not subject to change in line with popular consensus) that are not always compatible with the tenets and requirements of liberal democracy.

Before reflecting further on the political impact of an at least partially "restored" Latin American Church, it might be appropriate to summarize what we have learned from the Chilean and Peruvian cases.

The Church and Chile's Transition to Democracy

The Catholic Church contributed significantly to Chile's transition to democracy at several junctures. Individual bishops and lay activists both played important roles. The hierarchy pressed for a peaceful and orderly transition, playing first a neutral, and then a mediatory, role. It was not of a single mind from the outset, however. A handful of progressive bishops favored persistent and forceful criticism of the military government. They wanted to show the young, the poor, and the oppressed that the Church was still "with them." Moderates, on the other hand, wanted to avoid breaking with the regime completely, to disavow those advocating "all means of struggle" against it, and to deemphasize political issues because of their divisive internal effects. Conservative elements, finally, either liked the Pinochet regime or preferred it to a government dominated by the left. Like the moderates, they wanted relations with authorities that included ceremonial and other regular contacts with government authorities.

For its part, however, the Pinochet regime treated the Church as an adversary, allowing, if not directing, security and paramilitary forces to attack its pastoral agents and programs. This hostility led the Bishops' Conference to close ranks. When pastoral workers were harassed, foreign priests were expelled, and individual bishops were attacked, moderate and conservative bishops joined with progressives in defense of perennial core interests.[5]

Had the regime been less heavy-handed, moderate and conservative bishops might have remained on the sidelines. But with the Church as a whole threatened, they could not. With secular political forces unable to function, they stepped forward to challenge the regime during the late 1970s and early 1980s. With the renewal of political activity in 1983, they helped to develop and rally support for an alternative to the regime.

Their most significant achievements on behalf of a peaceful transition were the National Accord of 1985 and the Social Pact of 1988. The first of these was rejected by General Pinochet but brought opposition parties together around a program for the postmilitary period with which many regime people and their supporters could live. In a similar vein, in early 1988, the bishops helped to broker an agreement between labor and business groups that persuaded the latter they could survive, and even prosper, in a Chile run by opponents of the regime that they had supported to that point.[6]

It is ironic, perhaps, that the hierarchy was most influential when it was least prophetic, when it worked behind the scenes on behalf of compromise instead of denouncing evils and abuses publicly. In this respect, the relatively conservative Francisco Fresno, who succeeded the more outspoken Raúl Silva as head of the Santiago archdiocese in 1983, may have been the ideal Church leader for Chile at the time. Silva was a hero in the eyes of the world and the Chilean people generally. But he could never have allayed or disarmed the fears of either religious or political conservatives to the extent that Fresno did.

Other elements of the Catholic community also played important roles in the transition process. *Practicing Catholic elites*, prominent in business circles and active in parties of the left, center, and right, advised the bishops, facilitated the dialogues hosted by Fresno and other bishops, and generally embodied the openness to negotiation and compromise that the bishops were calling for. It is difficult to imagine the hierarchy playing as effective a role as mediators, in fact, without the assistance and initiatives of these elites. The unwillingness of Catholic business and political leaders to heed the bishops' call for negotiation and dialogue in 1972 and 1973 clearly undercut the Church's potential role in that crisis. During the mid and late 1980s, on the other hand, they served as carriers for Catholic values of moderation and compromise, and had precisely the opposite effect.

For their part, *Catholic activists*[7] were an important part of the broader opposition movement. As members of local Christian communities, as participants in youth groups and in classes preparing them for the sacraments, or as volunteers in local, Church-sponsored meal programs or human rights defense groups, Catholic activists took part in marches, demonstrations, work stoppages, protests, and other anti-regime activities. Their ties to the Church made them valued allies and occasional sources of protection for other regime opponents. For many people, they "embodied" the Church in their respective parish or zone.

Our survey data do not permit us to characterize Chile's Catholic activists or larger Catholic population generally.[8] It is conceivable that they included a substantial number of religious or political radicals who either supported or were active in groups like the Sebastian Acevedo Movement against Torture or the Santiago based Coordinadora de Comunidades Cristianas. Among the local-level Chilean Catholics we interviewed (many of whom came from Santiago's most progressive Catholic parishes or most militant neighborhoods), however, most *organizational* and *sacramental Catholics* were quite moderate. In particular, they displayed attitudes supportive of the transition to democracy, i.e., a willingness to work with leftists in a reconstituted democracy, a reluctance to consider violence an appropriate political method, and a preference for a regime that respected the political rights of all.

Among Catholics living in the parishes that we studied in greater depth, the religious and political views of Christian community members and other *organizational* Catholics were more progressive than their *sacramental* or *cultural* counterparts in two of the three cases. However, the most active and progressive among them came from poorer, more highly politicized, and traditionally combative areas, to which Church authorities generally assigned progressive priests and nuns, and where they generally received a favorable response. Other factors might have affected the attitudes and behavior of Catholics in other areas, but most of those living in Dolores (Quinta Normal), Parrales (La Granja), and Cristo Liberador (Villa Francia) were influenced by convergent religious and secular forces.

Finally, *cultural Catholics* played important roles in Chile's transition at both national and local levels. Most of the leaders and militants of opposition social movements and political parties came from their ranks. The assistance that many of these men and women received from Church agencies and Church-sponsored organizations during the years of military rule left them with a new appreciation for the values of freedom, human rights and dignity, community, and solidarity. Their contact with the Church left them more favorably disposed to its perspective on social and political matters, and to the need for moderation and reconciliation at this juncture in Chilean history. Independent and entirely secular intellectual currents were pushing them in the same direction, but the Church provided important reinforcement and legitimation.

Their experience with the Church during these years made it easier for cultural Catholics to work with Catholic political elites. The latter were not just representatives of rival parties (the Christian Democrats, for example); they were fellow Catholics. They were the personal friends of a bishop, or political advisors or liaisons of a Christian community, to whom one could extend trust or make concessions with greater confidence. They were people with whom they had shared joys and sorrows in difficult times, with whom they now shared much stronger bonds.

In sum, in the transition process, Chilean Catholics at various levels of the Church were more than the secondary allies of which Drake and Jaksic speak.[9] The support for democracy offered by the Church did not cease in 1982. It resurfaced at key junctures in the mid and late 1980s, helping to bridge the gaps among opposition groups and to ease the anxieties of former regime supporters beginning to look beyond Pinochet. And it remained significant through the Social Pact and the plebiscite of 1988.

The Church and Chile's Consolidation Process

Conversely, however, the Church's impact on consolidation has been modest and at times obstructive. The armed forces and civilian rightists continue to oppose elimination of military autonomy and other antidemocratic features left over from the transition, and neither the Catholic bishops nor Catholic activists have been willing to press these issues with a forcefulness likely to produce positive results. Mindful of the trauma of more than sixteen years of military rule, they appear to prefer partially democratic civilian rule to no civilian rule at all. Both Church authorities and local activists have been more outspoken with respect to economic policy but apparently have come to similar or analogous conclusions there as well: they prefer the current stability, with only modest advances in equity, to redistributive policies that might reduce growth, unsettle economic confidence, or arouse political adversaries on the right.

Individual bishops and some Catholics active in base communities have been less patient regarding civil-military relations, justice for the victims of human rights abuses, and the rate at which poverty and gross income disparity are being reduced. But the public demands of the Episcopal Conference as a whole, and of most activist and rank-and-file Catholics as well, have been much more moderate with respect to these matters than many observers expected. The period since 1990 has hardly been tension-free. Periodically, military leaders engage in shows of force to underscore their unwillingness to surrender the quasi-blanket amnesty and other restrictions they imposed on the political system before leaving. An uneasy truce persists. Elected civilian authorities are "free" to pursue their interests and those of their constituents but within political limits seldom experienced in the country's pre-1973 history.

This state of affairs underscores the principal dilemma of Chile's consolidation process: the circumspection and restraint required for survival in the short run are obstacles to full democratization in the long run. In effect, there has been no progress beyond the original, partial transition because the center-left and democratic right have been sensitive to the fears and concerns of the armed forces and their right-wing civilian supporters. Unfortunately, neither the military nor these civilian supporters have reciprocated. In other words, one side is accommodating the other at the expense of the emergence and consolidation of truly democratic institutions and processes.

The ambivalence of Church authorities with respect to democratic principles is more apparent and costly in this context. While older and more progressive bishops would like to encourage and abet fuller democratization, the ever larger contingent of theological conservatives has made it impossible for them to do so. In fact, these latter bishops (most of them named since 1978) have forced the conference to adopt unyielding positions on personal moral issues such as divorce, birth control, abortion, premarital sex, and the like. In so doing, they have been unwilling to accommodate contrary opinions and concerns. Indeed, they have alienated once supportive elements of the general public, and offer a precedent for antidemocratic sectors to invoke in justifying similarly uncompromising positions on issues of interest to them.

In the earlier transition process, the Episcopal Conference's increasingly conservative composition did not prevent it from playing a constructive role. Continued attacks on Church people helped the bishops to close ranks and convinced even ecclesially conservative prelates that a return to civilian rule was in the Church's best interests. Theological conservatives like Cardinal Fresno, in promoting dialogue and compromise among the civilian opposition, helped to win support for the transition from rightists who trusted his instincts and integrity.

A very different context persisted during the consolidation period. There were no external attacks on the Church to precipitate unified episcopal support for a fully democratic regime. Most bishops were satisfied with the reductions in hostility toward the Church and in human rights abuses. In the absence of stronger determination on the part of secular political leaders to roll back the restrictions on democratic rule, few prelates were willing to push on alone for the changes that consolidation required. Nor would they have been successful if they had. The secular carriers for their views that had been available during transition were no longer there.

At the national level, most Catholic lay activists took the same positions as the bishops on political and moral issues alike. Groups like the Association of Relatives of the Disappeared Detained, whose members included a significant number of Catholic social and political activists, kept steady pressure on both government and Church officials, demanding information on the fate of their loved ones and punishments for those determined to be responsible. But while highly visible, they were not representative of the broader Catholic community. In fact, the reluctance of most rank-and-file Catholics and Catholic activists to push more aggressively on behalf of the victims of human rights abuses or for additional democratic reforms enabled the bishops not to press more forcefully themselves on these same issues.

The Catholic activists that we interviewed at the local level exhibited the same moderate political views on consolidation that they had with respect to transition. Their attitudes were characterized by a high degree of tolerance for pluralism, optimism regarding political institutions and leaders, modest expectations about the economy and issues of distribution, and support for a more restricted role for the military. While not as active in political movements as either cultural Catholics or secular Chileans (those with no religious affiliations), those closest to the Church were utterly unsympathetic to extreme elements (of left and right), and provided at least residual support for consolidation.

Thus, during the first several years of civilian government, at no level of the Catholic community was there forceful leadership on either removing the remaining restrictions on civilian rule or implementing a majoritarian secular democracy. Nor were there significant forces in Chilean society demanding such changes. The various parts of the Catholic community shared the satisfaction of most Chileans with what had been accomplished in the transition, and they were equally reluctant to jeopardize its continuance by pushing for additional concessions from either military hard-liners or their civilian allies.

The Church and the Peruvian Transition

The Peruvian Church did not have much of an impact on its country's transition to democracy. That process was largely a function of relations between institutionalist and *pinochetista*[10] elements within the military, and between the military and Peru's center and right-wing parties. Priests, nuns, and lay activists played significant roles in the popular opposition to the Morales Bermúdez government. They provided meeting space, good offices, and other forms of assistance to workers, neighborhood activists, and partisan political organizers. The protest activities they helped to sustain put additional pressure on the military to withdraw but had little effect on its economic policy or its political tactics once military leaders decided to relinquish power.

The bishops, on the other hand, were unable to speak clearly or effectively during the transition period (1976–80). Their earlier ambivalence toward Peruvian democracy (e.g., at the time of Belaúnde's overthrow in 1968) weakened their credentials as allies or advocates of transition. From the outset, their internal divisions were more pronounced than those of the Chilean bishops, and they grew more obstructive as the influence of progressives like Cardinal Landázuri of Lima and his auxiliary, Luís Bambarén, declined. Divided in their ecclesial and ideological views, the Peruvian bishops lost their collective voice on social and political matters. The terms of Catholic social thought prevented conservatives from endorsing or defending the existing order, the socioeconomic or political agenda of right-wing groups, and the human rights abuses of the Morales Bermúdez government (except initially and at the very end of its term), but they refused to attack it or challenge it in any way. In this connection, it is worth noting that the Peruvian bishops were treated with greater deference and respect by military authorities than their Chilean counterparts. For this reason, perhaps, they were less inclined to confront the regime and more willing to rely on behind-the-scenes contacts with its authorities.

In the absence of statements or pronouncements from the Episcopal Conference, those of individual bishops were often featured in the national media, and some had significant impacts. More often than not, however, they prompted clarifications or rejoinders from other bishops, the effect of which was to underscore conference divisions and dilute the impact of more outspoken elements (such as the Southern Andean bishops).[11]

Finally, in contrast to the Chilean case, issues such as educational reform, family values, and separation of church and state intruded on the transition process in Peru, reducing the time and energy available for socioeconomic and political concerns. This was due partly to timing, and partly to the character of Peru's transition, in which most military officers were in favor of relinquishing power. But it also reflected the more substantial foothold that conservative bishops had within the Episcopal Conference and their greater concern for core Catholic interests.

Unlike their Chilean colleagues, the Peruvian bishops were thus neither the articulators of a broad consensus supporting democratic rule nor a mediating force among key political actors during transition. The Peruvian Church's influence on transition was greater at the local level, where its pastoral agents worked closely and supportively with popular movements and political organizations. Priests, sisters, and laypeople "representing" the Church in popular neighborhoods actively supported the efforts of neighborhood organizations and trade unions, rallying opinion against the military government and its policies. But even here, their impact was less in terms of concrete achievements than in the bases laid down for future organizational activity.

Ironically, the modest social and political involvement of local church leaders and of some Peruvian bishops may have intensified conflict within the Church and thereby further reduced its ability to speak on social and political matters. Beginning with Velasco, and continuing under Morales Bermúdez, many priests, nuns and lay activists were caught up in the commitments and aspirations of their peasant, worker, and slum-dweller congregations. They thus brought "the Church" into conflict with political authorities and established social forces (landowners, factory owners, etc.), and were troubling to conference conservatives, who had reluctantly accepted the changes introduced by Landázuri and legitimated by Vatican II and the Medellín Bishops' Conference.

In addition, a substantial number of conservative Catholic laypeople supported the Morales government's austerity policies and repression of dissent. The statements and actions of local clergy and progressive bishops drew their wrath and moved some to attempt to discredit prominent progressives like Bishop Bambarén. In Chile, where right-wing influence within the Church was less substantial, Cardinal Fresno, a conservative himself, helped to allay conservative fears and thereby eased an otherwise difficult transition. Conservatives were more numerous in Peru, however, and people like Landázuri and Bambarén merely heightened their anxieties, in effect intensifying the conference's divisions and further reducing its capacity for influence.

Finally, Catholic political activists in Peru were also more fragmented than their Chilean counterparts, and their impact on political events was similarly diluted. Rather than acting as a moderating leaven for compromise and tolerance, Catholic laypeople could be found at virtually all points along the political spectrum, including its extremes.

The Church and Consolidation in Peru

Besieged by socioeconomic and political crises from the mid 1970s through the early 1990s, Peru's reestablished democracy never took root. The country's social and political forces continued to value their own interests and perspectives more than they did working together or defending democratic principles and procedures. When President Fujimori suspended the Congress and assumed emergency powers in April 1992, most Peruvians supported him. Along with opposition political elites who were the objects of Fujimori's assault, the Catholic bishops were among the few Peruvians to speak publicly in defense of democratic institutions.

They did so in the language of recent Vatican pronouncements stressing the congruence between democratic and gospel values. For most of the period since 1980, however, the bishops did little to arrest the deterioration of social and political conditions. They lamented the scourge of mass poverty, the ever wider recourse to violence (by both Sendero Luminoso and the military's counterinsurgency forces), and the increasingly blatant and pervasive corruption among public officials. But they never offered guidelines for addressing the political or structural issues that underlay these problems, and they never proposed criteria for choosing among alternative strategies or policies. Leaving such matters to politicians and state authorities, they focused instead on ecclesial and traditional moral issues.

The bishops' continuing retreat was a function of Cardinal Landázuri's declining influence and of the growing number of conservative bishops appointed by Pope John Paul II beginning in 1978. Landázuri had been one of the few Peruvian prelates to embrace social Christian perspectives. His more conservative colleagues deferred to his judgments and leadership through the mid 1970s. But with John Paul's election in 1978 and the subsequent appointment of additional conservatives, his influence began to wane. During the early and mid 1980s, he lost control of the conference's agenda, and by the time he stepped down as its president (1988), it had come to focus on very different issues, i.e., divorce, abortion, birth control, questions of religious orthodoxy, and the recent growth of Pentecostal and other fundamentalist denominations.

This new, more conservative Episcopal Conference was not always tactful or measured in dealing with these matters. Its positions alienated some in the Church and failed to persuade or effectively pressure policy-makers. In fact, the uncompromising spirit with which the bishops addressed many issues (leaving no room for differences of opinion on issues of sexual morality) added to the divisiveness in Peruvian society at a time when consensus-building was needed.

The bishops remained aloof from social and political issues through mid 1991, when the enormity of the country's many crises convinced them of the need to address economic and political problems more forthrightly. One of the more progressive bishops, José Dammert of Cajamarca, was elected conference president, but his impact was diluted greatly by the 1990 appointment of conservative Augusto Vargas Alzamora to succeed the retiring Landázuri as archbishop of Lima.

Moreover, by the time the bishops began addressing socioeconomic and political issues again, it was too late; their moment of opportunity had passed. The vacuum was filled by a political outsider, Alberto Fujimori, who was no friend of the Church. The bishops had no congenial secular carrier, e.g., a Christian reform or moderate, centrist party, to which people could turn, as they had done in Chile.

Church people in Peru were socially and politically active at the local level. CEAS[12] staffers, priests, nuns, and lay pastoral agents came to the aid of workers, slum dwellers, peasant groups, and religious communities threatened by guerrilla violence, counterinsurgency measures, and ever deepening poverty. They helped people survive difficult circumstances, and encouraged them to defend their interests. They were most successful when they were endorsed and materially supported by diocesan and archdiocesan authorities, as in the Southern Andean region, the departments of Cajamarca and Chimbote, and some *pueblos jóvenes* and popular neighborhoods in Greater Lima. Church support helped rural and urban *rondas*[13] and other grassroots organizations to survive efforts by local elites to neutralize or eliminate them.

In the long run, these popular organizations could help to strengthen and stabilize democratic institutions. But their impact in the shorter term was probably negative. They helped to generate political support for the United Left at a time when some of its factions were moving in a radical direction, producing alarm and hostility among established political (e.g., APRA, rightist parties, and military officers) and economic elites.[14]

In fact, the obstacles to democratic consolidation were far greater in Peru than in Chile, making it harder for Church leaders to positively affect the outcome. For one, the Peruvian economy failed to generate or sustain growth, and failed to attend even minimally to the employment, housing, and health-care needs of popular-sector elements. Second, a sizeable part of the political left in Peru remained committed to a radical socialist agenda, embracing democracy only to the extent that it was useful in advancing that project. Third, the governments of this period proved unable or unwilling to control the Peruvian military. Human rights violations by military units that ignored or defied civilian authorities escalated during the consolidation process, greatly undermining it.[15]

Finally, the Peruvian right (unlike its Chilean counterpart) wanted nothing to do with even the constitutional left. In effect, there was little or no interest in Peru in a moderately progressive alternative around which the Church might have helped to rally political forces, as it had in Chile. Instead, with centrist projects (Belaúnde's moderate neoliberal experiment and García's radical populism) having failed during the 1980s, political forces were moving toward the political extremes.

These obstacles might have been insurmountable whatever the Peruvian bishops or Catholic activists said or did. But there were things that each could have done that were crucial to successful consolidation. The bishops, for example, could have pushed more forcefully for dialogue and compromise among political factions. They could have devoted more time and effort to facilitating relations among grassroots activists, party leaders, and public officials. They could have insisted that "collective social actors" (unions, neighborhood associations, etc.) be recognized as the legitimate representatives of popular groups, and that political elites deal with rather than denounce them.[16] In short, they could have been much more active and resourceful as moral tutors for democracy and as mediators helping to strengthen institutions and processes at critical junctures.

The unwillingness of the bishops to address complex issues or to mediate among contending forces had a stifling effect on the Catholic activists at the local level. Without authorization or encouragement from the hierarchy, Catholics belonging to local, church-sponsored organizations were less inclined to consider political involvement than they would have been otherwise.[17] And thus yet another of the Church's resources, the leavening impact of committed Catholic laypeople, was significantly diluted.

By way of contrast, during the transition in Peru, grassroots Church groups played a significant role in opposition to the Morales government and its policies. They were able to do so with minimal support from higher Church authorities because coordination with others (at regional and national levels) was not as crucial in opposing a government willing to relinquish power. In the case of consolidation, however, the crucial factor was the capacity of democratic groups to work together at the national and local levels, and here the absence of Church encouragement and support was more costly.

What These Cases Tell Us

In the introduction, we laid out three possible scenarios for the Latin American Church as a result of Vatican efforts to regain control over its activities: the first was a continued pullback from social and political involvement as newly appointed conservative bishops imposed a predominantly spiritual agenda on local leaders and rank-and-file Catholics; the second was schism, with politically radicalized local-Church communities rebelling against such leaders and forming a separate Church identified with the social agenda of popular-sector groups; and the third was mutual adjustment in which bishops and local leaders and communities accommodated one another, producing a Church with a more spiritual focus but still committed to social justice and relatively free of serious internal conflict. Our study of the Chilean and Peruvian cases leads us to conclude that none of these scenarios are likely to materialize.

A continuation of the pullback from social and political involvements is unlikely under current circumstances for several reasons. Despite the return to partial democracy in Chile and Fujimori's April 1995 reelection to a second term in Peru, the plight of the poor has improved only marginally, if at all, in these countries. Demands for social services still are being made of local churches, and most Chilean and Peruvian bishops want the Church to continue providing these. More important, because they embrace and affirm the Church's social teaching, even conservative bishops are committed to maintaining its social mission. This is especially true of those appointed by (and intensely loyal to) the present pope, whose recent pronouncements have encouraged pastoral agents to continue promoting social justice.[18] Finally, none of the recent conservative appointees has questioned the affirmations of the Medellín (1968), Puebla (1979), and Santo Domingo (1992) Conferences that the poor are to be the Church's "preferential option" or that the Church is to minister to their material needs, as long as this does not detract from preaching of the gospel and administering sacraments.[19]

In fact, Catholic bishops, including the conservatives, have several reasons for avoiding confrontations with local clerics and activists over these issues. For one, they value the credibility that the Church has regained among lower-income Catholics, and are not about to risk it by pushing restoration too hard. To insist that parishes and communities become primarily or purely spiritual in focus would be to repudiate the insistence that spirituality includes a concern for justice, and might alienate Church activists and followers who try to combine the two dimensions. Moreover, Vatican and local Church authorities had little trouble curtailing or controlling popular Church groups in Nicaragua, where they were relatively weak,[20] but have met with stiffer resistance in El Salvador, Brazil, and urban Chile and Peru, where Catholic communities and organizations are established and respected fixtures in their respective locales. Finally, even conservative bishops are aware that alienation of either pastoral agents or lay activists under present conditions could be costly, with Protestants making inroads among lower-income Catholics and with the number of priests available for ministering to their needs already limited.[21]

With respect to the second scenario, our interview and survey data indicate that local leaders and activists are neither ideological nor theological radicals. They are moderately progressive Catholics who have assimilated the values of Vatican II, i.e., lay initiative, collegial decision-making, respect for conscience, and a faith that is active in the pursuit of social justice. They are also loyal to the institutional Church and seek neither confrontation with, nor autonomy from, its authorities.[22]

But like their bishops, most local Catholic activists can also think pragmatically. They know from experience that the hierarchy provides important protection for local communities under state attack. They also know that their links with Catholic authorities preserve their official Catholic status, keeping them eligible for financial assistance from the international Church. Accordingly, most base community activists would think twice before marching off to create a new "people's" Church, even were they so inclined. They would be cut off from resources needed to help the poor they wish to serve, and they would be vulnerable to any resumption of state repression. In our judgment, therefore, schism between conservative bishops and local Catholic activists over the Church's social mission is quite unlikely from the standpoint of local activists as well.

Given the shared interests of higher- and lower-Church leaders, the third scenario of mutual adjustment appears more probable than the first or second. But while only modest conflict is likely with respect to social issues, the area of moral teaching will be more troublesome. The Vatican and newly appointed conservative bishops are determined to defend traditional Catholic teachings on abortion, divorce, birth control, and Protestantism, and to retain laws that embody official Church positions in these matters. Data from Chile and Peru indicate that many Catholics and Catholic activists hold more liberal views than their bishops and defend their right of conscience when it conflicts with official Church teaching. With neither side (Rome and the conservative bishops, on the one hand, and increasingly assertive priests, nuns, and lay followers, on the other) willing to budge, the potential for open conflict is clearly present.

In fact, conflict has already arisen over how base communities are organized and the role of women in them. The Vatican has been adamant that neither lay nor religious women may exercise priestly or ministerial functions,[23] and yet the majority of activists and leaders of base communities in Chile and Peru are lay and religious women. Moreover, for the past two decades many of them have been performing functions (preaching, baptizing, officiating at weddings) traditionally reserved for priests. In fact, the incorporation of women into key pastoral roles has been widespread throughout Latin America and in North Atlantic countries as well, given the chronic global shortage of priests.[24]

Although not yet an issue of conflict between conservative bishops and local church people, the potential for contention is there if Rome insists on removing women from their current leadership positions.

If none of these scenarios is likely to materialize as described, what do the Chilean and Peruvian experiences suggest with respect to the future of Latin American Catholicism and its role in the region's newly (re)constituted democracies? Much will depend on political and economic developments in individual countries, and in this regard we would like to underscore several of our findings.

First, very few of the local Catholic activists that we surveyed have been radicalized by either liberation theology or their contacts with leftist groups. In both our Chilean and Peruvian samples, those involved in the organizational life of the Church were socially and politically less active than "sacramental" and "cultural" Catholics during transition, and after civilian rule had been restored.[25] Their attitudes, however, have been generally supportive of democratic institutions and practices. Peruvian Catholics shied away from partisan involvement at a time (1985–90) when it might have helped to strengthen democratic institutions. But Catholic activists in both countries strongly endorsed liberal forms of democracy, rejected insurrection and authoritarianism as options for redressing social problems, and harbored relatively modest expectations of their respective postmilitary governments.

Second, there is no indication, in either Chile or Peru, that the theological conservatism of newly appointed bishops will translate into political positions subversive of democracy. The era of Catholic integralism in Latin America seems to have passed, and no contemporary Catholic movements, not even Opus Dei,[26] have publicly challenged the legitimacy of the newly constituted democracies.[27] In fact, as the Peruvian hierarchy's public criticism of Fujimori's closing of the Congress in 1992 indicated, even conservative bishops are willing to stand publicly in support of democracy and constitutional government.

The old chestnut that Catholicism is antithetical to democracy is thus no longer, if it ever was, true.[28] The decentralization of its internal structures and the delegation of some ministerial responsibilities to the laity have made the Church more comfortable with democratic values. The sinister character of contemporary authoritarian regimes has pushed it in this direction as well. Their experiences under military governments convinced many conservative prelates that such governments were a threat to Catholic values for their assaults on the human rights of real and suspected political activists, and on those ministering to them. Neither Chilean nor Peruvian Church leaders regard authoritarian regimes as reliable or effective allies; they view constitutional democracies as both more congenial to Catholic values and more likely to defend them effectively.

Evidence from Chile and Peru also suggests that the contemporary Church is less inclined to identify with partisan groups of any sort. While clerics and lay activists still harbor personal sympathies for specific parties or candidates, they no longer impose them from the pulpit nor hint at them in pastoral letters. Not even Chilean Christian Democracy receives the kind of public endorsement from Church officials that it did in the early and mid 1960s. Collaboration between Christians and Marxists has taught clerics and lay activists in both countries that the Church can coincide work with the left in pursuit of some common goals but need not embrace Marxism or compromise their Catholic beliefs or commitments in the process.[29]

Clerics are thus unlikely, in the years ahead, to condemn or attempt to discredit parties or movements that are committed to democratic goals and procedures. The age of overt and politically reactionary clericalism in Latin American politics is over. Backed by the Vatican, and in the interest of maintaining spiritual unity, Church officials have committed themselves to neutrality in partisan battles. And whether intended or not, this shift enhances their potential for acting as a neutral guardian of values that all groups must embrace if democratic institutions are to survive.[30]

It is also clear that the Latin American Church is no longer tied to the rich and powerful as preferred social strata. In terms of adherents, resources, and pastoral ministries, the fulcrum point of the Catholic community is now in the popular or working classes, to which the vast majority of Latin Americans belong. The Church itself would explain this in moral and theological terms, arguing that its prophetic forces and impulses have brought it to reembrace a more biblical Christianity that shares the lot and the hopes of the poor. Others might ask what else the Church could have done to arrest the advance of secular reform, Marxism, or Protestantism among popular-sector Catholics in the twentieth century. In either case, not even conservative bishops want the Church to return to pre–Vatican II positions or resurrect its sociopolitical alliances with the rich and powerful. The Church's class composition has changed, and this has affected the clergy's makeup and the way in which the

Church sees and addresses social issues. Just as its historical alliance with the upper class conditioned the way in which bishops and priests viewed social problems in the nineteenth and early twentieth centuries, so too has its new rootedness in the poor now, despite the conservative turn that the Church has taken theologically.

In this connection, we note the opposition of Catholic bishops in Chile and Peru to the neoliberal or structural adjustment policies pursued by their governments.[31] Confronted with the testimony of pastoral agents in poor neighborhoods, they have been critical of the negative impact on the poor of privatization, cutbacks in public expenditures, elimination of subsidies for basic commodities and services, the lowering of tariffs, and the provision of investment incentives to large foreign corporations. In Chile, with the return to democracy, social spending has increased and poverty has been reduced, although not at the rate or to the extent that many had hoped. In Peru, on the other hand, Fujimori has contained inflation and has attracted significant foreign investment, but social spending has fallen dramatically and the lot of the poor continues to deteriorate. Church leaders in both countries have been critical of government economic policies and are likely to remain so unless conditions for the poor improve.

What effects they will have on the Chilean and Peruvian governments, or on the strengthening of democratic institutions and practices, is more difficult to determine. Democracies do not require full economic and social equality to survive. But representative regimes need to address social and economic inequities among classes if they wish to remain stable and if ordinary citizens are to develop confidence in them. Here, the voices of the bishops and the efforts of local Catholic activists can help to keep these issues before policy-makers, thereby strengthening long-term prospects for stable democratic government.

In contrast with these first two, largely positive findings, however, we believe that the theological conservatism of the bishops appointed by John Paul II has had an undermining effect on values (e.g., toleration, consultation, accommodation, due process, and respect for majority will) on which the development of liberal democracy depends in both Chile and Peru. Traditionally Roman Catholicism has taught that there are immutable moral precepts discoverable by human reason (natural law) which the Church is obliged to proclaim and defend absolutely. The Church also teaches that when there is doubt or disagreement as to the content of these principles or their application, it (alone) has the authority (from God) to resolve ambiguities. These views challenge the liberal-democratic assumption that multiple views on complex public moral issues are legitimate and that the state lacks the authority to prefer or impose any one view in particular.

The potential for conflict over these matters can be contained as long as an overwhelming majority of Latin Americans continue to consider themselves Catholics and to accept the hierarchy's views as both obligatory and properly enshrined in public law. But with the development of religious pluralism in society, and with the emergence of independent thinking among rank-and-file Catholics, the authority of the bishops to define law and acceptable practice has diminished, and democratic legislatures are under increased pressure to liberalize laws dealing with personal and family morality. From the standpoint of liberal democracy, of course, public rights and laws are subject to either majority rule or to previously drawn contracts or conventions. But for the Vatican and for local Catholic bishops, clear and compelling issues of right and wrong, among which abortion as well as torture are included, cannot depend on the public's opinion. The two ideologies, liberal democracy and Catholicism, are thus, still, at odds. They are more convergent than a century ago, but they still disagree fundamentally on how moral issues are to be defined and resolved in the public arena.

With Latin American bishops firmly opposed to liberalization of laws governing abortion, divorce, birth control, and equal rights for all religions, tensions between church and state, and between the hierarchy and local Church people, may become critical. If the bishops were to demand that laypeople publicly oppose initiatives challenging the Church's positions on birth control and abortion, activist and rank-and-file Catholics might be reluctant to comply, and some priests, nuns, and laypeople might even resist in an organized and public manner, thereby prompting the bishops to take disciplinary action to preserve Church authority. In any event, these moral and religious issues are more likely to precipitate internal problems for the Church in the years to come than are disagreements relating to the Church's social mission.

It is not clear to what extent such matters will affect democratic consolidation. Were the current center-left government in Chile to face increased opposition from either its left or right, or both, antidemocratic forces would find it easier to argue (as the bishops themselves have done) that the issues at stake (variously, criminal justice, military dignity, the need for greater equity) are too important to be left to the vagaries of public opinion, and thus require an authoritative solution. And, of course, the potential for such dynamics exists even more clearly and intensely in Peru and in other countries that do not enjoy Chile's solid economy and accommodating political elites.

Under no foreseeable circumstances, however, will the Church be able to sustain or preserve Latin American democracy by itself. During the last year of Allende's presidency, the Chilean bishops spoke frequently and forcefully in favor of dialogue and compromise but failed to carry the day. Neither the opposition nor those supporting Allende's government believed, at that point, that democracy still served their interests. And even in a long-standing democratic country such as Chile, if the country's principal political groups are not prepared to defend democracy and civilian rule at the expense of their partisan interests, there is little that an institution like the Church can do. The Church's support for constitutional government may well be a necessary condition in certain contexts, but it is not sufficient in itself; other forces must be willing to follow its advice (as a ready "secular ear") and join together in common commitment and action. At best, the moral voice of the Church provides a second line of defense. It can strengthen the resolve and impact of secular groups that are also committed to democratic values, but it cannot substitute for them. Others must hear its message and carry it into practice in social and political arenas.

Nor are the interests of the Church and liberal democrats identical or of equal intensity. The Church is first and foremost a moral and religious institution, whose mission is to preach the gospel, administer sacraments, and uphold Catholic ethical teachings. As the Chilean bishops indicated in their treatment of the Falange Nacional (during the 1940s), and of Christians for Socialism and Tradition as well as Family and Fatherland (in the 1970s), they will oppose movements that they consider threats to such core concerns as episcopal authority, the universal scope of Church membership, and proper specification of its social norms.

Their acceptance of the military coup of 1973, for example, enabled the Chilean bishops to maintain and expand Church ministries under a potentially hostile authoritarian government. Their decision was a forceful reminder that the Church will not stand by an embattled democracy at all costs. It will do what it can, but will then move on to defend and preserve its religious and moral missions. The Church has survived many such crises in its 2,000 years of existence largely because it has known how to blend (or balance) faithfulness to principle with the need for institutional preservation. Thus, its emerging preference for democratic institutions and practices, while sincere, is not likely to override its other concerns. Regimes of various sorts will rise and fall, some of them more, others less, congenial to Catholic values. But the Church will be permanently and irrevocably committed to none.

The Church's Role in Political Life

On the basis of the Chilean and Peruvian cases, we conclude with some final reflections on the Church's role in Latin American political life. The first is that its political influence is primarily a function of the moral authority of its bishops, priests, sisters, and lay activists. Their moral authority rests on their personal charisms and on their status as depositories of divine inspiration or favor. It also rests on the not always warranted assumptions that Church people have no vested political interests, and, in the case of Church authorities, that most Catholics will follow their lead. In addition, Catholic leaders and activists are more likely to retain their moral and political authority if they remain united and concern themselves with social issues that relate directly to the Church's religious mission.

In sum, if it is to be put to political use, the Church's moral authority must be exercised judiciously and infrequently. In times of crisis, the Church may serve as a surrogate political actor or social service provider. With secular forces or institutions unable to operate normally, Church authorities may take over their functions or launch initiatives of their own. In doing so, they may help to stabilize and sustain a country's social and political processes. They can also enhance the cause or credibility of specific groups for whom they choose to speak (the poor or the oppressed, for example). And, finally, the Church's programs and facilities can provide formative, life-sustaining places of gestation, from which repressed social and political groups may emerge revitalized.

Under normal circumstances, however, the Church's moral authority will have a more modest political impact. Presentation of the gospel in a new light can free people from religious inhibitions that might be limiting their political choices. The timely lending of encouragement and resources can help them to win concessions from employers or government officials. In some instances, the Church actually changes the way its people think. More often, however, it reinforces them in positions they have already embraced. In fact, despite their widely alleged "docility," most Latin Americans listen to their religious leaders selectively. They tend to focus on, retain, and assimilate those views that they find congenial or useful, and to reject or ignore those that they do not. When acting in pastoral capacities, for example, priests and bishops are rarely in a position to tell people what to think or do politically. They can help to affirm the direction in which people have begun to move for motives or reasons of their own, and to counter obstacles that others have placed in their paths. But they are unlikely to determine that direction on their own.

Moreover, the Church's influence is greatest when the issues are morally compelling, i.e., when human rights, sheer survival, democracy vs. dictatorship, and the right to fair wages and safe working conditions are involved. It is not as substantial or credible in terms of specific policies as it is at the level of general principles and objectives. Issues like human rights are issues to which Church people can speak forcefully and credibly, particularly when others (who might) cannot. They are issues that call to people's consciences but leave room for prudential and technical judgments when it comes to actual application.

Finally, the Church's moral authority is not something that its leaders possess and can cash in as they see fit. This would imply the existence of a passive, mass following whose conduct the Church can control, whatever the issue involved, as long as it does not attempt to do so too often. With a mass following that is no longer passive on either religious or political issues, the Church's politically usable moral authority is more limited; it lies in the consciousness of the Catholic community itself, to which, of course, Church leaders and programs may contribute significantly.

Appendix

Tables 4.1 through 4.4

Left Political Tendency. This open question asked: "Do you sympathize with a particular political tendency or political party? Which?" Very few people identified a particular party. Answers indicating the left (as opposed to the center-left, center, center-right, and right) are considered more progressive.

Best Government. This question asked respondents to choose the best government of the last thirty-five years (from among the Ibáñez, Alessandri, Frei, Allende, and Pinochet governments). Answers indicating the Allende government are considered the most progressive.

Best Regime. This closed question asked the respondent to choose the most appropriate regime for the country from among the following alternatives: a military regime, a civil government that excludes extremists and undemocratic groups, a civil government that excludes no one, or a civil government that excludes the rich. Answers preferring the last of these are considered the most progressive.

No Violence. This question asked: "Do you justify the use of violence as a means of political action?" The alternatives were "yes," "yes, as a last resort," "yes, when one's adversaries do," and "never." Answers indicating an unqualified yes are considered the most progressive.

Christian-Marxist Collaboration. This question asked: "Do you agree that there should be collaborative activities [*actividades conjuntas*] between Catholics and Marxists?"

Social Organization. This question asked: "Do you currently belong to a production workgroup, a meal program, a glass of milk program, or a health committee?"

Political Organization. This question asked: "Are you a member of any political party?" Which?

Tables 5.1 through 5.4

Interested in Politics. This closed question asked: "People have very different opinions about politics. Can you tell me if you agree or disagree with the opinion I am going to read you: all Chileans have the obligation to take an interest in politics?"

Politicians Work for Good of All. This closed question asked: "People have very different opinions about politics. Can you tell me if you agree or disagree with the opinion I am going to read you: most people who take an interest in politics do so for the good of others?"

Adequate Party/Social Organization Relations. This open question asked: "Do you think that the relations between social organizations (such as trade unions and popular social organizations) and political parties are adequate or not?" People could answer yes or no and support their choice with arguments or examples.

Parties Do Not Reflect Interests of Ordinary Chileans. This closed question asked: "Do you think that political leaders actually reflect the thinking of their bases or are just speaking for themselves?" The options included: "they reflect the thinking of their bases"; "some reflect the thinking of their bases, others speak for themselves"; "they speak for themselves"; and "other."

Leader-Base Relations. This question asked: "How, in your judgment, should the leadership and base of a political party be related?" Responses clustered around three options: a leadership better informed about constituent needs and interests, better communication between leaders and followers, and equal footing between the two.

Best Regime. This closed question asked the respondent to choose the most appropriate regime for the country from among the following alternatives: a military regime, a civil government that excluded violent or undemocratic groups, a pluralistic democratic government that excluded no one, or a popular or socialist government. Answers preferring the popular or socialist democracy are considered the most progressive.

Accommodate Class Interests. This question asked: "Do you think that one ought to defend the interests of one's own class or accommodate the interests of all social classes?" Answers indicating the former are considered the most progressive, those indicating the latter, the least.

Traditional Military Role. This open question asked: "What role would you assign to the Armed Forces in a democratic civilian government?" Responses clustered into 9 categories: "defending national sovereignty, territory, or security," "preserving order or domestic peace, and enforcing traffic laws," "helping in the event of natural disasters," "assisting judicial authorities," "defending the people, human rights," "an economic or productive role, national reconstruction," "serving the country, not its own interests," "no government role, no political role, back to their barracks, or to their role prior to the military coup," and "they should be dissolved." The figure in table 5.4 indicates the percentage choosing any of the first three.

Follow Conscience over Church Authorities. This question asked: "In the great moral decisions of your life, do you take more into account your own conscience or the position of the Catholic Church?"

Follow Priest or Bishop. This closed question asked: "If there were a difference in pastoral orientation between the priest in charge of your parish or community and the bishop, which of those orientations would you follow?"

Tables 7.1 through 7.7

Leftist Political Tendency. This open question asked: "With which political tendency do you sympathize the most?" Most people answered "the left" (some center or right), although a significant number specifically mentioned the APRA.

Structural/Class Problems. This open question asked "Does the Peruvian society in which we are living seem just to you?" Answers clustered into five general categories: "not so unjust," "unjust because too much poverty," "unjust because of oppressive state or oppressive government policies," "unjust because laws are not obeyed," and "unjust because of structures of domination or profound class differences." The last response was considered to be the most radical or progressive.

Ills Due to Nonstructural Problems. All respondents who gave any but the last response to the previous question.

Leftist Ideas Good/Important. This closed question asked: "In general, what is your opinion of leftist ideas or positions?" The four options included: "Good, in agreement with them," or "important"; "favorably disposed to some of them"; "not in agreement, but defend the left's right to offer a political alternative"; "opposed to the left, and to its offering of itself as a political alternative."

Differences Between Rich and Poor Will Disappear. This open question asked: "Will the differences between rich and poor ever disappear?"

Followers Behind Leaders. This closed question asked: "What is the most important thing in a social or political movement?" The three options included: "a leader or group of leaders"; "an organized base behind a leader or group of leaders"; and "an organized and autonomous base and leaders capable of serving and interpreting it." The figure appearing in table 7.6 indicates the percentage of respondents choosing the second of these options.

Residents, Residents/Authorities Responsible. This open question asked: "Who do you think should solve these local neighborhood problems?" Answers clustered into five alternatives: "the government," "the mayor," "the party," "the residents themselves," and "residents working with authorities." The figure given in table 7.6 indicates the percentage of respondents choosing either of the last two alternatives.

Interested in Politics. This open question asked: "Are you interested in social and/or political issues or problems?" The figure appearing in table 7.7 indicates the percentage of respondents answering: "yes, a lot," as opposed to "no," or "yes, but not very much."

Political Organization or *Party Member.* The question asked: "Do you presently belong to any political organization?"

Neighborhood Organization. The question asked: "Do you presently belong to any neighborhood [*barrial o vecinal*] organization?"

Partisan Involvement. This closed question asked: "What is your opinion regarding the participation of Christians in politics?" Responses clustered into four general categories: "they should avoid politics generally and party politics especially"; "they should participate politically, but not join any particular politial party"; they should support, and/or become an active member of, some party or front"; "it depends on each individual's personal option."

Single Party. This closed question asked: "On what problems or in what areas ought Christians and leftists to be collaborating?" Responses ranged among the following alternatives: "human rights," "neighborhood problems," "labor affairs," "the popular movement," "party politics," and "there should not be any." The number appearing under "single party" indicates the percentage reponding "party politics."

Table 7.8

Christ a Political Revolutionary. This open question asked: "Why do you think Christ was persecuted and killed?" Answers that referred to Christ's role as a political leader or as someone opposed to the system of domination of his time are considered as the most progressive, those referring to his desire to "save us" as the least.

Church's Economic Role. This closed question asked: "Are you more or less in agreement with the fact that the bishops of the Catholic Church take positions [*se pronuncian*] on the government's economic policy?" Answers indicating unqualified agreement are considered the most progressive.

Conscience over Authorities. This question asked: "In your opinion, should a person accept or comply with the Church's instructions even if one disagrees with them?"

Poor Need Not Love Rich. This open question asked: "Given that Christ said that we should love one another, do you think that the poor should love the rich?" Answers that said no, using ideological justification, are considered the most progressive, those saying yes, because God said so, the least.

Notes

Introduction

1. Fremantle, 1963, pp. 149, 152.

2. Leo XIII, 1942, p. 5.

3. Pius XII, n.d., pp. 2–4.

4. Gremillion, 1976, pp. 203, 205–6, 212; and Abbott, 1966, pp. 284, 399–400, 416–17, 420, 425–26, 513–15, and 679.

5. Leo XIII, 1942, p. 5; Pius XII, n.d., pp. 8 and 13; and Abbott, 1966, p. 289.

6. This possibility appeals to authors for whom the Church remains a deeply conservative institution whose preferred allies are still political, socioeconomic, and military elites. For Mutchler (1971), Langton (1986), Gismondi (1986), and Grigulevich (1984), for example, the reforms and changes of the 1960s and 1970s were defensive moves designed to preserve institutional interests during times of Marxist threat and were easily jettisoned once the danger had passed. Their argument is that the Church's internal dynamics remained largely unchanged, and the current emphasis on spiritual concerns would reflect the reassertion of control by dominant economic and political elites.

7. This outlook finds resonance with the position of those scholars who portray the Latin American Church as chronically polarized (Crahan, 1982 and 1991; and Sigmund, 1992), or who think some popular elements might be marching toward a radically new society or model of church (Berryman, 1984).

8. Levine (1981, 1992), Bruneau (1982), Mainwaring (1986), and Smith (1982) have argued that the theological changes since Vatican II have strengthened the commitment of Catholics at all levels to the "modern" values of social equity and popular participation in politics. This literature emphasizes the Church's overall coherence and the consensus in favor of religious and political reform. It downplays the significance of both radical and conservative or reactionary tendencies, arguing that they have engendered far more emotion and controversy than their actual numbers or influence warrant.

1. Church and Society in Theoretical Perspective

1. These four dynamics are discussed in detail in Smith, 1982, pp. 23–62.

2. Troeltsch, 1931, vol. 1, pp. 331–43.

3. An *ex cathedra* statement is one in which the pope speaks solemnly "from the chair [of Peter]," formally invoking his power to define a doctrine infallibly. The matter must be divinely revealed (in Scripture or natural law), or necessary to explain what is divinely revealed, taught for a long time with uniformity by previous popes and the body of bishops, and held to be universally true by the laity. Only one papal pronouncement in modern times has fulfilled these conditions since the doctrine was defined

by the pope and bishops at the First Vatican Council in 1870. This occurred in 1950, when Pope Pius XII declared as infallible the belief that Mary was assumed into heaven, body and soul, after her death. For an analysis of the levels of binding force of Catholic teaching, see Charles E. Curran, 1990, pp. 155–78 (esp. pp. 164–65).

4. From the time of Clement of Alexandria in the second century, the Church has used human reason to explicate and apply the values of Scripture to new and changing situations. The efforts of St. Augustine (an avid student of Platonic philosophy) and later St. Thomas Aquinas (an Aristotelian) helped to generate rationally deduced principles based on (or consistent with) these values, but whose practical application was always viewed as problematic. Cf. Thomas Aquinas, 1952, Ia IIae, quest. 94, article 4; McCormick, 1989, pp. 131–62; and Mahoney, 1987, pp. 77–83.

5. On the issues of birth control and abortion, for example, the Church's teaching has become more condemnatory, allowing for no exceptions, as developments in modern science (e.g., embryology) have provided more complete information on the processes of conception and fetal development. For the development of the Church's position on contraception and abortion over time, see Noonan, 1966, and Connery, 1977. The Church's opposition to the charging of interest through the Middle Ages was modified to allow interest, provided it was not excessive, once commercial capitalism replaced agrarian feudalism as the prevailing economic model in Europe in the seventeenth century. For a treatment of this evolution, see Connery, 1957. As it grew and became an accepted institution in the Roman Empire, the Church's early opposition to war also evolved. For a discussion of the shift in the Church's position from pacifism, during the first three centuries, to the doctrine of just war, see Johnson, 1987. In none of these areas, however, has the Church ever made an *ex cathedra* statement, and its teachings are subject to further development as it digests new information from changing human contexts.

6. For a detailed consideration of these models, see Dulles, 1978.

7. A glimpse of its structure is provided in both the Acts of the Apostles and the letters of St. Paul.

8. For a historical overview of the evolution of the institutional model of Church over time, see Congar, 1962, pp. 119–56.

9. As Vallier has argued, "the Church . . . is not conservative 'by nature.' Its tendencies and record in that direction, instead, are to be found mainly in the kinds of interdependencies it holds with society and how these either block or short-circuit its capacities to create and institutionalize new roles and structures." See Vallier, 1970, p. 14.

10. The Catholic Church still enjoys a privileged legal status among religions in most Latin American countries. Catholic schools continue to receive government subsidies, for example, and the Catholic catechism is taught in the public schools of most countries.

11. These notions are taken from the documents of the Second Vatican Council. See Abbott, 1966, pp. 15, 201, 237–38, and 232.

12. Ibid., pp. 213–14, and 283–85.

13. The Church's heralding of gospel values, its service to the poor, its promotion of faith-sharing communities, and its defense of the sacramental dignity

of all, have clashed with the religious tastes and ideological perspectives of these groups.

14. The institutional model (as practiced in the colonial and postcolonial eras) left it dependent and organizationally underdeveloped. The Church could not bite the hands (the state and landed interests) that fed it, and lacked the organizational resources to mount significant opposition to its patrons. It was, in the final analysis, an organization of ritual celebrated by clerics and limited to those educational and charitable activities deemed appropriate or useful by state and dominant class elites. Levine, 1981; Houtart and Rousseau, 1971; Dewart, 1963; Crahan, 1991; and Mabry, 1973.

15. Levine, 1981 and 1992.

16. Houtart and Rousseau, 1971; Dewart, 1963; Crahan, 1979; and Mabry, 1973. It could be argued that disestablishment really has not been complete in either Mexico or Cuba. The Mexican Constitution of 1917 placed severe legal controls on the Church by the government, including state ownership of Church buildings and prohibition of clerical voting in Mexico, and there have been real, if subtle, pressures on Catholics not to practice their faith if they get ahead in Cuban society. But there has been a relaxation of such restrictions in Cuba in recent years, and a removal of them in Mexico in 1992.

17. MAPU (Movimiento para la Acción Popular Unitario) was founded in 1969 by the dissident, *rebelde* faction of the PDC. It later (1972) split into two factions (MAPU and MAPU-Obrero y Campesino, or MOC) and spawned several other offshoots.

2. The Chilean Church: A Historical Overview

1. According to official Chilean Church sources, 1477 of Chile's 2161 secular and religious priests were native born. See Osore, n.d., p. 3. According to the Vatican's *Annuarium Statisticum Ecclesiae* (1991, p. 96), only Costa Rica, Uruguay, and Colombia had lower ratios during the early 1990s. In addition, Chile has 36 Catholic hospitals and 569 Catholic dispensaries, most of them owned and staffed by religious congregations. Finally, in 1990 Catholic schools enrolled 14.4 percent of the country's primary school population; in 1989 they enrolled 16.3 percent of its secondary population, and in 1988 they enrolled 6 percent of its university students. These figures were obtained by dividing the number of students enrolled in Catholic schools at each level (available from the volumes of *Annuarium Statisticum Ecclesiae* for 1988, 1989, and 1990) by the number of students indicated by the 1993 *U.N. Statistical Yearbook* (U.N., 1993, pp. 96, 109, and 120) as being enrolled in Chile as a whole at those levels for these years.

2. According to the Vatican's statistics, 80.4 percent of Chileans consider themselves Catholics (most Chilean surveys put the figure at 74 percent). Of these only 65 percent attend mass or some other ritual at least once a year. In addition, the number of baptisms (16), first communions (7.9), confirmations (5.4) and marriages (4.6) performed in 1991 (for every thousand Catholics) compares poorly with the numbers in other Latin American and European Catholic countries. The numbers for Peru, for example, are 16.9, 10.1, 6.8, and 3.2, respectively, and those for Colombia are 24.1, 16.3, 11.7, and 4.0, respectively. Poland's numbers are 15.4, 16.7, 14.7, and 6.2, and Ireland's 14.7, 16.7, 17.7, and 4.5. See *Annuarium Statisticum Ecclesiae*, 1991, pp. 299, 301, 306, 307, and 309.

3. Guerra, 1929, *La constitución de 1925*, p. 39.

4. Mecham, 1966, pp. 211–14.

5. Rome's willingness to accept separation in Chile was influenced by the bitter disputes surrounding church-state separations in France (1905), Portugal (1911), Mexico (1917), and Uruguay (1919). In each of these instances, the Church opposed separation strenuously and ended up having very unfavorable terms imposed on it. Alessandri's motivation was similarly self-interested. An amicable resolution of the church-state issue would deny his Conservative Party rivals full political use of the religious issue. See Smith, 1982, pp. 75–76.

6. He felt that identification with a particular party made priests and bishops appear to be adversaries of anyone with other ties, and made it more difficult for them to minister to them. Errázuriz developed these views in a 1922 pastoral letter. Cf. Araneda Bravo, 1956, pp. 208–10, and Stewart–Gambino, 1992, pp. 34–35.

7. A Nazi Party also emerged in Chile in 1932 and became one of the strongest in Latin America (with roughly 40,000 members) in the mid and late 1930s.

8. "La iglesia y los partidos políticos: documento pontificio dirigido a nuestros obispos," *Boletín de la Acción Católica de Chile* (Santiago) 2 (September 1934): 525–29. For another treatment of church-state and church-party relations during this period, see Stewart-Gambino 1992, chap. 3.

9. "Circular dirigida al clero y a nuestros amados diocesanos sobre la relación de la iglesia con la política," *Boletín de la Acción Católica de Chile* 3 (December 1935): 537–41.

10. These groups were called *patronatos*, but had nothing to do with the traditional church-state relationship of that name.

11. Aliaga, 1975, pp. 23–33.

12. See Larson, 1967, pp. 14–15.

13. Aliaga, 1975, p. 82.

14. Hurtado, 1941.

15. "Obediencia a los poderes legitimamente constituidos," *Estudios* (Santiago), no. 71 (October 1938): 39; "Normas del Episcopado sobre la acción política de los católicos," *Boletín de la Acción Católica*, no. 8, 1939, p. 273.

16. See Zañartu's analysis (1969, chap. 7) of the bishops' 1947 letter, "Deber social de los católicos," and their 1949 document *Instrucción pastoral acerca a los problemas sociales.*

17. Catholic Action flourished under Hurtado's leadership. By 1941 Young Catholic Action had more than 55,000 members, and was attracting even more to its public demonstrations and celebrations. See Aliaga, 1975, p. 85.

18. The communists, and later the socialists, dominated the industrial unions that emerged in Chile prior to World War I. Legislation passed in the 1920s forbade the creation of new unions in factories employing less than twenty-five workers (most of the country's establishments employed nine or less). It effectively precluded the formation of social Christian trade unions, and Catholic workers were forced to join the Marxist-dominated unions and work for change from within them. See Angell, 1972, pp. 12–31 and 42–56.

19. The exchange between bishops and the Falange is reproduced in *Política y Espíritu* 27–28 (November–December 1947): 124–71.

20. Tardini's letter, "Llamado a la unidad de los católicos chilenos," appears in *Boletín de la Acción Católica Chilena* 18 (1950): 1–2.

21. Again, the Vatican's strategy in Europe may have influenced its decisions with respect to Chile. In the late 1940s, Rome was promoting reformist Christian Democratic parties as an alternative to communism in Europe and probably did not want to see such a party condemned or discredited in Chile if it might play the same role there.

22. The development of functionally specific Catholic Action organizations, e.g., the JOC (Juventud Obrera Católica), the ACR (Acción Católica Rural), the JEC (Juventud Estudiantil Católica), and the AUC (Asociación Universitaria Católica), occurred at this time. See Aliaga, 1975, pp. 101–7; and Poggi, 1967.

23. Gibbons, (1955, pp. 22–23; 1956, pp. 22–23; and 1958, pp. 14–16) gives the total number of priests in Chile, the total number of Catholics, and the number of priests ordained for each of these years. He does not indicate how many Chilean and foreign priests retired during this period. We assume, however, that foreign priests are almost entirely responsible for the increase in the total number of priests (372 between 1954 and 1958,) since the number of new priests ordained (between 21 and 27 per year) is probably canceled out by the number of Chilean and foreign priests retiring. The number of religious women in the country increased by about 200. Although Chilean women made up a substantial proportion of the total numbers in previous years, we would place the number of foreign nuns arriving during this period at somewhere between 150 and 180.

24. Smith, 1982, pp. 112–13.

25. When Baggio arrived, the Chilean Episcopal Conference (the bishops' national coordinating body) had been functioning for about a year. According to Araneda Bravo (1988, pp. 107–8), president Jorge Alessandri lobbied to get Baggio recalled because of his pro-Christian Democratic (and falangista) sympathies.

26. In late July 1958, when the Law of Defense of Democracy (that had outlawed the Communist Party) was overturned, the Christian Democrats voted for its repeal. Prior to the vote, ecclesial and political conservatives had urged Cardinal Caro to excommunicate anyone voting to lift the ban on "collaborating with communists." Caro refused, citing papal norms that left the specific forms or methods for opposing communism to individual Catholics and adding that "it was unacceptable that documents of a religious nature be used for partisan political ends." In contrast with its earlier confrontation with the Falange, the hierarchy thus sided with social Christian, not conservative Catholics. Caro's refusal to condemn politically liberal Catholics was a signal that anticommunism could no longer be used to resist democratic reform.

27. According to a survey conducted by Professor Eduardo Hamuy's *Centro de Opinión Pública* a month before this election, 44.1 percent of regular mass-attenders defined themselves as rightists, and only 21.3 percent as centrists. Among these respondents, Alessandri outpolled Frei 37.1 percent to 23.5 percent. Cf. Smith, 1982, pp. 89–90.

28. Developments in the Chilean Church during the next two periods (1958–73 and 1973–82) have been treated elsewhere by one of us (Smith, 1982, chaps. 5–8). The rest of this chapter covers this material more summarily.

29. In "La Iglesia y el Problema del Campesinado Chileno" (*Mensaje*, no. 108 [May 1962]: 185–94A), the bishops denounced the underutilization of land by absentee landowners, and the dearth of technical assistance for small farmers. They

called for an agrarian reform that would keep production levels up, encourage more effective use of the land, and assure peasants access to property and related benefits through credit and training. To help accomplish these goals, they recommended higher taxes on land, regulation of farm prices, social security for agricultural workers, and the creation of teams of agricultural extensionists in each region.

30. In "El deber social y político en la hora presente" (*Mensaje,* no. 114 [November 1962]: 577–87), the bishops called for tax and industrial reforms, administrative reorganization of government, encouragement of investment in consumer- rather than luxury-goods production, and the cessation of capital flight.

31. Smith, 1982, p. 116.

32. Student leaders active in the AUC launched strikes in mid–1967 calling for democratic reforms and demanding more scholarships for working-class students at the Catholic universities of Santiago and Valparaíso.

33. Opus Dei was founded in Spain in the late 1920s by a conservative priest, José María Escrivá de Balaguer, offering a rigorous, quasi-monastic, and rigidly authoritarian religious formation to professional men who would build careers and influence in the worlds of politics and business.

34. Repression of worker and squatter demonstrations in 1967 and 1968 left more than thirty-seven dead and badly damaged the government's progressive image. At the same time, urban guerrilla violence protesting the slowness of reforms was being directed by the Movement of the Revolutionary Left (MIR), which was founded in 1966.

35. See the bishops' "Chile, Voluntad de Ser," *Mensaje,* no. 168 (May 1968): 190–97; "Declaración episcopal sobre la situación actual del país," December 12, 1970, *Mensaje,* no. 186 (January–February 1970): 77–79; Bishop José Manuel Santos (president of the Chilean Episcopal Conference), "Chile Exige el advenimiento de una sociedad más justa," September 4, 1970, in Oviedo Cavada, 1974, pp. 26–28.

36. Conferencia Episcopal de Chile, 1968 and 1969. It was during this period (August 1968) that the Latin American bishops met at Medellín, where they denounced the status quo's "institutionalized violence" against the poor, committed the Church's energies and resources to the struggle for social justice, and endorsed the formation of small Christian communities (*comunidades cristianas de base*) within larger parishes. Concerned with the growing polarization within the Church, the Chilean bishops encouraged these communities to work on behalf of social justice but insisted that they give equal time to spiritual activities and that they avoid political partisanship.

37. See Oviedo Cavada, 1974, pp. 23–25, and pp. 10–15.

38. While most favored Tómic, the bishops did not make an issue of Allende's Marxism or of the Marxist parties supporting him, as they had in 1964. They knew that many Catholics were disillusioned with Frei and were supporting Allende. They also did not want to be seen making common cause with anti-Marxists on the right. Following a show of force by a Santiago-based army regiment in October 1969, the bishops denounced the alternatives of a military coup and leftist terrorism (aimed at the MIR), warning that either would lead to a "reign of terror." See Conferencia Episcopal, 1969.

39. Oviedo Cavada, 1974, pp. 29–30.

40. *Ultima Hora* (Santiago), November 12, 1970.

41. The document, "The Gospel, Politics, and Various Types of Socialism," appears in Oviedo Cavada, 1974, pp. 58–100, and is analyzed in Smith 1982, pp. 174–77.

42. Paul's letter "Octogesima Adveniens" is reproduced in Gremillion, 1976, pp. 485–512.

43. Interview no. 17 (Chile), June 22, 1990, in Santiago.

44. See their October 1973 statement, "Christian Faith and Political Activity," in Eagleson, 1975, pp. 179–228.

45. Interview no. 11 (Chile), May 29, 1987, in Santiago.

46. Centro de Opinión Pública, 1973.

47. During interviews in 1975, Smith found that nearly nine out of ten bishops (24 of 27 polled) believed there was no alternative to a military coup in September 1973. Over three-fourths of the priests interviewed (55 of 72) and almost two-thirds of the lay leaders in the CEBs (33 of 51) felt the same. See Smith, 1982, p. 210.

48. "Declaración del Señor Cardenal y del Comité Permanente del Episcopado Chileno," September 13, 1973, *Mensaje,* no. 223 (October 1973): 509.

49. When asked in 1975 whether the Church should publicly confront authoritarian regimes whose practices were in conflict with Catholic doctrine, or adapt to such situations so as to maintain the structures of the Church and the possibilities of administering the sacraments, 27 of the 30 active bishops indicated that the second strategy was more prudent. See Smith, 1982, p. 302.

50. Reproduced in *Mensaje,* no. 228 (May 1974): 196–98.

51. "Evangelio y Paz," September 1975, reproduced in *Mensaje,* no. 243 (October 1975).

52. None of Chile's roughly one million Evangelical Protestants participated in efforts to blunt the effects of repression. In fact, in December 1974, 2,500 Evangelical pastors signed a statement characterizing the military's intervention as "the response of God to the prayers of all believers who see in Marxism satanic forces of darkness in their highest expression." See Puentes Oliva, 1975; and Cleary and Sepulveda, 1992.

53. Interview no. 1 (Chile), March 24, 1987, in Santiago.

54. At the time of the coup, 90 percent of the operating budget of the Catholic university system and approximately 50 percent of the costs of Catholic primary and secondary schools were subsidized by the state.

55. Enhancing Silva's influence were the new bishops appointed on the recommendation of Msgr. Sotero Sanz Villalba, a papal nuncio in the Baggio mold. Of the seven episcopal appointments that Paul VI made between 1974 and 1977, five (T. González, Ysern, Camus, Herrada, and Aristía) were solid progressives.

56. One of them was Jaime Castillo, a Christian Democrat, and the other was Eugenio Velasco, a Radical. Their offense was to have presented a letter to the foreign ministers of the Organization of American States (OAS) meeting in Santiago, in which they protested the continuing state of siege and the abdication of their responsibilities by civilian courts.

57. Sociedad Chilena de Defensa de la Tradición. . . , pp. 377–400.

58. "Declaración del Comité Permanente," *Mensaje,* no. 250 (July 1976): 316.

59. Bishop Carlos González of Talca, Auxiliary Bishop Enrique Alvear of Santiago, and Bishop Fernando Ariztía of Copiapo.

60. The bishops insisted that such attacks were not "isolated incidents" but "part of an overall process or system" that "subjugates citizens to a dreaded omnipotent police state." See the Permanent Committee's "Declaración," *Mensaje*, no. 252 (September 1976): 436–37.

61. "Nuestra convivencia nacional" *Mensaje*, no. 257 (April 1977): 166–69.

62. Statements referring to specific abuses of power mentioned above are reprinted in issues of *Mensaje* for 1978 and 1979.

63. "Declaración del Comité Permanente de la Conferencia Episcopal de Chile," *Mensaje*, no. 288 (May 1980): 228; "Declaración sobre el plebiscito," *Mensaje*, no. 292 (September 1980): 519–20.

64. A former Vicaría official told us that the VPO was instrumental in the development of the Central Unica de Trabajadores (CUT), a new labor confederation that brought Christian Democrat and leftist workers together under the name of the Communist-Socialist-dominated federation of the Frei and Allende years.

65. See Poblete, Galilea, and Van Dorp, 1980, pp. 76 and 32.

66. Interview no. 6 (Chile), May 19, 1987, in Santiago.

67. Interview no. 12 (Chile), July 31, 1987, in Santiago.

68. The "cutbacks" were limited to the area of peasant associations, which henceforth would operate independently. But these had never been a source of trouble with the government, and the shift was apparently undertaken for reasons of political convenience.

69. In an interview with an Italian news agency, Silva said that while he still hoped that Chile would find a good road out of its present impasse, he did not think that things were going very well at the moment. Pinochet took this to be "a direct attack," and lodged a formal complaint with the Vatican, forcing Silva into another round of public and private explanations.

70. Interview no. 2 (Chile), April 15, 1987, in Santiago.

71. Most Popular Unity activists saw them as another expression of bourgeois reformist sentiment, while their right-wing and Christian Democratic opponents seemed bent on getting rid of Allende at any price.

3. The Peruvian Church: A Historical Overview

1. According to the *Annuarium Statisticum Ecclesiae* (1991), Catholics made up 92 percent of the country's population. As late as 1962, less than 1 percent of the population declared themselves to be Protestant, although Protestant denominations and sects have grown (to roughly 5 percent) since then. See below, chapter 7, for further discussion of this phenomenon.

2. These figures have been calculated by using data presented in *Directorio Eclesiástico del Perú*, 1984 and the *Annuarium Statisticum Ecclesiae*, 1991. In Peru, Church schools enroll 5.9 percent of all primary school students, 11.3 percent of all secondary school students, and 13.9 percent of all university students. Ironically, however, the less adequately staffed Peruvian Church can claim higher levels of sacramental contact with its Catholic population than its Chilean counterpart. In 1991, for each 1,000 Peruvian Catholics, 16.9 were baptized before they were seven, 10.1 made their First Communion, 6.8 were confirmed, and 3.2 were married in the Church. See *Annuarium Statisticum Ecclesiae*, 1991, pp. 299 and 307.

3. Mecham (1966, pp. 164–65 and 171) says that the Church acquired a great deal of land during the colonial era, becoming one of Peru's major landholders following independence. It does not appear to have kept the land for very long, however. According to Klaiber (1988, pp. 248–49), much of it was confiscated in 1826 and 1911, although he concedes that Catholic associations such as *cofradías* often owned land and used the revenues to finance their apostolic work.

4. Klaiber (1983, pp. 153–54) argues that only one of the country's forty oligarchical families, whose fortunes were made from the sugar and cotton lands of the northern coast, was either actively Catholic or supportive of the Church's agenda. The rest, he claims, had become secularized, and viewed the Church as an obstacle to change and progress.

5. According to the *Annuarium Statisticum Ecclesiae* for 1991 (p. 97), the number of Peruvian Catholics per priest was 8,425. Comparable figures for Chile were 4,989, for Brazil 9,423, for Poland, 1,518, for Ireland 687, and for the United States, 971. The Catholics per women religious ratio can be calculated by dividing the number of Catholics (pp. 35–39) by the number of women religious (pp. 87–91) in these countries. The resulting figures are: Peru, 4,001; Chile, 1,549; Brazil, 3,693; Poland, 1,305; Ireland, 402; and the U.S., 543.

6. The number of foreign-born priests working in Peru jumped from 874 (49.2 percent of the total number of priests) in 1959, to 1,153 (53.4 percent) in 1964, 1,465 (61 percent) in 1974, and 1,381 (62.3 percent) in 1984, before falling to 1,239 (52.7 percent) in 1987 and 1,111 (51.8 percent) in 1993. See the *Anuario Eclesiástico del Perú* for 1959 (p. 36), 1964 (p. 38), and 1974, and the *Directorio Eclesiástico del Perú* for 1984, 1987, and 1993. These figures are calculated in separate tables for 1959 and 1964. Place-of-birth data for 1968 are available only for the diocesan (non religious-order) priests, but are provided for individual dioceses (albeit with some omissions) in the 1974 *Anuario* and the 1984, 1987, and 1993 editions of the *Directorio.*

7. Klaiber, 1983, pp. 165–67.

8. Klaiber (ibid., p. 153) bases this conclusion on records indicating the occupations of those attending a meeting of the organization's chapter in Cusco in 1894.

9. Many of the legislators were Catholics representing rural areas dominated by large landowners (*gamonales*).

10. Ibid., pp. 167, 175.

11. The two leading parties of the nineteenth and early-twentieth centuries were the *civilista* and Democratic parties, each of which had liberal and conservative elements.

12. Astiz (1969, pp. 172–73) argues that the Church had become more independent of the state, and was not worried by the rising anticlericalism. Romero (1987) says that the Church was once one of the country's "basic institutions" but then lost stature and influence before later recovering them. Pike (1964) argues that the Church was weak during this period.

13. Quoted in Pike, 1964, p. 309. The archbishop himself was among the principal offenders in this regard. His ties to dictator Augusto Leguía (1920–31) were so close that he was forced to resign and leave the country when Leguía was overthrown in 1931.

14. By 1940, it would triple to 521,000, and by 1950 it reached 835,000, almost five times the original level. All figures are taken from Mitchell, 1993, p. 54.

15. See Klaiber, 1977, pp. 141–43.

16. APRA's penchant for violence worried most military officers. In 1932 in the northern city of Trujillo, APRA militants killed some 25 hostages before abandoning a military post they had seized. The army retaliated by shooting 1,000 suspected APRA militants and sympathizers in the immediate area.

17. Mecham, 1966, p. 167.

18. When strongman Manuel Odría was inaugurated as president in 1948 (after overthrowning the freely elected social Christian José Luis Bustamante y Rivero), Archbishop Guevara publically expressed the hope that "the new government and the Church could work closely together, because only when Church and state were closely united could the public welfare be effectively served." See *National Catholic Welfare Conference, News Bulletin*, 6/17/48. In contrast, Klaiber thinks (1988, pp. 318–19), that Farfán was much more progressive than most analysts portray him. He cites him warning that "charity alone cannot solve everything. It must be preceded by justice," and insisting that "there are good Christians in all the parties: on the right and on the left."

19. Klaiber (1988, pp. 288–89) mentions the Catholic Extension Center, the Young People's Social Action, and the Fides Center in this connection.

20. Ibid., p 321.

21. Dammert was named archbishop of Cajamarca in 1962, while Vallejos became archbishop of Cusco after entering the seminary in his thirties. Father Oscar Larson, who had been influential in the formation of social Christian youth in Chile, worked with some of these people while he was in Lima between 1935 and 1939.

22. César Arróspide, Ernesto Alayza, and Antonio Espinosa Lanz.

23. Klaiber (1983, p. 165) reports that there were 918 members of Catholic Action in the entire country in 1936, that only 130 members and 33 (!) ecclesial advisors attended the organization's first national congress in 1955, and that total membership in 1959 was about 2,100. These numbers compare poorly with the 100,000 men and women active in Chilean Catholic Action during the 1940s and 1950s, as reported above, in chapter 2.

24. Klaiber (1988, p. 320) reports that attendance at the Lima Congress in 1936 was over 100,000, fully one-fifth of the city's population at the time.

25. *Tercera Reunión Latinoamericana*, (Chimbote, Peru), cited in Coleman, 1953.

26. Ibid., pp. 27–33; Mecham, 1966, p. 176.

27. Priests were reluctant to serve in these locales because their salaries and living standards, which depended largely on the support they could generate from their parish communities, would be much lower. MacEoin, 1962 (pp. 156–57) describes a parish outside of Lima whose pastor barely managed to rush through weekend masses and catechism classes in each of its several population clusters.

28. Named a cardinal in 1962, Landázuri described himself as: "the pastor of everyone, but especially . . . the pastor of the poor, the sick, and those who suffer, because I consider them the privileged portion of my flock." *National Catholic Welfare Conference News Service Bulletin*, May 28, 1962. As one veteran observer noted (interview no. 15 [Peru], July 3, 1990, in Lima), Landázuri "was never known to be anything like an intellectual leader. I know for a fact that almost all his talks in public were written by other people. . . . But Landázuri, with his con-

tacts with the Church in the United States and Europe, was very much open to the social message."

29. According to Klaiber (1983, p. 169), 200 of the 348 people attending the week-long event were members of Catholic Action.

30. The "ordinary," as distinct from the "auxiliary," bishop is the highest ecclesial authority in a territory. Prelates, vicars, and apostolic administrators may exercise "ordinary" authority even though they are not consecrated bishops, i.e., even though they do not have the authority to ordain priests.

31. Landázuri was elected president of the conference's Permanent Committee (a body of eight to twelve bishops that met four times a year) in 1957, and held the position until 1988. He was also the leading force in the conference's yearly assemblies, and was responsible for naming the heads of its various commissions. The influence of national episcopal conferences has been publicly lamented in recent years by Vatican officials. According to Cardinal Joseph Ratzinger (1985, pp. 59–60) of the Sacred Congregation for the Faith, for example, "the decisive new emphasis on the role of the bishops . . . actually risks being smothered by the insertion of bishops into episcopal conferences that are ever more organized, often with burdensome bureaucratic structures. We must not forget that the episcopal conferences have no theological basis, they do not belong to the structure of the Church, as willed by Christ, that cannot be eliminated; they have only a practical, concrete function." And "No episcopal conference, as such, has a teaching mission; its documents have no weight of their own save that of the consent given to them by the individual bishops."

32. The term literally meant young towns, or towns in formation, as opposed to the older term *barriada,* which meant shantytown or "neighborhoodized" area.

33. These figures are rough approximations. The *Anuario Eclesiástico del Perú* puts the foreign priests working in Peru at 874 (49.2 percent of the total) in 1959 (p. 36). According to the Directorio Eclesiástico del Perú, the number climbed to 1,153 (53.4 percent) by 1964 (p. 38) and even higher in 1974 and 1984 (see note 6, above). Gibbons (1955, p. 32; 1956, p. 34; 1958, p. 22, and 1960, p. 24), puts the total number of priests working in the country at 1,490 in 1954, at 1,581 in 1956, at 1,673 in 1958, and at 1,740 in 1960. It may be that the numbers of new priests ordained (which Gibbons provides) and older priests retiring (which he does not) cancel each other out. Neither the *Anuarios* nor the subsequent *Directorios* differentiate between foreign-born and Peruvian nuns. We know from Gibbons, however, that the biggest jump in the total number of women religious came between 1958 (1,609) and 1960 (3,097). We do not know the number of Peruvian women entering religious life in any given year, but the substantial numbers of Spanish, Italian, Canadian, and U.S. sisters who arrived in the late 1950s almost certainly make up the bulk of these differences.

34. Maryknoll priests (roughly 70 at their peak) settled in Puno and Juli (in the southern Andean region) in the 1940s, and only later (1954) did some (roughly 20 percent) move to Lima. The orders sending the largest numbers of priests to Peru during these years were the Franciscans (174), the Jesuits (113), and the Vincentians (75). The Columban Fathers (some 30 in all) were assigned in the early 1950s to *barriadas* on Lima's northern outskirts.

35. The *hermandad,* or brotherhood, performs both religious and social functions in Peru. Typically, it was organized around a patron saint or personality, such as the Lord of the Miracles, with whom people identified, and in whose

honor festivals and public processions and rituals are held. Leadership positions in such organizations were a source of social identity and prestige.

36. The popular religious beliefs and practices to which most subscribed seemed to breed fatalism in both religious and social relationships, and were difficult for pastoral agents to understand or accept. But they used them as contexts for evangelization, and popular participation at the parish level and in smaller groups increased markedly. Their efforts to reach out to nonpracticing Catholics were aided by workshops and reflection groups organized by Father Gutiérrez, the Alvarez Calderón brothers, and lay popular educators.

37. Not all cases ended this way. Some priests and sisters kept on with their original methods and congregations. Others became disillusioned and returned to their home countries. Between a quarter and a third of the membership of some congregations left religious life altogether. Costello (1979, chap. 11) reports a dramatic decline in the number of missionary priests working in Latin America during this period. According to O'Keefe (1976, p. 395), of the 84 Maryknoll priests and brothers assigned to Peru in 1966, 37 left the region between 1967 and 1970. Of the the 67 who left between 1963 and 1973, 42 left the society within three years of leaving the region.

38. The statement issued by the group at the end of its first meeting is reproduced in Peruvian Bishops Social Action Commission, 1969, pp. 74–80. Archbishop Landázuri quickly endorsed the organization, saying he shared its concern for "material realities" (Valda Palma 1971, p. 74).

39. Neither Father Gutiérrez nor the Alvarez Calderón brothers were at the group's first meeting (at Cieneguilla, near Lima), although they endorsed the statement issued at its conclusion. At the next meeting, in Chimbote, in July, Father Gutiérrez traced the outlines of what would become his *Teología de Liberación.* According to one long-time Catholic social activist (interview no. 19 [Peru], July 14, 1992), Father Gutiérrez and others believed that an organization consisting entirely of priests would have greater credibility with both the bishops and the general public.

40. See McCauley, 1972, pp. 54–82; and Maloney, 1978, pp. 374–416.

41. Fifteen of the nineteen priests who were still with the organization after the first year were Peruvian diocesan priests, most of whom were associated with Father Gutiérrez. As to the consistency of the group's positions over time, compare its initial statement, or its pronouncement on agrarian reform (Centro de Estudios y Publicaciones, 1973, pp. 98–101), with its 1977 statements "The Situation of the People and Christian Responsibility" (Centro de Estudios y Publicaciones, 1978, pp. 36–38) and "To the People Struggling for Their Rights" (1978, pp. 87–89).

42. This organization (CONFERE) was made up of men and women elected by their orders and congregations to work in pursuit of common goals.

43. Father Antoncich served as secretary from 1968 to 1977.

44. Foreign priests working in the popular sector were more likely to be assertive than their Peruvian counterparts because of their cultural formation, and because they were organizationally and financially less dependent on their Peruvian bishops.

45. Religious priests and sisters were at an advantage vis-à-vis diocesan priests in this regard. The orders and congregations to which they belonged had their own chains of command and resource bases. They also had access to higher

levels of ecclesial authority (national and international). Finally, they were not obliged to work in any particular diocese, and thus provided personnel and resources that a bishop would not like to give up.

46. Lay pastoral agents and laypeople generally (from both middle- and popular-class backgrounds) were more inclined to think for themselves and to act on their own than previously.

47. Klaiber, 1988, p. 350; NCWC, 1961, 1963.

48. In 1963, three of the thirty-six bishops were members of Opus Dei, the conservative Catholic movement that originated in Spain in the 1920s. Many others were traditional conservatives in the mold of Archbishop Guevara.

49. A former PPC member (interview no. 15 [Peru], July 15, 1988, in Lima) described the Christian Democrats in the following terms: "Christian Democracy was such a minimal force that it was known as the 'four cats.' And when it divided, and a second party emerged, the four cats formed two parties, each with two cats. . . . I remember the time that one-time Christian Democrats were leaving the party and going over to Bedoya's (the new party led by Luis Bedoya Reyes). At first there were 5, then 10, 15, and 20, and I kept track, waiting until the number reached 100, because to say that 17 had changed parties would not really be news. So I told Rodrigo (another party member): 'there's no point in publishing this until we reach 100.' And he told me 'Hey, you don't know what you're talking about. The entire party consists of about 100.'"

50. None of the three candidates won the necessary 33.3 percent to win the presidency outright. Haya de la Torre of APRA received 32.98 percent, Belaúnde Terry of AP 32.1 percent, and former dictator (1948–56) ex-general Manuel Odría of the Unión Nacional Odrista (ONU) 28.45 percent.

51. Rodríguez Beruff, 1983, pp. 108–9, 123–25 (cited in Saba, 1987, p. 39).

52. The agrarian reform legislation passed by Congress covered only idle or inefficient haciendas in the poorest and most sparsely populated areas of the sierra. The large, lucrative coastal sugar plantations (owned by ONU supporters, and worked by APRA unions) were left untouched.

53. Peru's annual inflation rates under Belaúnde were 10 percent, 17 percent, 9 percent, 10 percent and 19 percent. See Kuczynski, 1977, pp. 101, 255.

54. Articles appearing in the *Revista Militar del Perú* during this period stressed the need for more coherent planning by the state, a more radical agrarian reform, and nationalization of the strategic mining sector, most of which was in foreign hands. See Stepan 1978, pp. 135, and 136–41.

55. Valda Palma (1971, p. 72) reports of insider dollar trading, million-dollar contraband rings, and other scandals that benefited ministers, deputies and other Belaúnde government officials during 1967 and 1968.

56. Army generals Rodríguez and Fernández Maldonado had recently completed *cursillos de Cristiandad* courses. See below, chap. 6.

57. See Stepan, 1978, chaps. 1 and 5–8. The French Dominican economist Josef Lebret, who assisted in the drafting of Paul VI's 1967 encyclical, *Populorum Progressio*, had lectured at CAEM during the mid-1960s, while the author of a recent *Revista Militar del Perú* article actually cited *Quadragesimo Anno*'s call for a state in which social organizations, not political parties, represented citizens and worked harmoniously for the common good.

58. The agreement (the Act of Talara) exempted the IPC from all taxes for

forty years, but the page on which these terms were spelled out was missing from the version released in August.

59. See Valda Palma, 1971, p. 75. ONIS also issued a statement denouncing the deal that Belaúnde had arranged with the International Petroleum Company. See McCaulay, 1972, p. 24.

60. The views of other officers ran the gamut from traditional conservatism, through pragmatic technocracy, to authoritarian nationalism. In an interview with one of us (no. 20 [Lima], July 15, 1992, in Lima), Velasco's minister of treasury and the man who ousted and succeeded him as president in 1975, General Francisco Morales Bermúdez, conceded that the military had not known what it was going to do before it seized power, and that its revolutionary program was worked out "as they went along."

61. Archbishops Dammert and Rodríguez Ballón, and Bishops Bambarén and Metzinger.

62. In March 1969, Jurgens suspended three of his priests for making statements in support of the wives of miners who had occupied the cathedral and for marching in protest against a society dance being held at a local country club. Bishop de Orbegozo was one of the three Opus Dei bishops at the outset of the Velasco period.

63. Pásara, 1989, p. 285.

64. Archbishops Landázuri, Rodríguez Ballón (Arequipa), Dammert (Cajamarca), Calderón (Puno), and Dalle (Juli).

65. Msgrs. Dammert, Dalle, and Durand argued that agrarian reform officials were mistakenly using coastal (as against indigenous) concepts and methods that were bound to misfire (interview no. 18 [Peru], July 12, 1992, in Lima). In October, Dammert reminded the public that the Church had been urging an agrarian reform for some time, but was critical of the "obstructionist tactics" of some parties and of the lack of qualified, honest and responsible administrative personnel for carrying out the reform.

66. Interview 17 (Peru), July 10, 1992, in Lima, with a person who worked for CEAS during the late 1970s and early 1980s.

67. For a discussion by a former official of the Velasco government, see Franco, 1979, pp. 96ff.

68. Pease García and Verme eds., vol. 1, 1974, p. 188.

69. Interview no. 18 (Peru), July 17, 1992, in Lima with Archbishop Ricardo Durand, then of Cusco and later of Callao.

70. Stepan, 1978, pp. 162–65. The acronym SINAMOS literally meant "without masters," but the organization was widely regarded as a mechanism for channelling popular participation in support of government programs. According to Franco (1979, pp. 55–56), SINAMOS did not really get started until 1973.

71. Bambarén was no radical. A priest who worked for many years in the *pueblos jóvenes* described him as impatient with ideological types and theoretical discussions, and as "more comfortable working with military policy–makers than he would have been with a left-wing civilian government." Interview no. 6 (Peru), July 10, 1987, in Lima.

72. "Documento de la Asamblea General del Episcopado sobre la justicia en el mundo, para el Sínodo de los Obispos, Agosto de 1971," in Conferencia Episcopal Peruana, n.d., p. 151.

73. "Evangelio, política y socialismos," cited in chap. 2.

74. Interview no. 15 (Peru), July 3, 1990, in Lima.

75. Interview no. 9 (Peru), July 15, 1987, in Lima.

76. See Klaiber, 1988, p. 384. Within a year Faith and Solidary Action evolved into a series of locally based groups, each of which tended to structure itself in accord with the interests and needs of people in its particular sector.

77. Gutiérrez, 1971.

78. The Peruvian bishops gave these communities a strong endorsement in their January 1973 letter "Evangelización y Liberación." See Conferencia Episcopal Peruana (n.d.), p. 208.

79. Two such centers were DESCO (Centro de Estudios y Promoción de Desarrollo) and CIDAP (Centro de Desarrollo y Participación). Ironically, DESCO was established by the Belgian Jesuit sociologist Roger Vekemans, who raised funds in Europe to get DESCO started but lost control of the organization to Peruvians who were much more sympathetic to left-wing social and political forces.

80. ONIS spokesmen were pointedly critical of Bambarén's insider relationship with the military, arguing that even when put to good use it tended to reinforce relations of dependence within both Church and society. Calderón Cockburn, Filomeno, and Pease García, eds., 1975, vol. III, pp. 800 and 1109; and Pease García and Filomeno, eds., 1977, vol. IV, pp. 1164, 1250–51, and 1567 provide excerpts from public statements by ONIS in support of striking workers and/or peasants seizing lands to which they had claims.

81. The text of ONIS's communiqué on the proposed decree-law to create a social property sector appears in Centro de Estudios y Publicaciones, 1978, pp. 169–73.

82. Pease García and Verme, 1974, vol. II, p. 568.

83. According to the bishops, "Solidarity with the poor and the oppressed also carries with it action to change the unjust structures that maintain the situation of oppression. True Christian charity compels us *[nos urge]* to work truly and efficaciously to create a more just society." See Conferencia Episcopal Peruana, n.d., pp. 184, 196, and 205.

84. Pásara, 1989, p. 285. Most accounts of the period of military rule in Peru refer to its first and second phases, the first (1968–75) under Velasco, during which the government pursued social reforms and populist economic policies, and the second (1975–80) under General Francisco Morales Bermúdez, whose government abandoned most of Velasco's reforms and imposed policies that were more conservative (i.e., anti-inflationary) and more accommodative of international interests. We identify three periods, believing that the Velasco regime had two distinct phases.

85. The number of man-hours lost and workers involved in strikes more than doubled in relation to the previous year's totals. See Stephens, 1983, p. 62.

86. Growth rates for 1974 and 1975 were 6.2 percent and 6.9 percent respectively, but inflation leaped to 13.7 percent in 1974 and 24.8 percent in 1975.

87. The bishops' formal statement, "Declaración del Episcopal Peruano sobre el Estatuto de la Prensa," appears in Conferencia Episcopado Peruano, n.d., pp. 261–64. Calderón Cockburn, Filomeno, and Pease García, eds., 1975, vol. III, p. 949, provides excerpts of the generally favorable response issued by ONIS.

88. Calderón Cockburn, Filomeno, and Pease García, 1977, pp. 1404–10. Although it was drafted in May, the document was not released until December 1975, several months after Morales Bermúdez led a coup against Velasco.

89. "Mensaje del Consejo Permanente del Episcopado con Ocasión de la Jornada Mundial de la Paz," in Conferencia Episcopal Peruano (n.d.), pp. 268–269.

90. Langton (1986) and Pásara (1988) argue that the changes were self-interested and relatively insignificant "adjustments," and that the Church remains an essentially conservative force. Klaiber (1988), Romero (1987), and Cortázar (1988), on the other hand, insist that the changes were real, both in terms of episcopal leadership and at the intermediate and local levels. Klaiber thinks that the conservative restoration has the potential to reverse much of what has been achieved in view of the "docility" of many Peruvians. Romero and Cortázar think that developments at the intermediate and base levels are irreversible, and that conservative bishops should not be presumed hostile to them.

91. Not all Peruvian workers, peasants, and urban poor were organizationally or sacramentally active. Rather, a sufficient number of such people became active that they came to constitute the Church's principal social base, so that it was no longer dominated by either the middle or upper classes.

92. They said nothing, for example, when social Christian president Bustamante y Rivera was overthrown by Odría in 1948. Still committed to the strategy of seeking and retaining powerful protectors, they viewed Odría as more congenial and a better hedge against the future.

93. Belaúnde's uncle was the prominent Catholic intellectual and sometimes political figure Victor Andrés Belaúnde.

4. The Chilean Church and the Transition to Democracy

1. Most authors (see note 6 below) agree with Drake and Jaksic (1991, p. 10) that the influence of the Church declined as secular movements and parties reassumed their normal positions and functions in the political arena.

2. See Berryman, 1994; Mainwaring, 1986; Krischke, 1991; and Green, 1993.

3. Nielsen, 1991, and Wooster, 1994.

4. It might be difficult for churchmen who defended human rights under a repressive military regime (the Chileans, for example) to win acceptance as an impartial mediator. Conversely, a failure to condemn military intervention, or identification with the military while in power (in Peru from 1968 to 1973), might lead to a questioning of the commitment to democracy of other Church leaders.

5. This is the anticipated outcome of observers such as Lernoux (1980) and Berryman (1984), who focus on the emergent "popular" Church.

6. Remmer (1989) mentions the National Accord but not the Church's role in its development. Drake and Jaksic (1991, p. 10) describe the Church as a secondary "ally," once the country's political parties returned to near-normal activity (in 1982–83). Constable and Valenzuela (1991) acknowledge the Church's role in the accord, but consider it a failure; they view the regime as having manipulated the pope's April 1987 visit, and downplay Church involvement in the 1986–89 period. An exception to this view is Garretón (1987, pp. 118–23) for whom the Church's role in promoting the accord helped to bring the opposition together even though the government ended up rejecting it.

7. See, for example, Correa and Viera-Gallo (1986), Castillo (1986,), and De la Maza and Garcés (1985). An exception to the rule are Constable and Valenzuela (1991), who argue that most Christian communities were neither radical nor highly politicized.

8. A transition to democracy is a staged process in which an authoritarian regime gradually cedes, or prepares to cede, power to a more or less democratic successor.

9. People on the left would not have responded as positively to the Church and its initiatives were it not for the ideological renovation taking place among leftists throughout the world during the 1970s and 1980s. Another important factor were the resources and encouragement provided by the United States ambassador, Harry Barnes. See Sigmund, 1990.

10. Sodano was named nuncio in 1978 by Pope Paul VI. An ideological and theological conservative who valued cordial relations between the local Church and government, he was named a cardinal and Vatican secretary of state by Pope John Paul II in 1990.

11. The new appointees included a Schoenstatt priest, Camilo Vial (first as auxiliary of Santiago in 1980 and then as bishop of San Felipe in 1983), Alberto Jara (Chillán, 1982), Miguel Caviedes (Osorno, 1982), and José Joaquín Matte (Military Vicariate, 1983). Two additional appointees, Juan Francisco Fresno (Santiago, 1983) and Francisco de Borja Valenzuela (Valparaíso, 1983), were theological conservatives who were already bishops at the time. The remaining three, Bernardino Piñera (San Felipe, 1983), Alejandro Jiménez (Valdivia, 1983), and José Manuel Santos (Concepción, 1983), were liberals in the mold of Cardinal Silva and were transferred to new positions. For a discussion of these matters, see de Ferrari (1992).

12. Conservative Catholics were particularly happy that José Manuel Santos was named archbishop of Concepcion and not Santiago. They were critical of the Church's "preferential option for the poor" and of its adversarial relationship toward the government's policies. By late 1980, they were accusing the bishops of making "contingent" political judgments and claiming the right to dissent from them. *Mensaje*, no. 317 (August 1983): 95–102.

13. See above, chap. 2.

14. "Caminar Juntos en la Iglesia," in Conferencia Episcopal Chilena, 1984, pp. 77–84.

15. While insisting that Church personnel (i.e., bishops, priests, deacons, religious men and women, and Christian communities as such) avoid "partisan associations," the bishops affirmed the right of lay Catholics to identify with any political groups whose views were consistent with Church teachings. In fact, six months earlier they lamented that many Christian communities "remained unduly confined to religious, cultural, and catechetical activities," were "insufficiently open to their environment," and paid insufficient attention to "problems of work, the economic situation, unions, neighborhood organizations, etc." See ibid., p. 59.

16. Msgr. Cristián Precht, a Silva protégé who was the first head of the Vicaría de Solidaridad, was later replaced by the apparently more conservative Juan de Castro, who turned out to be an equally forceful defender of the poor and repressed. Precht became pastoral vicar of the archdiocese and remained one of the most visible and influential figures within the Chilean Church through 1990, when Fresno retired.

17. See ibid., pp. 107–9.

18. The recession was brought on by the downturn in the international economy, a sharp reduction in the flow of dollars to Chile (that had financed its

earlier expansion), a fixed exchange rate, and automatic currency adjustment (which magnified, rather than mitigated, cyclical tendencies).

19. De la Maza and Garcés (1985) describe and analyze this involvement extensively. Baño (1992) stresses the divergent styles and objectives of middle-class and popular-sector protestors. After the first several protests, middle-class protestors stayed home and banged pots and pans loudly at prearranged times. Workers participated occasionally but were difficult to mobilize given high unemployment and the desire of those who had jobs to keep them. The backbone for most of the protests were slum-dwellers (many of them active in their local parishes or Christian communities), the urban poor generally, women's groups, and secondary and university students. The residents of the more radical shantytowns usually set up burning barricades at the entrances to their *poblaciones* and on nearby highways and thoroughfares.

20. See their letter "Más Allá de la Protesta," issued in late June, and reproduced in Conferencia Episcopal Chilena, 1984, pp. 117–18.

21. In a study by the Center for the Study of Contemporary Reality (Huneeus, 1986), respondents rated the Church the country's most trusted and respected institution and political parties the least respected.

22. The opposition was split into two blocs: the Democratic Alliance (AD) and the Popular Democratic Movement (MDP), led by the communists and harder-line socialists. Both groups endorsed the protests. But the AD insisted that they be nonviolent, and a source of leverage in negotiating with the military government, while the MDP believed that Pinochet would abandon power only if "forced" to do so, and that negotiations would simply prolong his rule.

23. They blamed the "Chicago Boys," for turning the recession into a full-blown crisis. See Silva (1991, pp. 98–127), for a discussion of the politics of Pinochet's shift from radical to pragmatic neoliberalism.

24. The Democratic Alliance was established in early August 1983. Headed by the Christian Democrats and the moderate (initially Briones, later Nuñez) wing of the Socialist Party, and including the Liberal, Republican, Social Democratic, and Popular Socialist parties, it issued a *manifesto democrático* in March 1984 that called for Pinochet's resignation as part of a "national accord."

25. Hugo Zepeda attended for the Republican Party, Luis Bossay for the Social Democrats, Enrique Silva Cimma for the Radicals, Gabriel Valdés for the PDC, and Ramón Silva Ulloa for the Popular Socialists. A sixth member, the moderate Briones wing of the Socialist Party, pulled out at the last minute so as not to anger other Socialist factions with which it was exploring possible reunification. Also attending the meetings, but adopting a more flexible position than Jarpa, was rightist and one-time strong regime supporter Francisco Bulnes. See Cavallo et al. 1989, pp. 410–20.

26. Przeworski (Mainwaring, O'Donnell, and Valenzuela, 1992, pp. 68–69) uses these terms to discuss bargaining relations in processes of democratization.

27. See Renato Hevia's account of the attacks on Church facilities and personnel during this period in *Mensaje*, no. 352 (September 1986): 349–51.

28. *Miristas* were members of the Movimiento de Izquierda Revolucionaria (Movement of the Revolutionary Left), which had defended the need for armed struggle under Frei and Allende as well. The government insisted that they were common criminals who would be given a "fair trial." The nuncio and the Chilean bishops countered that the granting of asylum in no way precluded future legal

proceedings, that no charges had been issued at the time asylum was granted, that a fair trial under present circumstances was problematic, and that the government's failure to guarantee safe passage from an embassy was an affront to the pope and the Chilean Church. Safe passage was granted after more than two months of wrangling.

29. Conferencia Episcopal de Chile, 1988, "Declaración del Comité Permanente sobre Llamado a un Gesto de Entendimiento," p. 23.

30. The government argued that the state of siege and other repressive measures were needed to counter the sharp increase in the incidence of violent attacks by paramilitary groups on the left.

31. The bishops had been asked to attempt something similar by the National Labor Command in early 1984, but had declined, citing the impropriety of political involvement. See their "Respuesta al Comando Nacional de Trabajadores," in Conferencia Episcopal Chilena (1988, p. 21). Our discussion of the origins and subsequent evolution of the accord draws on Precht (1992) and Cavallo et al., (1989, pp. 456–66). According to Cavallo et al., the bishops agreed in December 1984 that the Church should help the parties overcome their differences, and that Fresno should host bilateral meetings with party leaders on the subjects of reconciliation and the transition to democracy. In a 1992 interview, however, one of Fresno's advisors told us that Fresno proceeded without informing his fellow bishops, believing that undue publicity and commentary had been fatal to the talks with Jarpa in 1983.

32. All three were personal friends of Fresno. On Zabala's advice, the first leader invited was Patricio Aylwin, a representative of the Christian Democratic Party's more conservative wing (the party's president, Gabriel Valdés, was invited later). From the right, he invited the National (PN) and National Union (MUN) parties, and several independent rightists, but not the Independent Democratic Union (UDI), which still supported the regime and was critical of liberal democratic institutions. From the left, he invited Social Democrats and moderate (Briones wing) Socialists who were willing to rule out the use of violence, but not the Communists or harder-line (Almeyda) Socialists (who were not).

33. Until their merger in 1989, the separate factions of the Socialist Party were identified with their respective secretaries general. The more radical of the two groups was led by Clodomiro Almeyda from the late 1970s on. The moderate faction was led by different people at different times and was referred to variously as the Briones, Nuñez, and Lagos faction.

34. A copy of the text appeared in *Hoy*, no. 424, September 2–8, 1985, and is excerpted in Garretón, 1989, pp. 441–44.

35. Fresno invited the signatories to occupy pews opposite those of government officials at the traditional *Te Deum* rite on September 18, but this gesture was ignored as well.

36. Merino is said to have thrown a copy of the accord into his wastebasket, saying "I don't read such things." Pinochet, on the other hand, dismissed Fresno and the signatories alike as "useful fools *[tontos útiles]* of the left."

37. According to Garretón (1987, p. 99, note 11; and 1989, pp. 420–21), the accord helped the opposition to become more cohesive and to broaden its support. It was followed, in September 1986, by the Bases of Support for a Democratic Order (Group of 13), from which the MUN and IC withdrew, but to which MAPU, the Partido Humanista, and various socialist parties were added, and, in

early 1988, by the Concertación por el No, which added the Almeyda Socialists, MAPU Obrero Campesino, the Party for Democracy (an electoral front headed by Briones Socialists), and a number of smaller parties.

38. With respect to the economy, the signatories agreed to give priority to overcoming extreme poverty and marginality, creating productive and stable job opportunities, and achieving a high and sustained rate of growth.

39. The Communists and Almeyda Socialists, who still endorsed armed opposition and a revolutionary platform for the postmilitary period, were not invited to either the early discussions or subsequent negotiations. Nor were the UDI and Avanzada Nacional, who continued to support the government.

40. Garretón (1989, p. 118) argues that in mediating the accord the Church went beyond its earlier denunciations of violations of human rights. At a time when opposition forces were badly divided, he contends, "the Catholic Church played a permanent, and at times central role, oriented toward the working out of agreements and the establishment of levels of mutual confidence that facilitate dialogue and increasingly powerful consensus favorable to the idea of moving toward a democratic regime." Parker (1990) makes a similar argument.

41. Cavallo et al., 1989, p. 488. According to another observer, "the interview became a 20 minute monologue in which the general lectured Fresno and enjoined him to stop playing politics. Fresno was reportedly dumbfounded and unable to respond. At the end of the interview Pinochet handed the confused cardinal an oversized, gift-wrapped box, threw open the doors of the conference room and heartily embraced him in the presence of reporters and television cameras specially convoked for the occasion." See *Latinamerica Press* 18, no. 5 (February 13, 1986): 3.

42. Interview no. 3 (Chile), April 15, 1987, in Santiago.

43. Interview no. 10 (Chile), May 28, 1987, in Santiago.

44. Interview no. 11 (Chile), May 29, 1987, in Santiago.

45. Silva once commented to a friend: "they say we [he and Msgr. Fresno] are very different, but we are actually quite a bit alike. It is just that he is easier to like *[simpático]*."

46. The Permanent Committee was more liberal on social and ecclesial issues than the conference as a whole, particularly with recent episcopal appointments. During this period, the committee's declarations attributed most of the the spiraling violence to socioeconomic conditions, the lack of opportunities for political participation, and the regime's repressive methods and tactics. See their declaration "Justicia y Violencia" of April 7, 1986, in Conferencia Episcopal de Chile, 1988, pp. 147–48.

47. Interview no. 13 (Chile), July 21, 1987, in Santiago.

48. The term *allegados* refers to family members or others who were welcomed into what was usually an already overcrowded apartment or single-family dwelling.

49. Interview no 4 (Chile), April 23, 1987, in Santiago.

50. Interview no. 9 (Chile), May 26, 1987, in Santiago.

51. Interview no. 8 (Chile), May 25, 1987, in Santiago.

52. Among them were bishops Prado, Medina, Infante, Moreno, Lizama, and Fuenzalida y Fuenzalida.

53. According to Cavallo (1992, p. 41), opposition leaders were stunned by claims that the weapons were wrapped to be stored for several years, and that the

communists apparently were planning to take up arms against the civilian government that would succeed Pinochet's.

54. The Papal Visit Committee was headed by Bishop Francisco Cox Huneeus, a moderate conservative, but included representatives from other tendencies. Among other things, it insisted on prior approval of the remarks to be made by the representatives of Christian communities and others who would be speaking at papal functions, although its efforts to control content were not always successful. See Cavallo et al., 1989, pp. 509–20.

55. Constable and Valenzuela (1991, p. 298) argue that the regime successfully manipulated the visit, and that its net effect was at best neutral, and more likely harmful, to the opposition.

56. See Precht 1992, pp. 9–11.

57. It was announced in June that the plebiscite would be held before February 1989 (the date most frequently mentioned was September 11, 1988). Leftist parties refused to have anything to do with any of the mechanisms (voter registration, party registration, plebiscite, etc.) established by the 1980 Constitution. The Christian Democrats were divided, with most preferring direct elections and holding out for them. In fact, many observers thought that Pinochet would be better off in an open presidential election, in which opposition votes might be split among several candidates. Pinochet preferred the plebiscite format, however, because it assured him of an either-or, us-and-them context, and because he thought that much of the left would refuse to participate.

58. Their decision was eminently political. They could argue, of course, that their ultimate concern was pastoral, i.e., the achievement of reconciliation and lasting peace, and that they were not promoting particular political groups or interests. But their judgment that it offered a reasonable opportunity for achieving peace rested on an assessment of the plebiscite's likely costs and consequences, and was both contingent and political.

59. The laymen were Alberto Etchegaray, Sergio Molina, José Aguilera, Javier Luis Egaña, and José Zabala.

60. CIEPLAN is a private economic research institute staffed by former professors of Santiago's Catholic University. Its president, Alejandro Foxley, was later named treasury minister of the Aylwin government. René Cortázar, who would be Aylwin's minister of labor, was particularly helpful in these discussions.

61. A member of the mediation committee told us (interview no. 20 [Chile], July 24, 1992, in Santiago) that some entrepreneurs thought the political moment was "too delicate" for them to sign such a document.

62. In August, the bishops issued a statement urging acceptance of the mechanisms provided by the 1980 Constitution (which called for a plebiscite before 1989), but asked for the designation of an official candidate other than Pinochet.

63. See Gárfias' statement in *Mensaje*, no. 373 (October 1988), pp. 463–64, in which he urged Catholics: (1) not to be paralyzed by fear, worries or threats; (2) to vote for the common good and not allow themselves to be swayed by either pressures or utilitarian personal interests; (3) to vote for what they really wanted lest they regret later not having done so; and (4) to remember that the act of voting was one that could be freely carried on in the complete privacy of the voting booth (the regime was promoting the idea that somehow Pinochet could and would find out if a person voted against him).

64. Interview no. 16 (Chile), June 1, 1988, in Santiago.

65. See Cavallo et al., 1989, pp. 583–88.

66. In interview no. 18 (Chile), July 22, 1992, a retired army general told us that he didn't know what Pinochet's initial intentions might have been. But what was important, he said, was what he ended up doing, and that even had he wanted to set aside the results, the military's commitment to the Constitution was too strong to permit him to do so. He also implied that the military commanders took seriously both the insistence of right-wing forces (such as Andrés Allamand) that the results be respected and the relative calm with which many entrepreneurial elements viewed the postmilitary future.

67. They produced four basic changes: (1) The status of human rights was strengthened and Article 8 (which permitted the exclusion of groups for the ideas they espoused and not just their conduct) was eliminated. (2) The National Security Council was expanded to include civilian as well as military members, and was given advisory instead of veto power. (3) The number of elected senators (in addition to the 9 appointed senators) was increased from 28 to 39. (4) The process of amending the Constitution was made considerably easier.

68. Pre-election polling proved remarkably accurate. Most polls gave Aylwin between 50 and 57 percent of the vote, Büchi around 30 percent, and Errázuriz between 9 and 20 percent. Contrary to what might have been predicted several years earlier, opposition parties worked together much more effectively than did former regime supporters.

69. The law required that, for a single list to obtain both seats within any electoral district, it would have to obtain two-thirds of the votes, otherwise the second seat would go to the opposing list's highest vote-getter, even if (s)he had fewer votes than either of the top two finishers. When all was said and done, however, opposition candidates won 53 percent of the vote and 62 percent of the seats (74 of 120) in the Chamber of Deputies, and 55 percent of the vote and 58 percent of the seats (22 of 38) being contested in the Senate.

70. Three retired military officers that we interviewed in July 1992 (interviews nos. 17 and 18, July 22, in Santiago, and no. 19, July 23, in Valparaíso) thought that the Church was too supportive of the opposition, and too hostile to the government to serve as a genuine mediator or interlocutor.

71. With the bishops calling for mutual concessions, people in the armed forces and on the political right had a more difficult time dismissing them as simply an extension or mouthpiece of the opposition.

72. See Silva, 1991.

73. Manuel Antonio Garretón taught university-level courses in the Academia's night school, while Luis Razeto and Humberto Vega worked in its Economics and Work Project.

74. Leftist professionals and social activists like Luis Hermosilla, José Manuel Parada, José Zalaquett, and Germán Correa held staff positions with the Vicaría de Solidaridad or the Vicaría de Pastoral Obrera, both of which were official Church agencies or programs; others worked in Church-sponsored organizations such as the Agrupación de Familiares de Detenidos Desaparecidos, which had its offices in the Vicaría de Solidaridad, or for the Center for Pastoral Reflection, which operated independently under one of the zonal vicariates.

75. Ana González, a communist whose husband and three sons were murdered by security forces in 1976, was the subject of a video film (*La Comunión de las Manos*), and falls somewhere between these alternatives. Although "not a be-

liever," she was drawn to share in the Eucharist with her Catholic co-workers in the Agrupación de Familiares de Detenidos Desaparecidos. For an account of how cultural Catholics found a new expression of religious faith in terms of commitment to social justice and solidarity with the oppressed in Church-sponsored programs after the coup, see *Mensaje*, no. 239 (June 1975), pp. 246–51.

76. Van Dorp, 1985, p. 65.

77. Mutchler, Langton, Gismondi, and Grigulevich see the Church having a decidedly conservative impact on such people.

78. It is difficult to generalize about the development of these groups. As one of the priests with whom we spoke (interview no. 17 [Chile], June 20, 1990, in Santiago) put it, "the process of [development of] the communities . . . is very uneven and difficult. Their launching is difficult. Some begin, fail to catch on, and disappear. Others replace them, and some of them catch on, but even they have their ups and downs. But those that have kept going . . . generally pull together in line with a common affinity. It can also be because of their physical proximity, but most of all because of a certain affinity. [It's that], well, each of the communities has its own particular defining feature *[toque particular]*."

79. See De la Maza and Garcés (1985), for discussion of the participation of Church groups in the various protests.

80. The Christian Left and MAPU were both radical offshoots of the Christian Democratic Party. MAPU-Lautaro broke off from the latter in the late 1970s, opting for armed struggle. A priest who worked for years in a militant sector of Santiago's west side (interview no. 17 [Chile], June 20, 1990, in Santiago) characterized the experience of many young Christians in these terms: "For many people it was an opening up to an entirely new, distinct, and vital Christianity, that attracted a lot of people, and was something that they would never lose, although some of them did, later, leave the community. But there are lots of political leaders today, those around here, you know, who say that 'our commitment to the people, we acquired it as Christians. It was the community that gave us that, it was [a vision of] Christ committed to the people. . . .' [It's just that now] they simply have muffled their way of living the Christian life. . . . They know they are Christians, but they don't share a common Christian life, either sacramental or religious."

81. Ibid. Vergara was one of two brothers, sons of Catholic activists Manuel Vergara and María Luisa Toledo, who were killed by carabineros in what officials claimed was an attempted robbery of a bakery in 1985.

82. Interview no. 5 (Chile), May 8, 1987, in Santiago. "Miguel," one of the young leftist militants interviewed by Politzer (Politzer, 1988), provides a striking example of this phenomenon. After discovering politics in his First Communion class, he joined the Young Communists and later belonged to a group of young people that attacked police during the 1983 protests. In his words, "All of a sudden I feel I would like to devote myself completely to organizing actions [setting up barricades, throwing Molotov cocktails, etc.], but I would have to abandon my studies and be a full-time activist in the MIR or somewhere else. I would have to be like the Vergara brothers, and it is a real shit not to do it *[me caga la onda no hacerlo]*. Because as I understand the prophet Jeremiah, today we have to be Christians in deed and not just in words" (ibid., p. 106).

83. Those being radicalized (those joining the Communist Party, MIR, and MAPU-Lautaro, for example), would presumably be obstacles to a peaceful

transition, while those holding more moderate attitudes and favoring the negotiation of differences with the regime would enhance its prospects.

84. A 1987 survey of Christian communities in greater Santiago (Valdivieso, 1989) found that only a few communities had one or more members active in social and political organizations: 16 percent of them had members active in health organizations, 10 percent in unions, 8 percent in human rights organizations, 8 percent in meal programs, and 7 percent in political parties. Most respondents agreed that the country's socioeconomic conditions were bad and that changes were necessary, but they did not think that their communities were responsible for doing anything about this. Members in 39 percent of the communities conceded that most of them thought alike on these matters, but 20 percent insisted that they did not talk about social or political matters at their meetings, 11 percent said that they did, but that they generally disagreed on what to do, although without arguing about it, 16 percent said that they disagreed and did argue about it, and only 15 percent thought that it would be better if they did think alike politically.

85. Another survey (Zona Sur de Santiago, n.d.), focusing on communities in the politically active southern zone, found that only 5.6 percent of the communities had members who admitted belonging to a "political group," that the focus of their meetings were ecclesial (42.4 percent) as opposed to social or politi-cal (27 percent) or personal (22 percent) matters, and that their activities rarely (11.6 percent) involved social action. But it also discovered that most group reflections were a mixture of spiritual, ecclesial, and social matters, that they often (55 percent) discussed social and political realities, and that 76.9 percent of them had some members who participated in social or political organizations.

86. Warwick and Lininger (1975, pp. 75–77) describe a purposive sample as based on selection from significant groups chosen for their importance in testing hypotheses. Other recent books on religion and politics in Latin America (and in the U.S.) have constructed such purposive surveys. Levine (1992) and Burdick (1993) each interviewed large numbers of local-level Catholic activists participating in base communities. Both relied on contacts and advice from clerics to identify, gain access to, and acquire credibility with their respondents. In the United States, Jelen (1991) surveyed only church attenders in fifteen denominations in one Indiana county to generate new data on correlations between religious and political beliefs. Because they did not construct random samples, none of these studies can be used as a basis for generalizing about all activists in these countries. But they do provide valuable new insights into the interplay between religious and political beliefs and practices, and they have generated new hypotheses for further research. We have conducted our study in a similar spirit.

87. Until recently, social scientists studying rank-and-file Catholics focused almost exclusively on sacramentals, whose regular mass attendance presumably made them the more responsive to Church teachings and authorities, and therefore more "representative" of the Church in their dealings with the secular world. We differentiate organizationals from sacramentals, however, because we believe that the former have been more fully exposed to Church teaching.

88. In looking for "cultural Catholics," we asked informants (local priests and other pastoral agents) to identify politically active lapsed Catholics with whom they had regular contact. Not surprisingly, the "cultural" respondents they came up with included a disporportionately large number of leftist activists whose views were likely to be more radical than those of either organizational or sacra-

mental Catholics. To limit the possibly distorting effects of their leftist political ties, we decided to consider only "lapsed" Catholics, i.e., those who still considered themselves to be Catholics, but who only attended mass infrequently (less than once a month or only on religious feast days). We lumped the remainder (those who never go to mass) with the "fallen-away," i.e., those who once were, but no longer consider themselves to be, Catholics.

89. The wording for these and other survey questions are provided in the Appendix.

90. In terms of general ideological progressiveness, however, a person willing to consider recourse to violence (either generally or under certain circumstances) would appear to be more committed to change, and therefore to be more progressive. In terms of a peaceful transition to civilian rule, however, categorical rejection would be more functional than either qualified or unqualified endorsement.

91. I.e., those whom our informants identified as activists (and therefore presumably leftists), but who still considered themselves Catholic and had at least infrequent contact with the Church.

92. While our Chilean survey included more than 200 respondents from middle- and upper-class parishes, the one we conducted in Peru (chapter 7) was limited to people living in *pueblos jóvenes* and popular neighborhoods.

93. For class status, we took the respondent's occupation. If he or she did not work, we took the occupation of the head of the household in which he or she was living. Among our sample's Chilean Catholics living in *poblaciones* or lower-middle-class areas, blue-collar and independent workers made up 38.7 percent, white-collar workers 27.5 percent, and professionals, small businessmen, owners, and managers 32.7 percent. Independent and blue-collar workers were slightly overrepresented among organized Catholics, comprising 42.5 percent of those included in the sample.

94. If their answers to all four questions are combined, 44.8 percent of the independent workers chose the most progressive option. Corresponding figures for blue-collar workers, white-collar workers, and petite bourgeois elements were 50.6 percent, 42.5 percent, and 54.8 percent.

95. Spiritually oriented parishes stress sacramental ministry, encourage traditional (e.g., devotional) apostolic activities, and view social commitments and concerns as potentially distractive and / or divisive. Socially oriented parishes encourage the integration of social concerns and commitments into the life of the religious community. The religious sociologists of CISOC who assisted in the survey's design included an equal number of each parish type in the sample.

96. Santiago's averages were 58.7 percent for the plebescite, and 55.2 percent for the presidential election.

97. Most of the order's priests working in Chile were Chileans. Prominent theologians Pablo Fontaine and Ronaldo Muñoz, Rev. Felipe Barriga, formerly vicar of the archdiocese of Santiago's southern zone, and Father Juan Andres Peretiakowicz, the popular former-head of the Archdiocesan Youth Office, are among its most prominent members.

98. For liturgies, special occasions (like weddings and first communions), and even organizational activity, people were drawn to the parish church, and away from the local chapel, because of its location "on the avenue."

99. De la Maza and Garcés (1985) report high-level Christian community support during several of the protests in the general area served by the parish.

100. Father Bolton preferred to "accompany," as opposed to "lead" the community, and for years refused to serve as "pastor." He wanted people to minister to one another. Most people in the community, on the other hand, wanted a "real" pastor, and most of the sisters believed they would be better served by one as well. In 1990, Father Bolton was persuaded by archdiocesan officials to assume conventional pastoral responsibilities, which he did until 1992.

101. Typical issues include whether or not to allow a political group to use a community facility, whether or not to support as a community a candidate in neighborhood board elections, and so on.

102. Rev. Rafael Hernández, who was one of the seminary students who lived with Father Mariano Puga in the early 1970s, is currently vicar of the archdiocese's southern zone. Manuel Vergara and Luisa Toledo de Vergara, whose three sons were killed by government security forces during the mid-1980s, are no longer active in the community although Manuel works for the southern zone vicariate.

103. Support for the No vote in the Las Américas section of Estación Central was 66.1 percent. Support for Aylwin's presidential candidacy was 63.5 percent, and for opposition candidates for the Senate and Chamber of Deputies was 65.5 percent and 62.7 percent respectively.

104. If our concern were general ideological progressiveness, we would reverse the sign, thereby making Dolores organizationals even less progressive than other Catholics, while making organizationals in the other two parishes or communities more so.

105. Father Bolton told us (interview no. 17 [Chile], June 20, 1990, in Santiago) that although he has been arrested many times for protest and other antiregime activities, he was not the militant radical that many people (including his ecclesiastical superiors) assume him to be. "As people were often stoic and unable to move, unable to take action even in the face of mobilizations and protests, I would often preach in favor of participation in the people's struggle. And once I said this, I realized that many people had probably assumed that I was referring to armed struggle, when in fact I was not. In fact, I have always been skeptical of the armed path because things have turned out to be different, you know. There was a time when we did not think there was any other way out. But, well . . . I have abandoned that position."

106. Progressive responses included: (1) viewing Christ's death as a consequence of his role as a political revolutionary; (2) following one's conscience as against Church authorities, in great moral decisions; (3) approval of the bishops taking positions on economic issues; and (4) siding with one's priest, as against one's bishop, if their views differed.

107. The party with which people were affiliated may account for some of this disparity. The number of cases here is extremely small but may still be indicative. The number of party members in Cristo Liberador (2) is too small to be considered. But in Parrales, four of the five party members belonged to leftist parties, while in Dolores three of the nine were Christian Democrats, whose views were much less progressive (averaging 12.5 percent) than those of the other six (averaging 66.8 percent) who were leftists.

108. See Valdivieso, 1989.

5. Chile's Consolidation of Democracy

1. The rates of increase of GDP for 1991 and 1992 were 6.0 percent and 10.0 percent, respectively. Ruiz-Tagle (*Mensaje*, no. 415 [December 1992], pp. 480–83) attributes the reduction in the incidence of poverty under Aylwin to a reduction in the rate of inflation (and thus an increase in the value of wages), an increase in the number of jobs, and increased social benefits.

2. As late as October 1995, military authorities defended retired General Manuel Contreras' refusal to begin serving the seven-year sentence to which he was condemned for his role in the Letelier-Moffit murders. Rosenberg (1995, p. 49) emphasizes the strong support for democracy of the country's principal right-wing parties but quotes a leading businessman's opinion that the government should not press the military too hard.

3. See Mainwaring, O'Donnell, and Valenzuela, eds., 1992, p. 3, and Pérez-Díaz, 1993, p. 3.

4. Pacts like the one that the Church helped employers and workers reach in 1988 and the Aylwin government formalized in June 1990, reduced the extent and scope of bargaining, and could postpone or preclude social learning opportunities. Alternatively, the efforts of opposing sides to undermine or extend earlier arrangements might simply resurrect old controversies, sowing further mistrust and antagonism.

5. Principally industrial and financial elites and the Unión Democrática Independiente (UDI) and Renovación Nacional (RN) parties.

6. Most analysts (e.g., Guillermo O'Donnell and Samuel Valenzuela, both in Mainwaring, O'Donnell, and Valenzuela, 1992; and Reuschemeyer, Stephens, and Stephens, 1992) prefer the standard of "formal" as opposed to "substantive" democracy in assessing transition and consolidation. In their view, cultural and socioeconomic developments that would strengthen nonelite influence on, and support for, political institutions are not prerequisites for democratic status. Because formal and substantive democracies are not always related linearly (the former's stabilization can adversely affect the latter), and because transition and consolidation are not always chronologically or politically distinct, we follow Garretón (1987) in insisting on the need for nonpolitical supports (e.g., modernization, an economic model consistent with social reforms, and new, more responsive and accountable political parties) if institutions are to be fully democractic or enduring.

7. Modernization strengthens democratic institutions and practices by exposing people to citizens with different points of view, helping them to become more sympathetic to them, to view authorities in terms of performance (not status), and to develop appropriate standards for use in assessing institutions and policies. Social democratization, on the other hand, strengthens civil society, encourages organizations that are run along democratic lines, increases access to information and ideas with which citizens can form better judgments, and distributes resources and political influence more broadly and equitably. As it proceeds, "popular" or subordinate classes are able to promote their interests and concerns in the political process more effectively.

8. See Reuschemeyer, Stephens, and Stephens, 1992, chap. 2.

9. According to 1989 survey data, 42.5 percent of all Chileans expressed a high degree of confidence in the Catholic bishops, a figure 12 percentage points

higher than that for any other national institution or group. See Centro de Estudios Públicos, 1991a, cuadro 15, p. 46 and table 72.

10. The Concertation parties had a majority in the Chamber of Deputies, but the Senate (with its nine appointed senators) was under opposition control, and could block any legislation it wished.

11. González was bishop of Talca and president of the Episcopal Conference's Permanent Committee. His letter ("Truth and Reconciliation") was intended to help orient pastoral ministers in his own diocese but was reprinted and distributed throughout the country, effectively pressuring the Aylwin government to address the issue more forcefully.

12. See González, 1990, p. 22.

13. The grave contained twenty bodies. Each body was buried, in a separate gunny sack, with its hands tied behind the back.

14. "Asumiendo la Verdad," in *Mensaje*, no. 390 (1990): 260.

15. *Latinamerica Press*, vol. 22, no. 45 (December 6, 1990), p. 3.

16. The only exceptions to the amnesty were the Letelier-Moffit murders of 1976 and other overtly criminal acts.

17. *Latin American Weekly Report*, 91–24 (June 28, 1990): 10.

18. A case in point was the call in December by both the Vicaría de Sodalidaridad and Archbishop Jiménez of Valdivia urging the exhumation of eighteen bodies in the south.

19. Pinochet was concerned about a possible congressional investigation of his son's involvement in a $3 million arms procurement scheme, while senior army and navy people were afraid that they too might be the targets of inquiries or judicial proceedings.

20. González criticized Pinochet's order as lacking in proper respect toward civilian leadership and as contributing to a climate of fear. See *Latinamerica Press*, vol. 22, no. 48 (December 27, 1990): 2.

21. In early 1991, government and right-wing members of the Chamber of Deputies agreed to a multipartisan "peace proposal." It called for: (1) public acknowledgment of the truth concerning abuses; (2) extension (not abrogation) of the 1978 amnesty law; (3) acceptance by all groups of at least some responsibility for the 1973 coup; and (4) commitment to a process of mutual forgiveness and reconciliation among all. But before the military could respond, human rights organizations and groups representing the relatives of victims dismissed it as a "deal" between an overly timid government and rightist elites.

22. It came to be known as the Rettig Report, after commission chairman, Raul Rettig, a moderate Radical Party jurist. Other members included Ricardo Martín, a lawyer and one-time Pinochet supporter, Professor Gonzalo Vial, a conservative (and promilitary) historian, José Zalaquett, who worked for the Comité para la Paz and the Vicaría de Solidaridad, before being forced into exile (and working for Amnesty International), Christian Democratic lawyer Jaime Castillo of the Chilean Committee for Human Rights, Mónica Jiménez, a lawyer for the archdiocese of Santiago, Laura Jiménez, a corporate lawyer, and José Luis Cea, a moderate constitutional lawyer. The commission conducted hearings and interviews over a period of nine months, and had full access to records compiled since 1973 by the Vicaría de Solidaridad and the Chilean Human Rights Commission.

23. Some names had been published elsewhere, however.

24. Had the commission tried to look at all cases of torture and mistreat-

ment, its time frame (six months, renewable for another three) and staff resources would have been inadequate.

25. See "Con los Criterios del Evangelio" ("From the Standpoint of Gospel Values"), reproduced in *Mensaje,* no. 397 (March–April, 1991): 103–4.

26. See *Latin America Weekly Report,* 91–14 (April 18, 1991): 2.

27. According to a poll taken in the wake of the assassination, the government's rating for its efforts in the area of human rights went down slightly, the percentage of people who believed that the release of the Rettig Commission's report was a good idea fell from 63.4 percent to 52.6 percent, and the percentage of respondents who thought human rights issues were one of the areas to which the government should be directing its attention fell from 14.2 to 8.2 percent. See Centro de Estudios Públicos, 1991b, pp. 82 and 83.

28. "Día de Oración de Chile," *Mensaje,* no. 404 (November 1991): 468.

29. *Latin America Weekly Report,* 92–30 (August 6, 1992): 1.

30. Looking back on his efforts on behalf of human rights, Aylwin said: "The truth was established, and even if we might not have the truth in every case, we have the overall truth *[a verdad global]* and [in some instances] fairly specific truth, which half the country denied when I became president. Today, I think that everyone accepts it. I don't know if the entire military accepts it. I think the military's problem is that a formal acknowledgment would create big morale problems among its officers." See the interview with Aylwin in *Mensaje,* no. 426 (January 1994): 34.

31. "El Reencuentro que Anhelamos." in *Mensaje,* no. 423 (October 1993): 535–36.

32. See Vives Péres-Cotapos, Cristian, "¿Cómo disminuye la pobreza en Chile?" in *Mensaje,* no. 445 (December 1995): 29–33.

33. Vergara, 1994, p. 248.

34. Reuschemeyer, Stephens, and Stephens (1992, p. 222) explain the frequent acceptance of restricted democracy and the repeated support of antidemocratic coups by Latin America's middle classes as a result of the weakness of the region's working class. Barrett (forthcoming) characterizes the reformed labor code, to which the CUT leadership reluctantly agreed in June 1990, as severely limiting both collective bargaining and the right to strike.

35. The term was used to refer to the Christian Democratic economists (some of whom were devout Catholics) of CIEPLAN, an independent institute for economic research that they established (with funds from abroad) when they were dismissed from the Catholic University following the 1973 coup.

36. "Día de la Oración por Chile," in *Mensaje,* no. 404 (November 1991), p. 468.

37. *Latin America Weekly Report,* 91–42 (October 31, 1991): 2.

38. Oviedo's position harkened back to statements made by Chilean bishops in the 1930s and 1940s, in which they attributed economic ills to the acts and attitudes of individuals and called for a "change of heart" rather than for changes in social structures. In view of the pastoral statements cited in chapters 2 and 4, however, it is clear that Oviedo and most of his fellow bishops have been attentive to the structural dimension of socioeconomic problems for some time. But in keeping with the humanistic thrust of Catholic social teaching, they believe that structural initiatives must be accompanied by individual conversions if they are to be effective.

39. "El Reencuentro que Anhelamos." in *Mensaje*, no. 423 (October 1993): 534.

40. See *Mensaje*, no. 450 (July 1996), for coverage of the document.

41. *Latinamerica Press*, vol. 23, no. 38 (October 17, 1991): 7.

42. Rayo and Porath, 1990, pp. 37–39. Intensity of religious practice had a great deal to do with one's views in this regard: While only 34.9 percent of the practicing Catholics approved of legalization, 45.6 percent of the not very practicing Catholics, and 51.4 percent of the nonpracticing Catholics did. In 1985, only 57 percent of the Catholics interviewed in a comparable survey were willing to approve of divorce and remarriage in certain instances. See Van Dorp, 1985, p. 112.

43. Centro de Estudios Públicos, 1991a, p. 61.

44. Centro de Estudios Públicos, 1991b, p. 63.

45. *Latinamerica Press*, vol. 23, no. 26 (July 11, 1991): 4.

46. For Archbishop Piñera's views, see *Mensaje*, no. 450 (July 1996), pp. 32–35.

47. Their declaration *Unido para Siempre* appears in ibid, pp. 226–28. Part of it was published as "El Divorcio con Disolución del Vínculo" in *Mensaje*, no. 401 (August 1991): 293–94.

48. Ibid., p. 294. By mid-1996, several proposals legalizing divorce were circulating in the Congress, prompting another lengthy pastoral letter from the bishops, defending indissolubility. See Conferencia Episcopal de Chile, 1996a.

49. To date, Cuba is the only Latin American country in which abortion has been legalized, although the World Health Organization estimates that between 10 and 12 million Latin American women obtain illegal abortions each year, and almost one in ten of these requires hospital treatment for postabortion complications. Legislation legalizing abortion in extenuating circumstances was introduced in Argentina and in the state of Chiapas in Mexico in 1990, but was rejected in both cases. See *Latinamerica Press*, vol. 22, no. 35 (September 27, 1990): 5, and vol. 23, no. 14 (April 18, 1991): 7.

50. Much to the Vatican's displeasure, as many as a third of the Chilean bishops are said to be willing to support one or the other of these bills on grounds that they will produce less harm to families than existing laws. For a representative expression of their thinking, see the essay by retired Archbishop Bernardo Piñera, "Intervención de la iglesia en un Debate Público," in *Mensaje*, 450 (July 1996), pp. 32–35.

51. Practicing Catholics were less likely (70.9 percent) to approve than nonpracticing (79.3 percent) in the first of these cases. See Rayo and Porath, 1990, p. 40. These figures represent a substantial increase in liberal sentiment in relation to 1985, when only 42 percent were willing to justify abortion if the mother's life was at stake, and 32 percent if there had been a rape. See Van Dorp, 1985, p. 112.

52. See "AIDS in Latin America: Deception and Denial, An Epidemic Looms." *New York Times*, January 25, 1993, pp. A1 and A6.

53. The Chilean bishops had always opposed artificial means of contraception but had never targeted the government's family-planning programs specifically.

54. *Latinamerica Press*, vol. 25, no. 38 (October 21, 1993): 5.

55. According to a September–October 1991 CEP-ADIMARK study, 63.1 percent of these interviewed thought that it was all right for people who truly loved one another to have sexual relations before they were married. Significantly, 54.4 percent of the observant Protestants interviewed were opposed to the idea (Centro de Estudios Públicos, 1991c, p. 79). In 1985 (cf. Van Dorp, 1985, p. 101), 31 per-

cent of all Catholics were strongly in agreement, and 28 percent somewhat in agreement, with the view that "there is nothing wrong with an engaged couple having sexual relations before their marriage."

56. *El Mercurio*, June 11, 1992, p. A8.

57. Ibid.

58. Caro was backed by Antonio Moreno, newly appointed archbishop of Concepción, and his auxiliary bishop, Felipe Bacarrezza.

59. Bishop González was the seminary's rector and Bishop Caro was one of Correa's classmates.

60. Stoll, 1990, pp. 111 and 337, argues for 21 percent. CERC (Rayo and Porath, 1990, p. 3) puts the 1989 figure (including Mormons and Jehovah's Witnesses) at 15 percent, as does Van Dorp (1985).

61. Stoll, 1990, p. 316.

62. Conferencia Episcopal de Chile, 1992b, p. 28.

63. According to Rayo and Porath (1990, p. 3), Pentecostal evangelicals grew from 1 to 1.8 percent of all Chileans. Evangelicals as a whole remained about the same (10.5 as against 10.6 percent). Adventists fell from 1.7 to .5 percent while Jehovah's Witnesses fell from .8 to .7 percent, and Mormons grew from .9 to 1.3 percent during the same period.

64. Conferencia Episcopal de Chile, 1992b, p. 42.

65. Ibid., pp. 72–78.

66. *The Catholic Herald*, Milwaukee, October 22, 1992, pp. 1, 17.

67. Centro de Estudios Públicos, 1991b, pp. 45 and 73.

68. See "Porqué no se escucha a la Iglesia," in *Mensaje*, no. 413 (October 1993): 173–74.

69. According to the 1989 CERC study (Rayo and Porath, 1990, p.5), the percentage of Catholics describing themselves as "practicing" fell from 37.8 in November 1987 to 32.8 percent in October 1989, while the percentage describing themselves as "nonpracticing" rose from 8.7 to 16.7 percent.

70. Conservative in the sense of being more concerned with doctrine, clerical roles, orthodoxy, and discipline. According to one observer (*Latinamerica Press*, vol. 25, no. 38 [October 21, 1993]: 5): "Most of the younger clergy reportedly fit this (conservative) pattern. While there are notable exceptions, newly ordained priests tend to discourage lay participation and are more comfortable with parish structures than with base communities."

71. Interview no. 9 (Chile), May 26, 1987, in Santiago.

72. Interview no. 3 (Chile), April 15, 1987, in Santiago.

73. Interview no. 5 (Chile), May 8, 1987, in Santiago.

74. Interview no. 6 (Chile), May 18, 1987, in Santiago.

75. Interview no. 15 (Chile), May 30, 1987, in Santiago.

76. Reilly, 1986, p. 50, is representative of this line of thought.

77. Interview no. 2 (Chile), April 15, 1987, in Santiago.

78. See above, chapter 4

79. The questions were worded as follows: (1) "Of the following types of government, which do you consider the most appropriate for the country: (a) an authoritarian military government; (b) a civilian government that excluded extremist and nondemocratic groups; (c) a civilian government that excluded no one; (d) a civil government that excluded the rich [popular democracy]; (e) other?" (2) "In your opinion, what is most important in politics: to defend the

interests of one's own class or social group, or to try to harmonize them with those of other groups or social classes?" (3) "What role would you assign the Armed Forces in a civilian democratic government?" (open question).

80. Rayo and Porath, 1990.

81. It was based on a representative national sample that included more conservative small towns and provinces beyond Santiago.

82. They were less politically oriented than nonreligious Chileans (58.5 percent), however.

83. These Catholics correspond to the regular sacramentals in our 1987 survey. Unfortunately, the CERC survey did not identify Catholics who were active in Church-sponsored social or religious organizations, and we cannot know if they remained (as they were in 1987) more politically oriented than sacramentals.

84. Ibid., pp. 17, 13, and 33. It might be noted, however, that most young people (38.3 percent) identified with the political center as well.

85. The figures for all Chileans were 37 percent, 22.2 percent, and 26.2 percent respectively. The figures for those living in Santiago were 38.5 percent, 23.3 percent and 28.1 percent respectively. Evangelicals were less centrist (29.7 percent) and both more rightist (24.1 percent) and more leftist (29.9 percent). Ibid., pp. 24 and 26.

86. Sixty-one percent of evangelicals and 68.1 percent of those having no religion at all held such views. Ibid., p. 29.

87. Almost sixty-seven percent of evangelicals and 72.8 percent of those having no religion at all thought so. Ibid., p. 30.

88. The CERC survey asked respondents how frequently they attended mass, but not (as did our 1987 survey) whether or not they belonged to a Church-sponsored social or religious organization.

89. Support for dissent among middle- and upper-income Catholics in our survey was lower. Only a small portion of the religious organizationals (13.3 percent) and the CC and social organizationals (18 percent) chose conscience over Church authority, and only 16.9 percent of the latter would support their priest over the bishop. One-half of the religious organizationals, however, said they would side with their pastor over against the bishops in a disagreement between the two. Middle- and upper-income Catholics only account for about 20 percent of the overall Catholic population in the country, whereas working-class Catholics make up well over two-thirds. Hence, the loyalty to the Church by more wealthy Catholics does not offset the potential dissent by greater numbers of poor.

90. Centro de Estudios Públicos, 1991a, pp. 61 and 63.

91. Gilfeather, 1976, pp. 191–94; 1980, pp. 198–224, and 1985, pp. 58–73.

6. The Church and the Transition to Civilian Rule in Peru

1. These figures are taken from Wilkie, ed., 1983, vol. 22: pp. 21, 179, and 348; and Wilkie, ed., 1984, vol. 23: p. 286.

2. The document was entitled "Declaración de la Comisión Episcopal de Acción Social sobre Participación Política." It was an abbreviated, pamphlet version of a report written earlier in 1975 by a special commission appointed by the Permanent Council of Bishops in response to a request by the (then) Velasco government. It was "reactualized" by CEAS for distribution to Christian communi-

ties in early December, 1975, and was picked up by the press in February, 1976. It appears in Conferencia Episcopal Peruana, n.d., pp. 387–401, and in Pease Garcia and Filomeno, 1977, vol. V, pp. 1404–10.

3. See Conferencia Episcopal Peruana, n.d., pp. 323, 325, and 326.

4. In their October 1976 statement, the bishops said that the Church's social teaching did not offer a "model that simply needed to be carried out or applied." It rather provided a "light, a dynamism that is compatible with creativity, with the advances of the sciences, and with flexible responses that might be given different concrete circumstances." Ibid., p 321.

5. Interview no. 11 (Peru), July 27, 1987 in Lima. Following his talk, Pease was subjected to hostile questions by Fernando Vargas Somocurcio, the conservative archbishop of Arequipa. When the meeting adjourned, Vargas approached Pease to suggest that he, Landázuri, and Pease have lunch. During lunch, Pease took Landázuri aside, and asked him: "Didn't you tell me that it was Fernando that had objected most strongly to my speaking?" To which Landázuri responded, "Who told you that?" and Pease replied, "You just have."

6. One of the bishops we interviewed (interview no. 18 [Peru], 1992, in Lima) told us that Morales always intended to relinquish power to civilians at some point, and that he confirmed this in an informal meeting in August 1976. Rumors that he would be stepping down began to circulate in September, although Morales said nothing publicly for several months.

7. Morales told us (interview no. 20 [Peru], July 15, 1992, in Lima) that the return of power to civilian hands was as important an objective of his government as economic revitalization.

8. The Southern Andean region comprised of archdiocese of Cusco, the diocese of Puno, and the prelatures of Sicuani, Ayaviri, Juli, and Chuquibambilla. Local lawyers, merchants, and both civil and military officials were all critical of the assistance that Church people provided the peasant organizations. They viewed the latter as "terrorists" in their own right and willing collaborators of the Shining Path when, in fact, they were often the only forces actively resisting the organization.

9. Centro de Estudios y Publicaciones, 1978, p. 39.

10. Father Klaiber told us (interview no. 15 [Peru], July 3, 1990, in Lima) that "In 1978, I think it was, at a celebration of Landázuri's 25 years as priest (bishop), . . . I remember that I was at the cathedral. The Plaza de Armas was full, the bishops were there, and a statement of the Southern Andean bishops had just come out supporting the teachers' strike, and it was circulating throughout Lima like wildfire. And as the bishops came out of the cathedral the entire Plaza de Armas broke out in applause. There was unexplicit political symbolism there, all over the place."

11. Interview no. 20 (Peru), July 15, 1992, in Lima.

12. The three generals had completed *cursillos de cristiandad* (i.e., little courses in Christianity). These were spiritual renewal retreats consisting of two or three days of prayer, reflection, and personal discussion with a priest director and in small groups. They promoted lay Catholic spirituality and involved laypeople more actively in the life of the Church. Respect for the hierarchy and compliance with official Church teachings were constant themes, and the focus of most groups was on charity toward the poor rather than social justice.

13. Garland, 1978, pp. 182–83.

14. A Jesuit who was his superior when he (Bambarén) was a seminarian told us (interview no. 18 [Peru], July 13, 1992) that Bambarén's ring was as described, and that he wore it as an expression of solidarity with workers and of faith that Christ would defeat Marxism in the end.

15. The document went on to question whether an order that precluded brotherhood could be termed "Christian." See Pease García and Filomeno, 1977, pp. 1938–40, which reproduces the document in its entirety.

16. This term was used by a CEAS functionary (interview no. 17 [Peru], July 14, 1992, in Lima) who worked with Antoncich for several years. She thought that by the time he left the commission in late 1977 it had lost credibility with many bishops.

17. Among them were Ernesto Alayza, Teresa Tovar, Pilar Coll, José Burneo, and Imelda Vega Centeno.

18. Like their predecessors, the Velasco and Morales governments distributed patronage at strategic junctures and developed constituencies of sorts. But Velasco's was the first Peruvian government to encourage the development of regional, district, and block-level organizations to provide active support for its programs and initiatives. Most priests and nuns working in these areas sympathized with Velasco's objectives, and urged people to participate but to avoid co-optation.

19. San Marcos University was an APRA stronghold, but most of its student activists abandoned the party when (in the mid-1950s) Haya renounced its reform agenda in exchange for regaining legal status. At the Catholic University, where the Christian Democrats were dominant, many saw the move as further evidence of the bankruptcy of all existing parties. For a discussion of the radicalization of university students during this period, see Cotler, 1988, p. 161.

20. A former UNEC activist whom we interviewed (interview no. 4 [Peru], June 23, 1987) told us that she refused the demand of Vanguardia Revolucionaria leaders that she renounce Catholicism before being allowed to affiliate with that organization.

21. This distinction (Gutiérrez, 1970) enabled some of Gutiérrez's followers to reject Christian Democracy and to define themselves as leftists on strictly political grounds. Gutiérrez later abandoned the notion when he decided, in working out his theology of liberation, that religious and political commitments could not and should not be compartmentalized.

22. Interview no. 2 (Peru), April 13, 1987, in Lima.

23. A long-time lay activist told us (interview no. 4 [Peru], June 23, 1987, in Lima) that Gutiérrez openly pushed the PCR as "the party of preference" for progressive Catholics during this period. Gutiérrez himself (interview no. 14 [Peru], August 5, 1987, in Lima] denied this, as did several other Catholics politically active at the time.

24. Generals Rodríguez Figueroa, who had been head of SINAMOS under Velasco, and Fernández Maldonado, who had been his minister of mines, were the movement's leading figures. Morales forced Rodríguez to retire early, and had him expelled from the country (along with three associates) for publicly accusing Morales of having betrayed the military's revolution.

25. Interview no. 15 (Peru), July 3, 1990, in Lima.

26. They charged that the notebooks developed by the National Catholic

Education Office's Extension Department were Marxist in their orientation and "incited" readers to "class struggle." See Klaiber, 1986, p. 149.

27. Klaiber (ibid.) quotes Bishop Arboliú of Huanuco as saying the materials were an "impossible and monstrous symbiosis of Marxism and Christianity."

28. When the date for the Constituent Assembly elections was announced in October, the government urged citizens to unite "to create a climate of peace and mutual respect to make the transition possible."

29. Morales Bermúdez told us (interview no. 20 [Peru], July 15, 1992, in Lima) that most of the bishops were critical of his government's economic policy in their private communications and that some (he specifically mentioned the Andean bishops) were quite outspoken publicly, which he saw as an attempt to pressure him to abandon his policies. He complained that the bishops in general had no "experience" with "adjustment" policies, and that they were much more demanding of him than of the Belaúnde and García governments that followed his, even though their policies were much more drastic in this respect.

30. Interview no. 17 (Peru), July 10, 1992, in Lima.

31. Interview (no. 17 [Peru], July 10, 1992) with one of the two CEAS staffers assigned responsibility for this work during the late 1970s.

32. This view was expressed by a long-term CEAS staff-person in interview no. 21 (Peru), July 16, 1992, in Lima.

33. By 1979, the rate of economic growth had rebounded to 4 percent, real wages rose by 1.8 percent in 1979 and 11.7 percent in 1980, the country's balance of payments showed surpluses of $1,556 million for each of those years, and foreign exchange reserves of $1,295 million and $1,600 million were accumulated during the same period, thanks to increased oil exports.

34. Vatican permission was necessary since the existing concordat was an agreement between Rome and the Peruvian state, and any changes would have to be approved by the two parties.

35. Cardinal Landázuri told us (interview no. 16 [Peru], July 7, 1992) that in the 1930s the government took a more active role, submitting a list of three candidates from which the Vatican was asked to choose, but that in more recent times (since the 1960s) the Church simply submitted its choice to the government before announcing it, to see if the government had serious reservations.

36. The most complete study of Protestantism in Peru is Marzal, 1989.

37. Several days later the concordat (*el patronato nacional*), which gave the Peruvian government a hand in the naming of new bishops, was formally abolished. See Marzal, 1989, p. 268, and Klaiber, 1988, pp. 474–75.

38. ONIS issued several statements in late 1978 and early 1979, but then gradually receded from the public and political arenas.

39. In one section, the bishops wrote: "Because of its agonizing urgency, the sorrowful situation in which most of our people are living ought to be the first order of business of the government assuming power following the elections. The problems of hunger, health care, housing, work, and education are of such a magnitude that all citizens must assume responsibility for them and join together to overcome them. The gravity of these problems demand of the various parties and their leaders an intense search for development alternatives [models] that are appropriate for our reality." See Conferencia Episcopal Peruana, 1989, p. 47.

40. Some leftist leaders urged support of Belaúnde, arguing that none of the

leftists had a chance, and that he was the least reactionary of the remaining candidates.

41. See Cotler, 1988, p. 172.

42. Cotler (1988, p. 166) defines these as institutionalist and *pinochetista* elements, the latter being those who wanted to remain in power.

43. Cortázar (1989, pp. 54–55) disputes Klaiber's assertion (1988, p. 411) that the bishops chose not to criticize the government during the late 1970s for fear of jeopardizing the return to civilian rule. In his view (1989, p. 53), the hierarchy "put aside its earlier caution and began openly to support such [popular] struggles." But he includes in "the hierarchy" the Episcopal Conference's Permanent Council, its secretary general, the Social Action Commission (CEAS), and regional groups of bishops like those from the Southern Andes; and he later concedes (ibid., p. 54) that they were impelled by "a dynamism that originated in intermediate and base-level groups." We think Cortázar understates the Episcopal Conference's retreat from organized labor and other anti-Morales initiatives following the Constituent Assembly election of 1978.

7. The Church and the Consolidation of Democracy in Peru

1. Measured in terms of government intervention, tax burdens, and overall regulation, Belaúnde's policies were much more "liberal" or "neoliberal" than those of Morales Bermúdez, of whom the bishops had been quite critical.

2. According to *Latin America Weekly Report* (80, no. 35 [September 5, 1980]: 6), APRA "somewhat surprisingly" accepted the Belaúnde-Ulloa "program," but Cotler (1988, p. 174) says that, APRA refused Belaúnde's offer to join his government, opting instead for the "loyal opposition." It is not clear from his account how APRA reacted to his economic policies initially. Apparently, the party's conservative and populist wings disagreed on what its position should be.

3. Conferencia Episcopal Peruana, 1989, pp. 55, and 57–58.

4. These figures are drawn from Ramírez, 1993, pp. 61–77.

5. Conferencia Episcopal Peruana, 1989, pp. 90–92. The blatant misuse of public positions for personal gain was deeply resented by ordinary Peruvians and made them cynical regarding newly restored democratic institutions.

6. A missionary priest who had worked in the *pueblos jóvenes* of Lima for twenty-five years described (interview no. 7 [Peru], July 11, 1987, in Lima) the Velasco period as "an important era with a great deal of social enthusiasm, an era that opened up new possibilities for the popular sectors, for the poor. It was a difficult time. . . . But in social terms it was the last era of good news the country has had. Nothing has gone right in Peru since then." In contrast, many middle- and upper-class Peruvians blamed the country's current difficulties on Velasco and openly questioned the political judgments of churchmen (like Landázuri, Dammert, and Bambarén) who had supported him.

7. Thirteen of the seventeen bishops appointed during the next five years were theological and ecclesial conservatives. Five of the thirteen were Spaniards and a sixth was a Peruvian who had studied in Spain. Their selections were the work of a new papal nuncio, Msgr. Mario Tagliaferri, a conservative diplomat who shared the new Pope's determination to reduce the Church's political exposure and to emphasize more traditional moral and ecclesial concerns. Only three of the seventeen (one of whom had been a bishop in all but name since the 1950s)

were moderate progressives who thought the Church should address social and political matters such as development models. These characterizations are based on a review of the Peruvian press of the period and on discussions with Father Jeffrey Klaiber, the author of two books on the Peruvian Church (Klaiber, 1977 and 1988) and a professor of history at Lima's Catholic University for more than twenty years.

8. In an interview (no. 15 [Peru], July 3, 1990, in Lima), Klaiber told us that "the conservative vote dominates the conference, there is no doubt about that. But the other problem is that it's a weak conference anyway, because Peru is such a highly missionary pastoral country, that about half the bishops are missionaries who really don't want to talk about ideology at all. And so they just want to go back and do things in their dioceses. So the conservatives dominate but everyone is uncomfortable about explicit battles. So they back off of a lot of issues."

9. Ramírez, 1993, pp. 61, 69, and 73.

10. Michel Ascueta in Villa El Salvador, José Távora in Comas, and Arnoldo Medina in San Juan de Miraflores, for example.

11. Church officials and activists in the Southern Andean region were an exception to the general tendency not to challenge the organization.

12. Sendero confined itself to Ayacucho during most of 1980 and 1981. In 1982, it began making forays into neighboring departments, and the number of civilians ("suspected terrorists") killed increased dramatically (particularly once the army assumed responsibility for counterinsurgency efforts from the police. Deaths per year jumped from 200 or 300 the first two years to 2,000 over the next several. They were reported regularly in Desco's *Resumen Semanal* and summarized from time to time in *Latin American Weekly Report.* A priest (interview no. 21 [Peru], July 16, 1992, in Lima) who worked for eight years in a Sendero-controlled area told us that the organization "took as its enemy those who took it as theirs."

13. Critiques of Gutiérrez's work began to appear in the publications of CEDIAL, a Catholic research institute directed by the Belgian Jesuit Roger Vekemans, following the Medellín Conference (at which Gutiérrez played an influential role). CEDIAL worked with Archbishop Alfonso López Trujillo and other neoconservatives in combatting "misreadings" and "misapplications" of the Medellín documents. Conservative Peruvian Catholics (Garland, 1978) picked these criticisms up in the mid-1970s when they launched their attack on progressive forces.

14. In an interview (no. 16 [Peru], July 7, 1992, in Lima), Landázuri recalled that a theologian friend of his scrutinized Gutiérrez's work, and reported that it needed some "clarification" but contained no heresy.

15. The first document was published in August 1984 as *Instruction on Certain Aspects of the Theology of Liberation.* See Sacred Congregation for the Doctrine of the Faith, 1984.

16. This story was related by a priest (interview no. 7 [Peru], July 11, 1987, in Lima) who had accompanied the Peruvian bishops to Rome. While there, he persuaded a high-ranking Vatican official who had visited his Lima slum parish a year earlier to approach John Paul II on Gutiérrez's behalf. Apparently, the official did, defending the theologian as Church treasure rather than enemy, and persuading the pope to examine his writings personally.

17. Conferencia Episcopal Peruana, 1989, pp. 270–71. David Molyneaux

(*Latinamerica Press,* vol. 16, no. 45 [December 6, 1984]: 1–2) thinks that the document was more progressive than we are suggesting here. He stresses the concessionary paragraphs that Father Gutiérrez's allies managed to have included, but we see these as symbolic gestures that leave the document's otherwise conservative tone intact.

18. Conferencia Episcopal Peruana, 1989, pp. 281–82.

19. In 1981, the conference secretary Bishop Vargas Alzamora urged regional bishops to refer to their gatherings as "meetings" and not "assemblies." See Conferencia Episcopal Peruana, 1989, pp. 61–62.

20. Tovar's (1986b, p. 131) figures put unemployment (outside agriculture) at 13.9 percent in 1973 and 16.4 percent in 1984. Underemployment for those years was 51.5 percent and 53 percent. Unemployment and underemployment levels for people living in *pueblos jóvenes* (more than half of Lima's population) would be much higher.

21. According to Tovar (ibid., pp. 135–39), the economic difficulties of the mid and late 1970s followed upon a period of prosperity under Velasco. In that context, people were willing to set their sights high (on wage increases, structural changes, the fall of the government, etc.), and to run greater risks in pursuing their goals. After rebounding modestly in the late 1970s and early 1980s, the Peruvian economy fell into another recession, but with civilian rule newly restored there was no political crisis and no dramatic political alternative in sight. Accordingly, people lowered their expectations, hoping merely to hold on to the little they had, and to help one another survive.

22. Interview no. 15 (Peru), July 3, 1990, in Lima.

23. IU candidates won nine of the eleven poorest districts in greater Lima in the 1980 municipal elections, and all twelve in 1983 elections. See Tuesta Soldevilla, 1989, p. 51.

24. García won more than 50 percent of the vote in each of the twelve districts that Barrantes had won in 1983, while Barrantes managed between 20 percent and 36 percent of the vote in these areas. Apparently, pastoral agents (who were openly sympathetic toward the IU) were unable to shake people from their traditional political attachments or logic. They were unable, in effect, to project the Church's considerable local influence into national political leverage.

25. It was clear by late 1986 that García would not compromise with the left and was unlikely to abandon policies that were either economically risky or likely to further alienate key economic and political groups. See Poole and Rénique, 1992, pp. 7–11.

26. The growth rate was 8.6 percent, and the rate of inflation fell from 163.4 percent to 77.9 percent. But the idle industrial capacity (the economy was said to be at 64 percent capacity at the start of García's term) was quickly absorbed, and the lack of new investment signalled trouble ahead.

27. Unemployment fell from 9.1 percent (1985) to 3.4 percent (1986).

28. Poole and Rénique, 1992, p. 11.

29. Several months earlier, the bishops had withdrawn their representative from the National Peace Commission because García refused to order the military to accept the commission's recommendations regarding counterinsurgency methods.

30. Conferencia Episcopal Peruana, 1989, pp. 297–99.

31. A conservative observer (interview no. 10 [Peru], July 21, 1988, in Lima)

viewed Landázuri's endorsement as a way of staying on good terms with García: "The fact that the Cardinal has been in office since the oligarchic government of Prado, that he was chosen Archbishop of Lima as the result of pressure from General Odría, who was a dictator, that he was archbishop under the populist Belaúnde, that he had no difficulty maintaining his position during the Velasco government, since his brother-in-law was Velasco's prime minister, that he continued on under the Morales Bermúdez government, which tried to undo what Velasco did, that he got along well with Fernando Belaúnde in his second government, when he was not so populist, and that today he is close to Alán García, seem to me to reflect an impressive political capacity, wouldn't you say?"

32. As long, the bishops said, as these activities appeared to be consistent with efficient and socially responsible use of resources. See Conferencia Episcopal Peruana, 1989, pp. 325–28.

33. The Southern Andean region suffered a significant turnover in leadership in recent years. Msgr. Dalle of the Ayaviri prelature died in an automobile accident in 1982. His replacement, a French Dominican named D'Alterroche, was not as conservative as Mendoza, but was much less progressive than Dalle. Peruvian Jesus Calderón remained bishop of Puno. U.S. Maryknoller Al Koenigsknecht headed the Juli prelature until his death in 1986. Another Maryknoll priest, Michael Briggs, served for four years as apostolic administrator until a permanent successor, Bishop Raimundo Revoredo, was named in 1990. Canadian Albano Quinn remained in the Sicuani prelature.

34. The fact that his brother was an army general may have influenced the archbishop in this matter.

35. Three new auxiliary bishops (Sala in Cuzco and Ugarte in Abancay in 1986, and Cipriani in Ayacucho in 1988) were members of the conservative secular institute Opus Dei.

36. See Conferencia Episcopal Peruana, 1989, p. 327.

37. The second instruction was issued in March 1986 as *Instruction on Christian Freedom and Liberation*. See Congregation for the Doctrine of the Faith, 1986.

38. Interview no. 15 (Peru), July 3, 1990, in Lima.

39. Conferencia Episcopal Peruana, 1989, p. 334.

40. Pope John Paul II, 1988, p. 5.

41. In contrast, another liberation theologian, the Uruguayan Jesuit Juan Luis Segundo, conceded that the Instructions probably were referring to him but said they were wrong to do so. See Segundo 1985, p.14.

42. According to Klaiber (interview no. 15, July 3, 1990, in Lima), Durand defeated progressive Jose Dammert of Cajamarca by a margin of two votes.

43. Bishops identified with Sodalitium and Opus Dei assumed the presidencies of key conference commissions, and conservative Jesuits were named to other posts. See McCoy, 1989, p. 528.

44. Conferencia Episcopal Peruana, 1989, pp. 368 and 370.

45. Ibid., pp. 372–73.

46. Sendero militants would steal cattle and give them to peasant communities, hoping to incriminate them and turn them against the government. With the exception of Mendoza, however, the Southern Andean bishops continued to defend the peasant communities.

47. Dammert argued that the *rondas* were necessary as long as the government was incapable of dealing with problems. Local elites and authorities argued

precisely the reverse, with some, if not equal, plausibility. Dammert strongly opposed García's efforts in 1987 to take control of the *rondas* by subordinating them to local military authorities.

48. Dammert and other bishops tried to persuade *ronda* leaders to be more conciliatory toward local authorities. But they could not always constrain them and were even less successful in convincing officials to be more reasonable. In fact, conflict intensified, showing again that the politics were complicated, that the bishops were not always as politically sophisticated or effective as they needed to be, and that the "right" thing was not always easy or possible.

49. With floods in the north, the drought in the Puno area, and the spread of guerrilla activity and counterinsurgency operations beyond Ayacucho, the need for the social services that CEAS provided increased dramatically in 1987.

50. According to one local Church observer (interview no. 8 [Peru], July 14, 1987, in Lima), conservative bishops just wanted CEAS to hand out used clothing.

51. A CEAS staffperson told us (interview no. 21 [Peru], July 16, 1992, in Lima) that Misereor was contributing about U.S. $350,000 yearly in the late 1980s and early 1990s, but was willing to give more if the Peruvians could come up with matching funds from another source.

52. CEAS hired eight to ten lawyers (on contract, not permanent staff) to handle cases involving people accused of belonging to Sendero. Card. Landázuri defended CEAS and occasionally followed the counsel of its lay and clerical advisors, but he never gave it the active leadership or attention that Cardinal Silva provided the Vicaría de Solidaridad.

53. We based our survey on a purposive sample with quota controls for two reasons. A representative sample of all Catholics would not have yielded enough Catholic activists on which to generalize (as in Chile, Christian community members made up less than 1 percent of the nominal Catholics in greater Lima). In addition, the academic and Church people with whom we spoke expressed doubt that Catholics (activists or not) in popular neighborhoods or *pueblos jóvenes* would answer questions candidly unless given prior assurances by a trusted priest or nun.

54. Ninety-one percent supported the idea of Christian-Marxist collaboration within the same popular movement, but only 5.6 percent suggested forming or joining a single political party.

55. Peruvian respondents were asked a number of political questions that the Chileans were asked (political tendency, whether one favored "popular" or nonexclusionary democracy, the propriety of defending one's own class interests (as against reconciling them with those of other classes), and whether one favored collaboration with leftists (Marxists) and, if so, of what sort (a single party, a united popular movement, or something less committal). Two other questions were added: the terms in which one defined Peru's social problems (structural, class, or other) and what one thought of leftist (Marxist) ideas (good or important, valuable, etc.). Peruvians were also asked three of the religious questions put to Chileans (the reasons for Christ's death, economic involvement, and Church authority), although the latter two were phrased somewhat differently, and a fourth question: whether or not the poor were obliged to love the rich (poor love rich). For the precise wording of these questions, see Appendix 1.

56. In an interview appearing in the August 1987 issue of *Páginas*, a nun who has worked in Villa since 1977, Sister Nelcida Solozano, describes these communi-

ties (similar to the Christian Base Communities in Chile) as coming together as church, maintaining a relationship with one another, and working with cultural or Christian community groups in the popular communities. Typically, she said, their members would be block leaders in their neighborhoods, religious educators, or leaders of reflection groups.

57. Villa residents shifted to Fujimori and his Cambio 90 Party in the 1990 and 1995 elections.

58. One priest we interviewed (interview no. 6 [Peru], July 10, 1987, in Lima) stressed the overlap between ecclesial and social activists in his area. At one recent meeting of sector Christians, he said, all but one of the eighteen people attending were community leaders and activists and veterans of more than ten years of pastoral group activity. The tendency in most *pueblos jóvenes*, however, was for organizational Catholics to steer clear of partisan political involvement.

59. A massive procession, attracting hundreds of thousands of people, is held in October in honor of the Lord of Miracles, a mural depiction of Christ painted by a slave on a Lima prison wall that remained intact through an earthquake that destroyed much of the city in the sixteenth century.

60. Tuesta Soldevilla (1989, p. 17) cites García (1985, p. 127), who gives Comas a relative poverty rating of 5.05 (based on a scale of 8 indicators) as compared to 9.55 for Villa El Salvador, and 4.88 for San Juan de Miraflores.

61. The questionnaires did not identify the respondent as a member of a particular parish, although the priests and nuns in each of these parishes helped the interviewers to identify and approach the respondents.

62. Villa organizationals were more apt to mention Velasco than other Catholics, although they were also more supportive (13 percent) of García (no sacramentals or culturals mentioned his government) as well, even though the economy was beginning to break down as the survey was being administered.

63. Radical IU elements might have rejected such a program and bolted the coalition, but the candidate of the parties and sectors that remained might have made the runoff in 1990 and could have defeated a conservative opponent like Vargas Llosa in the general election.

64. The Revolutionary Left Movement, the Revolutionary Communist Party, and the Democratic Union of the People, respectively.

65. Interview no. 3 (Peru), June 22, 1987, in Lima.

66. Interview no. 11 (Peru), July 27, 1988, in Lima.

67. The constitution was drafted by an elected Constituent Assembly dominated by Fujimori's supporters. It was approved in a national referendum by a margin of 52 to 48 percent.

68. Stoll, 1990, pp. 337–38.

69. Jesuit anthropologist Manuel Marzal, who has studied Protestant communities in Lima, notes their success in rehabilitating alcoholics, offering opportunities for religious leadership (neither celibacy nor education are required of their ministers), and helping to keep the personal lives of migrants together in a new, strange environment. Their stress on frugality and the avoidance of alcohol also helps them improve the living standards of some of their converts. See Marzal, 1989, pp. 400–403, 410–11 416–20.

70. Vargas had been secretary-general of the Episcopal Conference since the early 1980s, and was reputedly close to Sodalitium Christinae. Some observers expected Rome to name Bishop Bambarén, who was said to be keenly interested in

the position. In fact, the trouble he was having with the progressive priests in his diocese (Chimbote), and his public criticism of Protestant sects, might have been part of an effort to position himself in the center of the ideological and ecclesial spectra, thereby improving his prospects for nomination. In the end, however, he was passed over for his fellow Jesuit, Msgr. Vargas.

71. The papal nuncio, Bishop Dossena reputedly urged that the bishops openly support Vargas Llosa. Bishop Bambarén was quoted as saying: "If at this point they [the evangelicals] are already showing their aggressiveness against the Catholic Church, what will they do if they have power in their hands?" *Latinamerica Press,* vol. 22, no. 20, (May 31, 1990): 1.

72. Fujimori was born in Peru of Japanese immigrant parents. During the campaign, he was widely referred to as *el chinito,* literally the little Chinaman, an affectionate term used to refer to both Asiatic and assimilated Andean people. When, as president, he began making "tough" decisions and treating his opponents harshly, the press dropped the endearing diminutive, referring to him simply as *el chino.*

73. The subsidies on food and other basic items were removed, and prices jumped from 200 to 300 percent. Ten days later, rates for public services such as telephones and electricity increased between 15 and 20 times. Fujimori's popularity rating went from 56 percent in June to 37 percent following the price increases. But by November it was back at 59 percent despite the prospect of tariff cuts, new (more accommodating) rules for foreign investment, and reductions in state investment.

74. The bishops did not comment on Fujimori's willingness to adopt policies that in his campaign he said should never be adopted. They simply said: "our people continue to bear increasing poverty and misery. Today, they feel defeated by strong economic measures, which, in principle, look to end the country's grave economic crisis." See *Latinamerica Press,* vol. 22, no. 30 (August 23, 1990): 2.

75. In announcing the birth control program, Fujimori told reporters: "We don't want a country populated by children feeding themselves from garbage dumps" (*New York Times,* November 16, 1990). The bishops had persuaded García to reject the decriminalization of abortion in the latter stages of his presidency.

76. Some bishops had other thoughts with respect to birth control but were not prepared to dissent at the time.

77. The Partido Unificado Mariateguista (a successor to the UDP), the Peruvian Communist Party, Bandera Roja (Red Flag) faction, and the Peruvian Communist Party, Patria Roja (Red Fatherland) faction, respectively.

78. Bishops Miguel Irízar, José Ramón Gurruchaga, Emilio Vallebuona, and J. L. Martín turned out to be much less theologically and ecclesialogically conservative than reputed, while other, true conservatives supported Dammert in exchange for concessions on specific issues.

79. A former head of Caritas was killed in Ayacucho in December 1987. Two German priests were assassinated in December 1989. A seventy-year-old nun, María Agustina Rivas, was murdered in September 1990, an Australian nun was murdered in May 1991, an Italian missionary priest and two Polish priests were killed in August 1991, and other priests and nuns were the objects of unsuccessful assassination attempts during these years as well. See *Latinamerica Press,* vol. 23, no. 32, (September 5, 1991). These acts were part of a guerilla offensive targeting foreign "social workers." It was designed to isolate the people they wished to at-

tract, diminish their options, and force them to choose between Sendero and the government. It did not appear to be particularly antireligious, although senderista ideology was explicitly disdainful of religious faith and sentiment.

80. Conferencia Episcopal Peruana, 1991, p. 14. Reaffirming one of the Medellín Conference's hallmark themes, the bishops also stated: "there will only be a stable and lasting peace in our country to the extent that we manage to forge economic and political structures that are more just and more respectful of the dignity and rights of all. Therefore, a strategy of pacification has to rest not only on the struggle against subversion and terrorism but, and above all, on a strategy of development and social transformation designed to overcome marginalization and injustice."

81. According to opinion polls cited in *Latinamerica Press*, vol. 24, no. 13 (April 9, 1992): 2, 80 percent of Lima's population supported Fujimori's action generally, 73 percent favored dissolution of the Congress, and 80 percent supported restructuring of the judicial branch. Of course, ex post facto support is easy to come by, and at this point, relatively early in Fujimori's term, most Peruvians had little stomach for abandoning his government. But in any event, confidence in the country's "democratic institutions" was clearly limited.

82. Interview, no. 16 (Peru), July 7, 1992, in Lima.

83. Consejo Permanente del Episcapado Peruana, 1992, pp. 10–11.

84. Ibid., p. 5.

85. Ibid., p. 4.

86. Enrique Obando, a researcher for the Peruvian Center for International Studies, quoted in the *New York Times*, July 28, 1993, p. A3.

87. *Latinamerica Press*, vol. 25, no. 28 (July 29, 1993): 8; vol. 25, no. 38 (October 21, 1993): 8; and vol. 25, no. 40 (November 4, 1993): 1–2.

88. According to a long-time CEAS staff associate (interview no. 21 [Peru], July 16, 1992), in Lima], conservative bishops were fearful of CEAS because of the close relationship that CEAS people have maintained with Gutiérrez over the years. He told us that Opus Dei Bishop Cipriani (of Ayacucho) forbade the organization from entering his diocese under any circumstances, and that other bishops were willing to assess its programs on a case-by-case basis, but only if tangible material benefits (e.g., job-training programs) were involved.

89. *Iglesia en el Perú*, no. 174, January, 1991, p. 1.

90. Interview no. 13 (Peru), August 3, 1988, in Lima.

91. Such an organization would help Father Gutiérrez, now in his sixties, to convey to younger priests the experience and perspectives of priests of his generation.

92. According to *Latinamerica Press*, vol. 27, no. 24 (June 29, 1995), 87 percent of all Peruvians disagreed with the amnesty, 77 percent thought its purpose was to cover up the involvement of senior officers, and Fujimori's approval rating fell to 8 percent following the bill's signing.

93. *Caretas*, no. 1381 (September 21, 1995), p. 12.

94. Vargas responded, accusing Fujimori of "declaring war on the Church, on the Pope, and on God." Ibid.

95. For Fujimori, easy access to birth control was an important part of his overall approach to economic growth, while for most of Peru's conservative bishops it was a typical expression of the self-indulgence of contemporary culture.

96. Some observers think that Moyano, a former schoolteacher and catechist,

was killed because of an interview, published several weeks earlier, in which she claimed that Sendero's following in Villa was minuscule.

97. In the districts encompassed by our 1987 survey, Belmont's Obras movement garnered over 50 percent of the vote, while the IU polled only 16.7 percent. See Tuesta Soldevilla, 1994, p. 168. We have argued above that Catholic activists were influenced by the political climate of the neighborhoods or districts in which they were located. In 1987, Catholic activists in these districts supported the left by an even larger margin than other residents. It would be reasonable to assume, although we cannot know for sure, that their support for the left fell off later as did that of the population generally.

98. See ibid., pp. 161–62.

99. Ibid., pp. 134–35.

100. The figures in the column headed "Poor Need Not Love the Rich" indicate the percentage of respondents answering that the poor had no obligation to love rich people, citing either ideological or religious reasons, or saying that it depended on the attitude or behavior of the rich.

101. Almost a fifth of the religious organizationals, a quarter of those belonging to Christian base communities (CEBs), and almost 35 percent of the social organizationals had this view.

102. *Latinamerica Press,* vol. 27, no. 35 (September 28, 1995): 445.

103. The lone exception to this was the offer to mediate with Sendero that Arch. Vargas Alzamora made to the Fujimori government in 1992 following the dissolution of Congress and the judiciary.

104. See the essay by O'Donnell in Mainwaring, O'Donnell, and Valenzuela, eds., 1992, pp. 17–56.

8. Conclusions

1. It is not clear that Protestant gains have been among previously active Roman Catholics. Guillermoprieto (1995, pp. 166–67) thinks that most Brazilian converts come from *umbanda, candomblé,* or other spiritist sects. According to a 1985 study of Santiago (Van Dorp, 1985, p. 41), evangelical (i.e., Evangelical, Adventist, Baptist, Jehovah's Witnesses, and Mormon) growth in Chile has been greater among those having no religion (54 percent) than among Catholics (31 percent), although the extent to which the latter "practiced" their religion is not known. Surveys conducted by Hamuy and the Bellarmine Center over the years show Catholic affiliations in Santiago remaining stable from 1958 (83.5 percent) through 1980 (82.8 percent), while evangelical Protestant figures rose from 3.9 percent to 8.5 percent. During the next five years, Catholic affiliations fell off dramatically (to 65 percent in 1985), while those of evangelicals rose to 15 percent. According to Rayo and Porath's study of the entire country (1990, p. 3), both Catholic and evangelical affiliations held steady, at 74 percent and 15 percent respectively, between 1986 and 1989.

2. This scenario is consistent with the positions of Mutchler, 1971; Langton, 1986; Gismondi, 1986; and Grigulevich, 1984. They see the Church's "reforms" as defensive moves designed to preserve institutional interests, and for them the Church remains a deeply conservative institution whose natural allies are still political, socioeconomic, and military elites.

3. This option finds resonance with the position of scholars who have por-

trayed the Latin American Church as chronically polarized (Crahan, 1982 and 1991; and Sigmund 1992), or who thought some popular elements might be marching toward a radically new society or model of Church (Berryman, 1984).

4. Levine, 1981 and 1992; Bruneau, 1982; Mainwaring, 1986; and Smith, 1982 have emphasized the Church's overall coherence and the consensus in favor of religious and political reform. They downplay the significance of both radical and conservative or reactionary tendencies, arguing that they have engendered far more emotion and controversy than their numbers or actual influence warrant.

5. Specifically, its authority, its ability to function ritually, and its religious and moral teachings.

6. In each instance, the bishops brought social and political leaders together in an atmosphere of confidence and trust in which they could talk and see the extent of their common concerns. In the end, they gave valuable moral legitimation to the final terms of the agreements. In short, they helped opposition forces to unite around moderate and broadly appealing alternatives to Pinochet that robbed him of much of his earlier support.

7. Again, Catholic activists (or "organizational Catholics") are those involved in local Church-based or Church-sponsored activities of either a social or religious nature. We distinguish them from "sacramental Catholics" (those only participating liturgically) and nonpracticing ("lapsed") or "cultural" Catholics." See above, pp. 9, 16, and 144.

8. Valdivieso's 1989 study looked at most of the roughly 750 Christian communities in the Santiago area, and argues on that basis that radicals and politically active militants were a very small minority.

9. Drake and Jaksic, 1991, p. 10.

10. I.e., military elements who wanted to remain in power.

11. Cortázar (1989, pp. 54–55) rejects Klaiber's assertion (1989, p. 411) that with the election of the Constituent Assembly in 1978 the bishops were more reluctant to challenge the Morales Bermúdez government and its policies for fear of jeopardizing the return to civilian rule. According to Cortázar (1989, p. 53), the hierarchy "put aside its earlier caution and began openly to support such (popular) struggles." But the hierarchy to which he is referring consists primarily of the Secretary General of the Episcopal Conference, CEAS, and the Southern Andean bishops, and it is being pressured, he concedes (p. 54) by "a dynamism that originated in intermediate and base-level groups." We believe that Cortázar understates the retreat by the Episcopal Conference itself (i.e., the bishops speaking as a whole) from both organized labor and other anti-Morales activities.

12. CEAS, the Social Action Commission of the Episcopal Conference, was headed for many years by Bishop Bambarén. Conservative bishops were skeptical of its various programs and documents, seeing in them the hand of liberation theologian Father Gustavo Gutiérrez.

13. See the description of the rondas in chap. 7, pp. 235–37.

14. A number of these elites developed second thoughts about "democratic" forces and processes that they could not control.

15. In contrast, Chilean civilians were fighting to gain a measure of justice for those abused by the military in the past. This was a formidable task but less difficult than curbing ongoing violations by a similarly unaccountable military.

16. See O'Donnell's essay in Mainwaring, O'Donnell, and Valenzuela, 1992, pp. 17–56.

17. The positions and actions of party leaders affected them negatively as well.

18. John Paul's statements on social and economic issues get much less media coverage than his more conservative pronouncements on moral theory (e.g., *Veritatis Splendor*) in 1993 and his May 1994 letter ruling out the ordination of women.

19. As Puga (1985a) makes clear, however, Chilean bishops differ in their understanding of the nature of this commitment and its implications for the Church's ministry.

20. See Castañeda, 1994, p. 209.

21. The number of priests being ordained and seminarians (some of them future priests) being trained has risen in both Chile and Peru in recent years but is still nowhere near being adequate to the number of Catholics to be ministered to. In Chile, there is 1 priest for approximately 5,000 Catholics; in Peru the ratio is roughly 1 to 8,500.

22. These findings parallel Levine's recent study (Levine, 1992, pp. 199–212) of Catholic activists in base communities in Venezuela and Colombia. He finds participants to be egalitarian in their social beliefs but respectful of the hierarchy. While not likely to wait for episcopal direction in matters affecting them, they neither openly oppose the bishops nor seek to operate independently of them.

23. The Pope's May 1994 letter is reported in the *New York Times*, May 31, p. 8.

24. Drogus (in Brazil) and Levine (in Colombia and Venezuela) also find women exercising leadership roles, even in religious activities, in most base communities. See Drogus, 1987; and Levine, 1992, pp. 205, 214–30. For evidence of the growing number of women exercising leadership in pastoral ministries of the Church in the United States, see Wallace, 1992.

25. Castañeda's (1994, p. 201) observations, drawn largely from Central America, confirm our findings: "as the fight to overthrow the dictatorship triumphed, these movements inevitably lost some of their revolutionary appeal and vigor, while at the same time settling into more traditional roles. They could not maintain in a democracy the drive and symbolism they had generated under authoritarian conditions." Hewitt (1991) found similar patterns among local Catholic activists in Brazil.

26. The organization is stronger in Peru, where seven of the fifty-plus bishops are members and overall membership is estimated at 2,000. Several years ago, however, in connection with efforts to promote the beatification of its founder, José Escrivá de Balaguer, 3,500 letters were written attesting to miracles in which allegedly he interceded. In Chile, on the other hand, only two of the thirty-plus bishops are affiliated with Opus Dei.

27. One prominent Chilean member of Opus Dei, the theologian José Miguel Ibáñez Langlois, was favorably disposed toward Pinochet but spent most of his energies attacking liberation theology (Ibáñez, 1989) rather than democracy. One of the conservative Chilean Catholic intellectuals with whom we spoke (interview no. 13 [Chile], June 21, 1987, in Santiago), who reputedly belonged to the organization, rejected democracy on philosophical grounds but conceded that "good" Catholics could think otherwise.

28. See Vallier, 1970, p. 14; and Reuschemeyer, Stephens, and Stephens, 1992, pp. 152–53.

29. The experience of the Church in Central America confirms what we have seen in Chile and Peru. Although not enthusiastic about Marxist movements, at no point did the Central American bishops condemn them outright. Some Catholic activists collaborated with Marxists in opposing repressive military regimes (e.g., Somoza's Nicaragua and the military regime in El Salvador). And while the Nicaraguan bishops were persistently critical of *sandinista* policies throughout the 1980s, they never condemned the regime as such. In fact, Archbishop Ovando y Bravo, the most staunch of its critics, was appointed by Daniel Ortega to head a commission of reconciliation charged with working out differences between the government and the *contras* in the late 1980s.

30. This was a role the Chilean and (to a lesser extent) Peruvian hierarchies played in the transitions to democratic rule in their respective countries. Catholic bishops in El Salvador, Guatemala, and Mexico mediated between governments and armed revolutionaries during the 1980s and early 1990s, an indication that other national Catholic churches in Latin America are playing important roles as spokespersons for values and processes that span political differences and further the cause of democratic rule. See Berryman, 1994.

31. The Latin American bishops as a whole (CELAM) took a similarly critical position at their May 1995 meeting. See "The Bishops vs. the Market," in *Latinamerica Press*, vol. 27, no. 18 (May 18, 1995).

Bibliography

Abbott, Walter M., S.J. 1966. *The Documents of Vatican II.* New York: Guild Press.

Aliaga R., Fernando. 1975. *Itinerario Histórico.* Santiago: Equipo de Servicios de la Juventud.

Allou, Sergio. 1989. *Lima en Cifras.* Lima, CIDAP.

El Amigo del Clero, Lima.

Angell, Alan. 1972. *Politics and the Labour Movement in Chile.* London: Oxford University Press.

Annuarium Statisticium Ecclesiae. 1987, 1988, 1989, 1990, and 1991. Vatican City: Typis Polyglottis Vaticanis.

Araneda Bravo, Fidel. 1956. *El Monseñor Errázuriz.* Santiago: Editorial Jurídica.

——. 1986. *Historia de la Iglesia en Chile.* Santiago: Ediciones Paulinas.

——. 1988. *El Clero en el Acontecer Político Chileno, 1935–1960.* Santiago: Editorial Emision.

Arzobispado de Santiago. 1949. *Instruccion pastoral acerca de los problemas sociales.* Santiago: San Pancracio.

Astiz, Carlos. 1969. *Pressure Groups and Power Elites in Peruvian Politics.* Ithaca: Cornell University Press.

Ballón, Eduardo, ed. 1986. *Movimientos Sociales y Democracia: La Fundación de un Nuevo Orden.* Lima: Desco (Centro de Estudios y Promoción del Desarrollo).

Baño, Rodrigo. 1992. *De Augustus a Patricios, la última (do)cena política.* Santiago: Editorial Amerinda.

Barrantes, Alfonso. 1985. *Sus Propias Palabras.* Lima: Mosca Azul Editores.

Barrett, Patrick. Forthcoming. "Forging Compromise: Business, Parties, and Regime Change in Chile." Ph.D. dissertation, Department of Political Science, University of Wisconsin, Madison.

Berryman, Phillip. 1984. *The Religious Roots of Rebellion: Christians in the Central American Revolution.* Maryknoll, N.Y.: Orbis.

——. 1994. *Stubborn Hope: Religion, Politics and Revolution in Central America.* Maryknoll, N.Y.: Orbis.

Boletin de la Acción Católica de Chile, Santiago.

Booth, David, and Sorj, Bernardo, eds. 1983. *Military Reform and Social Class.* New York: St. Martin's Press.

Bourricaud, Francois. 1970. *Power and Society in Contemporary Peru.* New York: Praeger.

Bruneau, Thomas. 1982. *The Church in Brazil.* Austin: University of Texas Press.

Burdick, John, 1993. *Looking for God in Brazil.* Berkeley: University of California Press.

Calderón Cockburn, Julio; Filomeno, Alfredo; and Pease García, Henry, eds. 1975. *Perú, 1974, Cronología Política.* Vol. III. Lima: Desco (Centro de Estudios y Promoción del Desarrollo).

——. 1977. *Perú, 1975, Cronología Política.* Vol. IV. Lima: Desco (Centro de Estudios y Promoción del Desarrollo).

Caretas. no. 1381. Lima. (September 21, 1995): 12.

Castañeda, Jorge. 1994. *Utopia Unarmed.* New York: Knopf.

Castillo, Fernando. 1986. *Iglesia Liberadora y Política.* Santiago: ECO.

Cavallo, Ascanio. 1988. *Los Te Deum del Cardenal Raul Silva Henríquez en el Régimen Militar.* Santiago: Ediciones Copygraph.

——. 1992. *Los Hombres de la Transición.* Santiago: Editorial Andrés Bello.

Cavallo, Ascanio, et al. 1989. *Chile, 1973–1988: La Historia Secreta del Régimen Militar.* Santiago: La Epoca.

Centro de Estudios Públicos. 1991a. *Estudio Social de Opinion Pública, Diciembre 1990,* Documento de Trabajo 151 (February).

——. 1991b. *Estudio Social de Opinión Pública, Marzo 1991,* Documento de Trabajo 156 (June).

——. 1991c. *Estudio Social de Opinión Pública, October–November 1991,* Documento de Trabajo 170 (December).

Centro de Estudios y Publicaciones. 1973. *Signos de Liberación: Testimonios de la Iglesia en América Latina, 1969–1973.* Lima: CEP.

——. 1978. *Signos de Lucha y Esperanza: Testimonio de la Iglesia en América Latina, 1973–1978.* Lima: CEP.

——. 1982. *Acompañando la Comunidad.* Lima, CEP.

——. 1983 *Signos de Vida y Fidelidad: Testimonios de la Iglesia en América Latina, 1978–1982.* Lima: CEP.

——. 1988. *Signos de Nueva Evangelización: Testimonio de la Iglesia en América Latina, 1983–1987.* Lima: CEP.

Cleary, Edward, O.P., ed. 1989. *Path From Puebla: Significant Documents of the Latin American Bishops Since 1979.* Washington, D.C.: U.S. Catholic Conference.

Cleary, Edward L., and Sepulveda, Juan. 1992. "Chilean Pentecostalism: Coming of Age." In Cleary, Edward L., and Stewart-Gambino, Hannah, eds., *Power, Politics and Pentecostals in Latin America.* Boulder: Lynne Rienner Publishers.

Coleman, William, M. M. 1953. *Latin-American Catholicism: A Self Evaluation.* World Horizon Reports. Maryknoll, N.Y.: Maryknoll Publications.

Collier, David. 1978. *Barriadas y élites de Odría a Velasco.* Lima: Instituto de Estudios Peruanos.

Comité Familiares de los Obreros de CROMOTEX. 1980. *Compañeros, Tomen Nuestra Sangre: La Lucha de los Obreros de Cromotex.* Lima: Tarea.

Conferencia Episcopal Chilena. 1982. *Documentos del Episcopado, Chile, 1974–1980.* Santiago: Mundo Ltda.

——. 1984. *Documentos del Episcopado, Chile, 1981–1983.* Santiago: Ediciones Mundo.

Conferencia Episcopal de Chile. 1968a. "Declaración de la Conferencia Episcopal de su reunión plenaria del 4 del octubre de 1968." Mimeographed. Santiago.

——. 1968b. *Orientaciones Pastorales I.* Santiago: Imprenta Alfonsiana.

——. 1969a. "Declaración episcopal sobre la situación actual del país," December 12.

——. 1969a. *Orientaciones Pastorales II.* Santiago: Tipografía San Pablo.

——. 1971. "Evangelio, política y socialismos." In Oviedo Cavada, Carlos, ed., *Documentos del Episcopado: Chile, 1970–1973.* Santiago: Ediciones Mundo: 58–100.

——. 1988. *Documentos del Episcopado Chileno, 1984–1987.* Santiago: Area de Comunicaciones de la Conferencia Episcopal.
——. 1992a. *Documentos del Episcopado Chileno, 1988–1991.* Santiago: CENCOSEP.
——. 1992b. *Evangélicos y sectas: propuestas pastorales.* Santiago: CENCOSEP.
Conferencia Episcopal Peruana. n.d. *Documentos del Episcopado: La Pastoral Conciliar en el Peru, en la Iglesia, 1968–1977.* Lima: EAPSA.
——. 1989. *Documentos de la Conferencia Episcopal Peruana, 1979–1989.* Lima: VE.
——. 1991. *Paz en la Tierra.* Lima: CEP.
Congar, Yves, O.P. 1962. "The Historical Development of Authority in the Church: Points for Christian Reflection." In Todd, John M., ed., *Problems of Authority.* Baltimore: Helicon Press.
Congregation for the Doctrine of the Faith. 1986. *Instruction on Christian Freedom and Liberation.* Washington, D.C.: United States Catholic Conference.
Connery, John, S.J. 1957. *Scholastic Analysis of Usury.* Cambridge: Harvard University Press.
——. 1977. *Abortion: The Development of the Roman Catholic Perspective.* Chicago: Loyola University Press.
Consejo Permanente del Episcopado Peruano, Conferencia Episcopal Peruana. 1992. *Un Nuevo Peru, Tarea de Todos.* Lima.
Constable, Pamela, and Valenzuela, Arturo. 1991. *A Nation of Enemies.* New York: Norton.
Correa, Enrique, and Viera-Gallo, Antonio. 1986. *Iglesia y Dictadura.* Santiago: CESOC.
Cortázar, Juan Carlos. 1988. "Institucionalidad y Proyectos en la Iglesia." *Allpanchis* XX, no. 32.
——. 1989. "La Iglesia Toma la Palabra." *Páginas* 96 (April).
Costello, Gerald M. 1979. *Mission to Latin America: The Successes and Failures of a 20th Century Crusade.* Maryknoll, N.Y.: Orbis.
Cotler, Julio. 1988. "Los Partidos Políticos y la Democracia en el Perú." In Pásara, Luis and Parodi, Jorge, eds. *Democracia, Sociedad y Gobierno en el Perú.* Lima: CEDYS.
Crahan, Margaret. 1979. "Salvation Through Christ or Marx: Religion in Revolutionary Cuba." *Journal of Interamerican Studies and World Affairs* 21: 156–84.
——. 1982. "International Aspects of the Role of the Catholic Church in Central America." In Feinberg, Richard, ed., *Central America: International Dimensions of the Crisis.* New York: Holmes Meier.
——. 1991. "Church and State in Latin America: Assassinating Some Old and New Stereotypes." *Daedalus* 20, no. 3: 131–58.
Curran, Charles. 1990. *Moral Theology: Challenges for the Future. Essays in Honor of Richard McCormick.* New York: Paulist Press.
De Ferrari, José Manuel. 1992. "10 Años de Cambios de Obispos en Chile, 1981–1991." *Reflexión y Liberación* 12.
De la Maza, G.; and Garcés, M. 1985. *La Explosión de las Mayorías Protesta Nacional, 1983–1984.* Santiago: Educación y Comunicaciones (ECO).
Dewart, Leslie. 1963. *Christianity and Revolution: The Lesson of Cuba.* New York: Herder and Herder.
Dooner, Patricio, ed. 1988. *La Iglesia Católica y el Futuro Político de Chile.* Santiago: CISOC-Bellarmino.

Drake, Paul, and Jaksic, Ivan. 1991. *The Struggle for Democracy in Chile.* Omaha: University of Nebraska Press.

Drogus, Carol Ann. 1987. "We Are Women Making History: Political Participation in São Paulo's CEBs." University of Wisconsin-Milwaukee Center for Latin America Discussion Paper #81.

Dulles, Avery, S.J. 1978. *Models of the Church.* New York: Doubleday Image.

Eagleson, John, ed. 1975. *Christians and Socialism: Documentation of the Christians for Socialism Movement in Latin America.* Maryknoll, N.Y.: Orbis.

Estudios, Santiago.

Fleet, Michael. 1985. *The Rise and Fall of Chilean Christian Democracy.* Princeton: Princeton University Press.

——. 1987a. *Encuesta Cristianos y Marxistas.* Santiago: CISOC-Bellarmino.

——. 1987b. *Estudio de Opiniones y Actitudes Religiosas.* Lima: Pontificia Universidad Católica del Perú, Facultad de Ciencias Sociales.

Franco, Carlos. 1975. *La revolución participatoria.* Lima: Mosca Azul Editores.

——. 1979. *Perú: participación popular.* Lima: Ediciones CEDEP.

Fremantle, Anne, ed. 1963. *The Papal Encyclicals in Their Historical Context.* New York: The New American Library.

Frühling, Hugo. 1985a. "Non-Profit Organizations as Opposition to Authoritarian Rule: the Case of Human Rights Organizations and Private Research Centers in Chile." Unpublished paper.

——. 1985b. "Reproducción y Socialización de Núcleos de Resistencia: La Experiencia de la Vicaria de la Solidaridad en Chile." Unpublished paper.

García, José María. 1985. "Pobreza, población y vivienda en distritos de Lima Metropolitana, 1981." In Henríquez, Narda and Ponce, Ana. *Lima: población, trabajo y política.* Lima: Universidad Católica.

Garland, Alfredo. 1978. *Como Lobos Rapaces.* Lima: Sapei.

Garretón, Manuel Antonio. 1987. *Reconstruir la Democracia, Transición y Consolidación Democrática en Chile.* Santiago: Editorial Andante.

——. 1989. "La Oposición Política Partidaria en el Régimen Militar Chileno, Un Proceso de Aprendizaje para la Transición. In Cavarrozzi, Marcelo and Garretón, Manuel Antonio, eds., *Muerte y Resurrección, Los Partidos Políticos en el Autoritarismo y las Transiciones en el Cono Sur.* Santiago: Flacso.

Gibbons, William J., S.J., comp. 1955. *Basic Ecclesiastical Statistics, 1954.* Baltimore, Loyola College.

——. 1956. *Basic Ecclesiastical Statistics, 1956.* World Horizon Reports, 12. Maryknoll, N.Y.

——. 1958. *Basic Ecclesiastical Statistics, 1958.* World Horizon Reports, 24. Maryknoll, N.Y.

——. 1960. *Basic Ecclesiastical Statistics, 1960.* World Horizon Reports, 25. Maryknoll, N.Y.

Gilfeather, Katherine Ann, M. M. 1976. "Women and Ministry." *America,* (October).

——. 1980. "Religious, the Poor and the Institutional Church in Chile." In Levine, Daniel H., ed., *Churches and Politics in Latin America.* Beverly Hills, Calif.: Sage Publications.

——. 1985. "Coming of Age in a Latin Church." In John, C. B. and Webster, Ellen Loro, eds., *The Church and Women in the Third World.* Philadelphia: Webster Press.

Gismondi, Michael. 1986. "Transformations of the Holy." *Latin American Perspectives* 13, no. 3.

González, Bishop Carlos. 1990. *Verdad y Reconciliación.* Talca: Marana-tha Ltda.
Gorman, Stephen M. 1982. *Post-Revolutionary Peru: The Politics of Transformation.* Boulder: Westview Press.
Green, Anne. 1993. *The Catholic Church in Haiti: Political and Social Change.* East Lansing: Michigan State University Press.
Gremillion, Joseph, ed. 1976. *Gospel of Justice and Peace: Catholic Social Teaching Since Pope John.* Maryknoll, N.Y.: Orbis.
Grigulevich, José (Josef). 1984. *La Iglesia Católica y el movimiento de liberación en América Latina.* Moscow: Editorial Progreso.
Guerra, José Guillermo. 1929. *La Constitución de 1925.* Santiago: Balcells.
Guerra García, Francisco. 1975. *El Peruano, Un Proceso Abierto.* Lima: Contratiempo.
——. 1983. *Velasco: Del Estado Oligárquico al Capitalismo de Estado.* Lima: Ediciones CEDEP.
Guillermoprieto, Alma. 1995. *The Heart that Bleeds.* New York: Vintage.
Gutiérrez, Gustavo. 1970. *Lineas Pastorales de la Iglesia en América Latina.* Lima: CEP.
——. 1971. *Teología de Liberación.* Lima: CEP.
Hansen, Laurie. 1992. "Pope Strikes Out Against Sects Wooing Latin American Catholics Away From Church," *Catholic Herald* (Milwaukee), 22 October 1992.
Hanson, Eric O. 1987. *The Catholic Church in World Politics.* Princeton: Princeton University Press.
Hewitt, W. E.. 1991. *Base Christian Communities and Social Change in Brazil.* Lincoln: University of Nebraska Press.
Hoy, Santiago.
Houtart, François, and Rousseau, André. 1971. *The Church and Revolution: From the French Revolution of 1789 to the Paris Riots of 1968; from Cuba to Southern Africa; from Vietnam to Latin America.* Maryknoll, N.Y.: Orbis.
Huerta, María Antonieta, and Pacheco Pastene, Luis. 1986. *La Iglesia Chilena y los Cambios Sociopolíticos.* Santiago: Pehuen.
Huneeus, Carlos. 1986. *Cambios en la Opinión Pública: Una Aproximación de la Cultura Política en Chile.* Santiago: Centro de Estudios de la Realidad Contemporánea (CERC), Academia de Humanismo Cristiano.
Hurtado, Alberto, S.J. 1941. *¿Es Chile un País Católico?* Santiago: Ediciones Splendor.
Ibáñez Langlois, José Miguel. 1989. *Teología de Liberación y lucha de clases.* Santiago: Ediciones Universidad Católica de Chile.
Iglesia en el Peru, Lima.
Jelen, Ted G. 1991. *The Political Mobilization of Religious Beliefs.* New York: Praeger.
John Paul II, Pope. 1988. "Discurso al Episcopado" 37. Vaticano: Sala de Prensa Santa Sede.
Johnson, James Turner. 1987. *Quest for Peace: Three Moral Traditions in Western Cultural History.* Princeton: Princeton University Press.
Klaiber, Jeffrey, S.J. 1977. *Religion and Revolution in Peru, 1824–1976.* Notre Dame, Ind.: University of Notre Dame Press.
——. 1983. "The Catholic Lay Movement in Peru." *The Americas* 40, no. 2: 149–70.
——. 1986. "The Battle over Private Education in Peru, 1968–1980: An Aspect of the Internal Struggle in the Catholic Church." *The Americas* 43, no. 2: 137–58.

——. 1988. *La Iglesia en el Perú.* Lima: Pontificia Universidad Católica del Perú, Fondo Editorial.

Krischke, Paulo J. 1991. "Church Base Communities and Democratic Change in Brazilian Society." *Comparative Political Studies* 24, no. 2.

Kuczynski, Pedro Pablo. 1977. *Peruvian Democracy under Economic Stress: An Account of the Belaúnde Administration, 1963–1968.* Princeton: Princeton University Press.

Lagos Schuffeneger, Humberto, and Chacón Herrera, Arturo. 1987. *Los Evangélicos en Chile: Una Lectura Sociológica.* Santiago: Ediciones Literatura Americana Reunida, Presor.

Lalive d'Epinay, Christian. 1969. *Haven of the Masses: A Study of the Pentecostal Movement in Chile.* London: Lutterworth.

Langton, Kenneth P. 1986. "Church, Protest, and Rebellion." *Comparative Political Studies* 19, no. 3: 317–355.

Larson, Rev. Oscar. 1967. *La ANEC y la Democracia Cristiana.* Santiago: Ediciones Ráfaga.

Latinamerica Press, Lima.

Latin America Weekly Report, London.

Latin American Episcopal Conference (CELAM). 1970. "Joint Pastoral Planning." In CELAM, *The Church in the Present-day Transformation of Latin America in the Light of the Council.* 2 vols. Bogotá: General Secretariat of CELAM.

Leo XIII, Pope. 1942. *On Civil Government.* (Orig. 1881). New York: Paulist Press.

Lernoux, Penny. 1980. *The Cry of the People: United States Involvement in the Rise of Fascism, Torture and Murder, and the Persecution of the Catholic Church in Latin America.* New York: Doubleday.

Levine, Daniel H. 1981. *Religion and Politics in Latin America: The Catholic Church in Venezuela and Colombia.* Princeton: Princeton University Press.

——. 1992. *Popular Voices in Latin American Catholicism.* Princeton: Princeton University Press.

Mabry, Donald. 1973. *Mexico's Acción Nacional: A Catholic Alternative to Revolution.* Syracuse: Syracuse University Press.

Macaulay, Michael G. 1972. *Ideological Change and Internal Cleavages in the Peruvian Church: Change, the Status Quo and the Priest; The Case of ONIS.* Ph.D. dissertation, University of Notre Dame.

MacEoin, Gary. 1962. *Latin America: the Eleventh Hour.* New York: P. J. Kennedy.

Mahoney, John. 1987. *The Making of Moral Theology.* New York: Oxford University Press.

Mainwaring, Scott, 1986. *The Catholic Church and Politics in Brazil, 1916–1985.* Stanford: Stanford University Press.

Mainwaring, Scott; O'Donnell, Guillermo; and Valenzuela, Samuel, eds. 1992. *Issues in Democratic Consolidation.* Notre Dame, Ind.: University of Notre Dame Press.

Maloney, Thomas J. 1978. "The Catholic Church and the Peruvian Revolution: Resource Exchange in an Authoritarian Setting." Ph.D. dissertation. University of Texas.

Marins, José; Trevison, Teolide M.; and Chanona, Carolee. 1978. *Praxis de los Padres de América Latina: Documentos de las Conferencias Episcopales de Medellín a Puebla, 1968–1978.* Bogotá: Ediciones Paulinas.

Maritain, Jacques. 1960. "Catholic Action and Political Action." In Maritain, Jacques, *Scholasticism and Politics*. Garden City, N.Y.: Doubleday.

Martin, David. 1990. *Tongues of Fire: The Explosion of Protestantism in Latin America*. Cambridge: Basil Blackwell.

Marzal, Manuel M., S.J. 1989. *Los Caminos Religiosos de los Immigrantes en la Gran Lima*. Lima: Pontificia Universidad Católica del Perú, Fondo Editorial.

McCormick, Richard, S.J. 1989. *The Critical Calling: Reflection on Moral Dilemmas since Vatican II*. Washington, D.C.: Georgetown University Press.

McCoy, John. 1989. "Liberation Theology and the Peruvian Church." *America*, 160, no. 21 : 526–30.

Mecham, J. Lloyd. 1966. *Church and State in Latin America: A History of Politico-ecclesiastical Relations*. 2nd edition, revised. Chapel Hill: University of North Carolina Press.

Mensaje, Santiago.

El Mercurio, Santiago.

Mitchell, C. M., ed. 1993. *International Historical Statistics: The Americas and Australia*. Detroit: Gale Research.

Mutchler, David. 1971. *The Church as a Political Factor in Latin America, with Particular Reference to Colombia and Chile*. New York: Praeger.

National Catholic Welfare Conference News Service Bulletin (NCWC), Washington, D.C.

New York Times.

Nielsen, Niels C. 1991. *Revolutions in Eastern Europe: The Religious Roots*. Maryknoll, N.Y.: Orbis.

Noonan, John T. 1966. *Contraception: A History of Its Treatment by the Catholic Theologians and Canonists*. Cambridge: Harvard University Press.

O'Keefe, Mary Early. 1976. *The Peruvian Polity and the Church: The Adaptive Role of Maryknoll, 1963–1973*. Ph.D. dissertation. Department of Political Science. University of California, Santa Barbara.

OSORE, n.d. *Datos Estadísticos 1995, Clero Secular, Congregaciones Religiosas, Sacramentación en Chile*. Santiago: Oficina de Sociología Religiosa, Secretaría General del Episcopado.

Oviedo Cavada, Carlos, ed. 1974. *Documentos del Episcopado: Chile, 1970–73*. Santiago: Ediciones Mundo.

Parker Gumucio, Cristian. 1985. "Cristianismo y Movimiento Popular en Chile." In *Plural* 4 (Primer Semestre).

——. 1990. "El aporte de la Iglesia a la sociedad chilena bajo el régimen militar." *Cuadernos Hispanoamericanos* 482–83.

Pásara, Luis. 1986. *Radicalización y Conflicto en la Iglesia Peruana*. Lima: Ediciones El Virrey.

——. 1989. "Peru: The Leftist Angels." In Mainwaring, Scott, and Wilde, Alexander, eds. *The Progressive Church in Latin America*. Notre Dame, Ind.: University of Notre Dame Press.

Pease García, Henry, and Verme, Olga, eds. 1974. *Perú, 1968–1973, Cronología Política*. Vols. I and II. Lima: Desco (Centro de Estudios y Promoción del Desarrollo).

Pease García, Henry, and Filomeno, Alfredo, eds. 1977. *Perú, 1976, Cronología Política*. Vol. V. Lima: Desco (Centro de Estudios y Promoción del Desarrollo).

Pérez-Díaz, Victor. 1993. *The Return of Civil Society*. Cambridge: Harvard University Press.

Peruvian Bishops Commission for Social Action. 1969. *Between Honesty and Hope*. Maryknoll, N.Y.: Maryknoll Publications.

Pike, Frederick B. 1964. "The Modernized Church in Peru: Two Aspects." *Review of Politics* 26, no. 3.

Pius XII, Pope. n.d. *1944 Christmas Message of His Holiness Pope Pius XII*. Washington, D.C.: National Catholic Welfare Conference.

Poblete B., Renato, S.J., Galilea W., Carmen, and Van Dorp P., Patricia, 1980. *Imagen de la Iglesia de Hoy y Religiosidad de los Chilenos*. Santiago: Centro Bellarmino, Departamento de Investigaciones Sociologicas.

Poggi, Gianfranco. 1967. *Catholic Action in Italy: The Sociology of a Sponsored Organization*. Stanford: Stanford University Press.

Política y Espíritu, Santiago.

Politzer, Patricia. 1985. *Miedo en Chile*. Santiago: Ediciones Chile y América (CESOC).

——. 1988. *La Ira de Pedro y los Otros*. Santiago: Planeta.

Poole, Deborah, and Rénique, Gerardo. 1992. *Peru, Time of Fear*. London: Latin American Bureau.

Precht, P. Cristián. 1992. "Del Acuerdo a La Reconciliación: La Iglesia de Chile y El Camino a la Democracia." Paper presented to the 17th International Congress of the Latin American Studies Association, Los Angeles (September).

Preston, Paul. 1993. "Brother Votes for Brother: The New Politics of Protestantism in Brazil." In Garrard-Burnett, Virginia, and Stoll, David, eds., *Rethinking Protestantism in Latin America*. Philadelphia: Temple University Press.

Puentes Oliva, Pedro. 1975. *Posición evangélica*. Santiago: Editora Nacional Gabriela Mistral.

Puga, Josefina. 1985a. *El Desafío de la Opción Preferencial por los Pobres: Pensamiento de Obispos y Párrocos*. Santiago: Centro Bellarmino.

——. 1985b. *La Iglesia Católica Hoy: Su Imagen en el Gran Santiago*. Santiago: Centro Bellarmino, Departamento de Investigaciones Sociológicas.

Ramírez, Miguel. 1993. "The Impact of Austerity in Latin America, 1983–1989: A Critical Assessment." *Comparative Economic Studies* 33, no. 1.

Ratzinger, Joseph. 1985. *The Ratzinger Report*. San Francisco: Ignatian Press.

Rayo, Gustavo, and Porath, William. 1990. *Perfil y Opciones Sociales de los Católicos Chilenos*. Santiago: CERC.

Reilly, Charles. 1986. "Latin America's Religious Populists." In Levine, Daniel H., ed., *Religion and Political Conflict in Latin America*. Chapel Hill: University of North Carolina Press.

Remmer, Karen. 1989. *Military Rule in Latin America*. Boston: Unwin Hyman.

Reuschemeyer, Dietrich; Stephens, Evelyn Huber; and Stephens, John. 1992. *Capitalist Development and Democracy*. Chicago: University of Chicago Press.

Richard, Pablo. 1978. *Mort de Chrétien y Naissance de l'eglíse*. Paris: Centre Lebret.

Rodríguez Beruff, Jorge. 1983. *Los Militares y el Poder: un Ensayo sobre la Doctrina Militar en el Perú, 1948–1968*. Lima: Mosca Azul.

Romero, Catalina. 1987. *Iglesia en el Perú, Compromiso y Renovación (1958–1984)*. Lima: Instituto Bartolomé de las Casas-Rimac.

Rosenberg, Tina. 1995. "Force is Forever," *New York Times Magazine* (September 24).

Saba, Raul. 1985. *Political Development and Democracy in Peru.* Boulder: Westview.

Sacred Congregation for the Doctrine of the Faith. 1984. *Instruction on Certain Aspects of the Theology of Liberation.* Washington, D.C.: United States Catholic Conference.

Santos, José Manuel. 1974. "Carta de Mons. José Manuel Santos, Presidente de la Conferencia Episocopal de Chile (CECH), a algunos dirigentes campesinos de Linares." In Oviedo Cavada, Carlos. *Documentos del Episcopado: Chile, 1970–73.* Santiago: Ediciones Mundo.

Secretaría Nacional del Episcopado Chileno. 1949. *Instrucción pastoral acerca a los problemas sociales.* Santiago: San Pancracio.

——. 1964. *Anuario Eclesiástico del Perú.* Lima: Editorial Salesiana.

——. 1974. *Anuario Eclesiástico del Perú.* Lima: Departamento de Estadística, Arzobispado de Lima.

Secretariado del Episcopado Peruano. 1947. *Anuario Eclesiástico del Perú.* Lima: no publisher.

——. 1949, 1951. *Anuario Eclesiástico del Perú.* Lima: Imprenta Santa María.

——. 1959. *Anuario Eclesiástico del Perú.* Lima: no publisher.

——. 1969. *Anuario Eclesiástico del Perú.* Lima: no publisher.

——. 1983. *Directorio Eclesiástico del Perú.* Lima: Centro de Proyección Cristiana.

——. 1984. *Directorio Eclesiástico del Perú.* Lima: Centro de Proyección Cristiana.

——. 1987. *Directorio Eclesiástico del Perú.* Lima: Centro de Proyección Cristiana.

Segundo, Juan Luis, S.J. 1976. *The Liberation of Theology.* Maryknoll, N.Y.: Orbis.

——. 1985. *Theology and the Church: A Response to Cardinal Ratzinger.* Minneapolis: Winston Press.

Sigmund, Paul. 1990. *Liberation Theology at the Crossroads.* New York: Oxford University Press.

Silva, Eduardo. 1991. "The Political Economy of Chile's Regime Transition: From Radical to 'Pragmatic' Neoliberalism." In Drake, Paul, and Jaksic, Ivan, eds., *The Struggle for Democracy in Chile, 1982–1990.* Lincoln: University of Nebraska Press.

Silva Henríquez, Arch. Raul. 1970. "Iglesia, sacerdocio y politica." In Oviedo Cavada, Carlos. *Documentos del Episcopado: Chile, 1970–73.* Santiago: Ediciones Mundo.

Smith, Brian H. 1982. *The Church and Politics in Chile: Challenges to Modern Catholicism.* Princeton: Princeton University Press.

——. 1990. *More Than Altruism: The Politics of Private Foreign Aid.* Princeton: Princeton University Press.

Sociedad Chilena de Defense de la Tradición, Familia y Propiedad. 1976. *La Iglesia del silencio.* Santiago.

Stepan, Alfred. 1978. *The State and Society: Peru in Comparative Perspective.* Princeton: Princeton University Press.

Stephens, Evelyn Huber. 1983. "The Peruvian Military Government, Labor Mobilization, and the Political Strength of the Left." *Latin American Research Review* 18, no. 2: 57–93.

Stewart-Gambino, Hannah. 1992. *The Church and Politics in the Chilean Countryside.* Boulder: Westview Press.

Stoll, David. 1990. *Is Latin America Turning Protestant?* Berkeley: University of California Press.

Thomas Aquinas. 1952. *The Summa Theologica of St. Thomas Aquinas*. Chicago: Encyclopedia Britannica.

Tovar, Teresa. 1985. *Velasquismo y Movimiento Popular, Otra Historia Prohibida*. Lima: Desco (Centro de Estudios y Promoción del Desarrollo).

——. 1986a. "Barrios, Ciudad, Democracia y Política." In Ballón, Eduardo, ed., *Movimientos Sociales y Democracia: La Fundación de Un Nuevo Orden*. Lima: Desco (Centro de Estudios y Promoción del Desarrollo).

——. 1986b. "Vecinos y Pobladores en la Crisis (1980–1984)." In Eduardo Ballón, ed., *Movimientos Sociales y Crisis: El Caso Peruano*. Lima: Desco (Centro de Estudios y Promoción del Desarrollo).

Troelstch, Ernst. 1931. *The Social Teachings of the Christian Churches*. 2 vols. New York: Macmillan.

Tuesta Soldevilla, Fernando. 1983. *Elecciones Municipales: Cifras y Escenario Político*. Lima: Desco (Centro de Estudios y Promoción del Desarrollo).

——. 1985. *El Nuevo Rostro Electoral, Las Municipales de 1983*. Lima: Desco (Centro de Estudios y Promoción del Desarrollo).

——. 1986. *PERU 1985: el derrotero de una nueva elección*, Lima: Centro de Investigación de la Universidad del Pacífico.

——. 1989. *Pobreza Urbana y Cambios Electorales en Lima*. Lima: Desco (Centro de Estudios y Promoción del Desarrollo).

——. 1994. *Perú Político en Cifras: Elite Política y Elecciones*. 2nd Edition. Lima: Fundación Friedrich Ebert.

Ultima Hora, Santiago.

United Nations. *Statistical Yearbook, 1993*. N.Y.: United Nations.

U.S. Catholic Conference. 1978. *LADOC Keyhole Series*, Washington, D.C.

Valda Palma, Roberto. 1971. *Los Obispos Rojos de Latinoamérica*. Lima: Studium.

Valdivieso, Gabriel. 1989. *Comunidades Cristianas de Base, Su Inserción en la Iglesia y en la Sociedad*. Santiago: Centro Bellarmino-CISOC.

Vallier, Ivan. 1970. "Extrication, Insulation, and Re-entry: Toward a Theory of Religious Change." In Landsberger, Henry A., ed., *The Church and Social Change in Latin America*. Notre Dame, Ind.: University of Notre Dame Press.

Van Dorp, Patricia. 1985. *Religiosidad en el Gran Santiago, 1985*. Santiago: Centro Bellarmino-CISOC.

Vergara, Pilar. 1994. "Market Economy, Social Welfare, and Democratic Consolidation in Chile." In Smith, William C.; Acuña, Carlos H.; and Gamarra, Eduardo A., eds., *Democracy, Markets, and Structural Reform in Contemporary Latin America: Argentina, Bolivia, Brazil, Chile, and Mexico*. New Brunswick: Transaction Publishers.

Vicaría de la Solidaridad. 1980. *Vicaría de la Solidaridad: cuarto ano de labor*. Santiago: Arzbispado de Santiago.

Vives, Cristián. 1978. "La solidaridad: una forma de evangelizar y de participar en la iglesia." Mimeographed. Santiago: Centro Bellarmino, 1978.

Wallace, Ruth A. 1992. *They Call Her Pastor: A New Role for Catholic Women*. Albany: State University of New York Press.

Warwick, Donald P., and Lininger, Charles A. 1975. *The Sample Survey: Theory and Practice*. New York: McGraw Hill.

Wilkie, James W., ed. 1983. *Statistical Abstract of Latin America* vol. 22 (1982). Los Angeles: UCLA Latin American Center.

——. 1984. *Statistical Abstract of Latin America* vol. 23 (1983). Los Angeles: UCLA Latin American Center.

Willems, Emilio. 1967. *Followers of a New Faith: Culture Change and the Rise of Protestantism in Brazil and Chile.* Nashville: Vanderbilt University Press.

Wooster, Henry. 1994. "Faith and Ramparts: the Phillippine Catholic Church and the 1986 Revolution." In Johnson, Douglas, and Sampson, Cynthia, eds., *Religion: The Missing Dimension of Statecraft.* New York: Oxford University Press.

Zañartu, Mario. 1969. *Desarrollo económico y moral católica.* Cuernavaca: CIDOC.

Zona Sur de Santiago. n.d. *Caracterización comunidades Eclesiales de Base.* Mimeographed. Santiago.

Index